Benjamin Williamson

An Elementary Treatise on the Differential Calculus Containing the Theory of Plane Curves

Benjamin Williamson

An Elementary Treatise on the Differential Calculus Containing the Theory of Plane Curves

ISBN/EAN: 9783337811419

Printed in Europe, USA, Canada, Australia, Japan

Cover: Foto ©berggeist007 / pixelio.de

More available books at **www.hansebooks.com**

AN ELEMENTARY TREATISE

ON

THE DIFFERENTIAL CALCULUS.

FOURTH EDITION.

AN ELEMENTARY TREATISE

ON

THE INTEGRAL CALCULUS,

CONTAINING

APPLICATIONS TO PLANE CURVES AND SURFACES.

BY

BENJAMIN WILLIAMSON, F.R.S.

IN THE PRESS.

AN ELEMENTARY TREATISE ON DYNAMICS.

BY

BENJAMIN WILLIAMSON, F.R.S.,

AND

FRANCIS A. TARLETON, LL.D.,

Fellows of Trinity College, Dublin.

AN ELEMENTARY TREATISE

ON

THE DIFFERENTIAL CALCULUS,

CONTAINING

THE THEORY OF PLANE CURVES,

WITH

NUMEROUS EXAMPLES.

BY

BENJAMIN WILLIAMSON, M.A., F.R.S.,

FELLOW OF TRINITY COLLEGE, AND PROFESSOR OF NATURAL PHILOSOPHY
IN THE UNIVERSITY OF DUBLIN.

Fifth Edition, Revised and Enlarged.

NEW YORK:

APPLETON & CO.

1884.

[ALL RIGHTS RESERVED.]

DUBLIN:
PRINTED AT THE UNIVERSITY PRESS,
BY PONSONBY AND WELDRICK.

PREFACE.

In the following Treatise I have adopted the method of Limiting Ratios as my basis; at the same time the co-ordinate method of Infinitesimals or Differentials has been largely employed. In this latter respect I have followed in the steps of all the great writers on the Calculus, from Newton and Leibnitz, its inventors, down to Bertrand, the author of the latest great treatise on the subject. An exclusive adherence to the method of Differential Coefficients is by no means necessary for clearness and simplicity; and, indeed, I have found by experience that many fundamental investigations in Mechanics and Geometry are made more intelligible to beginners by the method of Differentials than by that of Differential Coefficients. While in the more advanced applications of the Calculus, which we find in such works as the *Mécanique Celeste* of Laplace, and the *Mécanique Analytique* of Lagrange, the investigations are all conducted on the method of Infinitesimals. The principles on which this method is founded are given in a concise form in Arts. 38 and 39.

In the portion of the Book devoted to the discussion of Curves, I have not confined myself exclusively to the application of the Differential Calculus to the subject; but have availed myself of the methods of Pure and Analytic

Geometry, whenever it appeared that simplicity would be gained thereby.

In the discussion of Multiple Points I have adopted the simple and General Method given by Dr. Salmon in his *Higher Plane Curves*. It is hoped that by this means the present treatise will be found to be a useful introduction to the more complete investigations contained in that work.

As this Book is principally intended for the use of beginners, I have purposely omitted all metaphysical discussions, from a conviction that they are more calculated to perplex the beginner than to assist him in forming clear conceptions. The student of the Differential Calculus (or of any other branch of Mathematics) cannot expect to master at once all the difficulties which meet him at the outset; indeed it is only after considerable acquaintance with the Science of Geometry that correct notions of angles, areas, and ratios are formed. Such notions in any science can be acquired only after practice in the application of its principles, and after patient study.

The more advanced student may read with advantage the *Réflexions sur la Métaphysique du Calcul Infinitésimal* of the illustrious Carnot: in which, after giving a complete *resumé* of the different points of view under which the principles of the Calculus may be regarded, he concludes as follows:—

"Le mérite essentiel, le sublime, on peut le dire, de la méthode infinitésimale, est de réunir la facilité des procédés ordinaires d'un simple calcul d'approximation à l'exactitude des résultats de l'analyse ordinaire. Cet avantage immense serait perdu, ou du moins fort diminué, si à cette méthode pure et simple, telle que nous l'a donnée Leibnitz, on voulait, sous l'apparence d'une plus grande rigueur soutenue dans tout le cours de calcul, en substituer d'autres moins naturelles, moins commodes, moins conformes à la marche probable des inventeurs. Si cette méthode est exacte dans les résultats, comme personne n'en doute aujourd'hui, si c'est toujours à elle qu'il faut en revenir dans les questions difficiles, comme il paraît encore que tout le monde en convient, pourquoi recourir à des moyens détournés et compliqués pour la suppléer? Pourquoi se contenter de l'appuyer sur des inductions et sur la conformité de ses résultats avec ceux que fournissent les autres méthodes, lorsqu'on peut la démontrer directement et généralement, plus facilement peut-être qu'aucune de ces méthodes elles-mêmes ? Les objections que l'on a faites contre elle portent toutes sur cette fausse supposition, que les erreurs commises dans le cours du calcul, en y négligeant les quantités infiniment petites, sont demeurées dans le résultat de ce calcul, quelque petites qu'on les suppose; or c'est ce qui n'est point : l'élimination les emporte toutes nécessairement, et il est singulier qu'on n'ait pas aperçu d'abord dans cette condition indispensable de l'élimination le véritable caractère des quantités infinitésimales et la réponse dirimante a toutes les objections."

Many important portions of the Calculus have been omitted, as being of too advanced a character; however, within the limits proposed, I have endeavoured to make the Work as complete as the nature of an elementary treatise would allow.

I have illustrated each principle throughout by copious examples, chiefly selected from the Papers set at the various Examinations in Trinity College.

In the Chapter on Roulettes, in addition to the discussion of Cycloids and Epicycloids, I have given a tolerably complete treatment of the question of the Curvature of a Roulette, as also that of the Envelope of any Curve carried by a rolling

Curve. This discussion is based on the beautiful and general results known as Savary's Theorems; taken in conjunction with the properties of the Circle of Inflexions. I have also introduced the application of these theorems to the general case of the motion of any plane area supposed to move on a fixed Plane.

In this Edition I have made little alteration beyond the introduction of a short account of the principles of the determinant functions known under the name of Jacobians, which now hold so fundamental a place in analysis.

TRINITY COLLEGE,
   June, 1884.

TABLE OF CONTENTS.

CHAPTER I.

FIRST PRINCIPLES. DIFFERENTIATION.

	PAGE
Dependent and Independent Variables,	1
Increments, Differentials, Limiting Ratios, Derived Functions,	3
Differential Coefficients,	5
Geometrical Illustration,	6
Navier, on the Fundamental Principles of the Differential Calculus,	8
On Limits,	10
Differentiation of a Product,	13
Differentiation of a Quotient,	15
Differentiation of a Power,	16
Differentiation of a Function of a Function,	17
Differentiation of Circular Functions,	19
Geometrical Illustration of Differentiation of Circular Functions,	22
Differentiation of a Logarithm,	24
Differentiation of an Exponential,	26
Logarithmic Differentiation,	27
Examples,	30

CHAPTER II.

SUCCESSIVE DIFFERENTIATION.

Successive Differential Coefficients,	34
Infinitesimals,	36
Geometrical Illustrations of Infinitesimals,	37
Fundamental Principle of the Infinitesimal Calculus,	40
Subsidiary Principle,	41
Approximations,	42
Derived Functions of x^m,	46
Differential Coefficients of an Exponential,	48
Differential Coefficients of $\tan^{-1}x$, and $\tan^{-1}\frac{1}{x}$,	50
Theorem of Leibnitz,	51
Applications of Leibnitz's Theorem,	53
Examples,	57

CHAPTER III.

DEVELOPMENT OF FUNCTIONS.

	PAGE
Taylor's Expansion,	61
Binomial Theorem,	63
Logarithmic Series,	63
Maclaurin's Theorem,	64
Exponential Series,	65
Expansions of $\sin x$ and $\cos x$,	66
Huygens' Approximation to Length of Circular Arc,	66
Expansions of $\tan^{-1} x$ and $\sin^{-1} x$,	68
Euler's Expressions for $\sin x$ and $\cos x$,	69
John Bernoulli's Series,	70
Symbolic Form of Taylor's Series,	70
Convergent and Divergent Series,	73
Lagrange's Theorem on the Limits of Taylor's Series,	76
Geometrical Illustration,	78
Second Form of the Remainder,	79
General Form of Maclaurin's Series,	81
Binomial Theorem for Fractional and Negative Indices,	82
Expansions by aid of Differential Equations,	85
Expansion of $\sin mz$ and $\cos mz$,	87
Arbogast's Method of Derivation,	88
Examples,	91

CHAPTER IV.

INDETERMINATE FORMS.

Examples of Evaluating Indeterminate Forms without the Differential Calculus,	96
Method of Differential Calculus,	99
Form $0 \times \infty$,	102
Form $\frac{\infty}{\infty}$,	103
Forms 0^0, ∞^0, $1^{\pm \infty}$	105
Examples,	109

CHAPTER V.

PARTIAL DIFFERENTIAL COEFFICIENTS.

Partial Differentiation,	113
Total Differentiation of a Function of Two Variables,	115
Total Differentiation of a Function of Three or more Variables,	117
Differentiation of a Function of Differences,	119
Implicit Functions, Differentiation of an Implicit Function,	120
Euler's Theorem of Homogeneous Functions,	123
Examples in Plane Trigonometry,	130
Landen's Transformation,	133
Examples in Spherical Trigonometry,	133
Legendre's Theorem on the Comparison of Elliptic Functions,	137
Examples,	140

CHAPTER VI.

SUCCESSIVE PARTIAL DIFFERENTIATION.

	PAGE
The Order of Differentiation is indifferent in Independent Variables,	145
Condition that $Pdx + Qdy$ should be an exact Differential,	146
Euler's Theorem of Homogeneous Functions,	148
Successive Differential Coefficients of $\phi(x + \alpha t, y + \beta t)$,	148
Examples,	150

CHAPTER VII.

LAGRANGE'S THEOREM.

Lagrange's Theorem,	151
Laplace's Theorem,	154
Examples,	155

CHAPTER VIII.

EXTENSION OF TAYLOR'S THEOREM.

Expansion of $\phi(x + h, y + k)$,	156
Expansion of $\phi(x + h, y + k, z + l)$,	159
Symbolic Forms,	160
Euler's Theorem,	162

CHAPTER IX.

MAXIMA AND MINIMA FOR A SINGLE VARIABLE.

Geometrical Examples of Maxima and Minima,	164
Algebraic Examples,	165
Criterion for a Maximum or a Minimum,	169
Maxima and Minima occur alternately,	173
Maxima or Minima of a Quadratic Fraction,	177
Maximum or Minimum Section of a Right Cone,	181
Maxima or Minima of an Implicit Function,	185
Maximum or Minimum of a Function of Two Dependent Variables,	186
Examples,	188

CHAPTER X.

MAXIMA OR MINIMA OF FUNCTIONS OF TWO OR MORE VARIABLES.

Maxima and Minima for Two Variables,	191
Lagrange's Condition in the case of Two Independent Variables,	191

		PAGE
Maximum or Minimum of a Quadratic Fraction,		194
Application to Surfaces of Second Degree,		196
Maxima and Minima for Three Variables,		198
Lagrange's Conditions in the case of Three Variables,		199
Maximum or Minimum of a Quadratic Function of Three Variables,		200
Examples,		203

CHAPTER XI.

METHOD OF UNDETERMINED MULTIPLIERS APPLIED TO MAXIMA AND MINIMA.

Method of Undetermined Multipliers,	204
Application to find the principal Radii of Curvature on a Surface,	208
Examples,	210

CHAPTER XII.

ON TANGENTS AND NORMALS TO CURVES.

Equation of Tangent,	212
Equation of Normal,	215
Subtangent and Subnormal,	215
Number of Tangents from an External Point,	219
Number of Normals passing through a given Point,	220
Differential of an Arc,	220
Angle between Tangent and Radius Vector,	222
Polar Subtangent and Subnormal,	223
Inverse Curves,	225
Pedal Curves,	227
Reciprocal Polars,	228
Pedal and Reciprocal Polar of $r^m = a^m \cos m\theta$,	230
Intercept between point of Contact and foot of Perpendicular,	232
Direction of Tangent and Normal in Vectorial Coordinates,	233
Symmetrical Curves, and Central Curves,	236
Examples,	28

CHAPTER XIII.

ASYMPTOTES.

Points of Intersection of a Curve and a Right Line,	240
Method of Finding Asymptotes in Cartesian Coordinates,	242
Case where Asymptotes all pass through the Origin,	245
Asymptotes Parallel to Coordinate Axes,	245
Parabolic and Hyperbolic Branches,	246
Parallel Asymptotes,	247
The Points in which a Cubic is cut by its Asymptotes lie in a Right Line,	249
Asymptotes in Polar Curves,	250
Asymptotic Circles,	252
Examples,	254

CHAPTER XIV.

MULTIPLE POINTS ON CURVES.

	PAGE
Nodes, Cusps, Conjugate Points,	259
Method of Finding Double Points in general,	261
Parabolas of the Third Degree,	262
Double Points on a Cubic having three given lines for its Asymptotes,	264
Multiple Points of higher Orders,	265
Cusps, in general,	266
Multiple Points on Curves in Polar Coordinates,	267
Examples,	268

CHAPTER XV.

ENVELOPES.

Method of Envelopes,	270
Envelope of $La^2 + 2Ma + N = 0$,	272
Uundetermined Multipliers applied to Envelopes,	273
Examples,	276

CHAPTER XVI.

CONVEXITY, CONCAVITY, POINTS OF INFLEXION.

Convexity and Concavity,	278
Points of Inflexion,	279
Harmonic Polar of a Point of Inflexion on a Cubic,	281
Stationary Tangents,	282
Examples,	283

CHAPTER XVII.

RADIUS OF CURVATURE, EVOLUTES, CONTACT.

Curvature, Angle of Contingence,	285
Radius of Curvature,	286
Expressions for Radius of Curvature,	287
Newton's Method of considering Curvature,	291
Radii of Curvature of Inverse Curves,	295
Radius and Chord of Curvature in terms of r and p,	295
Chord of Curvature through Origin,	296
Evolutes and Involutes,	297
Evolute of Parabola,	298
Evolute of Ellipse,	299
Evolute of Equiangular Spiral,	300

Involute of a Circle,	300
Radius of Curvature and Points of Inflexion in Polar Coordinates,	301
Intrinsic Equation of a Curve,	304
Contact of Different Orders,	304
Centre of Curvature of an Ellipse,	307
Osculating Curves,	309
Radii of Curvature at a Node,	310
Radii of Curvature at a Cusp,	311
At a Cusp of the Second Species the two Radii of Curvature are equal,	312
General Discussion of Cusps,	315
Points on Evolute corresponding to Cusps on Curve,	316
Equation of Osculating Conic,	317
Examples,	319

CHAPTER XVIII.

ON TRACING OF CURVES.

Tracing Algebraic Curves,	322
Cubic with three real Asymptotes,	323
Each Asymptote corresponds to two Infinite Branches,	325
Tracing Curves in Polar Coordinates,	328
On the Curves $r^m = a^m \cos m\theta$,	328
The Limaçon,	331
The Conchoid,	332
Examples,	333

CHAPTER XIX.

ROULETTES.

Roulettes, Cycloid,	335
Tangent to Cycloid,	336
Radius of Curvature, Evolute,	337
Length of Cycloid,	338
Trochoids,	339
Epicycloids and Hypocycloids,	339
Radius of Curvature of Epicycloid,	342
Double Generation of Epicycloids and Hypocycloids,	343
Evolute of Epicycloid,	344
Pedal of Epicycloid,	346
Epitrochoid and Hypotrochoid,	347
Centre of Curvature of Epitrochoid,	351
Savary's Theorem on Centre of Curvature of a Roulette,	352
Geometrical Construction for Centre of Curvature,	352
Circle of Inflexions,	354
Envelope of a Carried Curve,	355
Centre of Curvature of the Envelope,	357

	PAGE
Radius of Curvature of Envelope of a Right Line,	358
On the Motion of a Plane Figure in its Plane,	359
Chasles' Method of Drawing Normals,	360
Motion of a Plane Figure reduced to Roulettes,	362
Epicyclics,	363
Properties of Circle of Inflexions,	367
Theorem of Bobilier,	368
Centre of Curvature of Conchoid,	370
Spherical Roulettes,	370
Examples,	372

CHAPTER XX.

ON THE CARTESIAN OVAL.

Equation of Cartesian Oval,	375
Construction for Third Focus,	376
Equation, referred to each pair of Foci,	377
Conjugate Ovals are Inverse Curves,	378
Construction for Tangent,	379
Confocal Curves cut Orthogonally,	381
Cartesian Oval as an Envelope,	382
Examples,	384

CHAPTER XXI.

ELIMINATION OF CONSTANTS AND FUNCTIONS.

Elimination of Constants,	384
Elimination of Transcendental Functions,	386
Elimination of Arbitrary Functions,	387
Condition that one expression should be a Function of another,	389
Elimination in the case of Arbitrary Functions of the same expression,	393
Examples,	397

CHAPTER XXII.

CHANGE OF INDEPENDENT VARIABLE.

Case of a Single Independent Variable,	399
Transformation from Rectangular to Polar Coordinates,	403
Transformation of $\dfrac{d^2 V}{dx^2} + \dfrac{d^2 V}{dy^2}$,	404
Transformation of $\dfrac{d^2 V}{dx^2} + \dfrac{d^2 V}{dy^2} + \dfrac{d^2 V}{dz^2}$,	405
Geometrical Illustration of Partial Differentiation,	407

	PAGE
Linear Transformations for Three Variables,	408
Case of Orthogonal Transformations,	409
General Case of Transformation for Two Independent Variables,	410
Functions unaltered by Linear Transformations,	411
Application to Geometry of Two Dimensions,	412
Application to Orthogonal Transformations,	414
Jacobians,	415
Case in which Functions are not Independent,	417
Jacobian of Implicit Functions,	420
Case where $J = 0$,	422
Examples,	427
MISCELLANEOUS EXAMPLES,	430
NOTE ON FAILURE OF TAYLOR'S THEOREM,	443
NOTE ON GENERAL CONDITIONS FOR A MAXIMUM OR MINIMUM,	445

The beginner is recommended to omit the following portions on the *first* reading:—Arts. 49, 50, 51, 52, 67–85, 88, 111, 114–116, 124, 125, Chap. VII., Chap. VIII., Arts. 159–163, 249–254, 261–269, 296–301, Chap. XXII.

CORRIGENDA.

Page 63, line 9, *for* $\dfrac{n(n-1)..(n-r-1)}{1.2...r} x^{n-r} y^r$, *read* $\dfrac{n(n-1)...(n-r+1)}{1.2...r} x^{n-r} y^r$.

,, 165, third line from bottom, *for* $\dfrac{u}{2} \pm \sqrt{\dfrac{a^2}{4} - u}$, *read* $\dfrac{a}{2} \pm \sqrt{\dfrac{a^2}{4} - u}$.

,, 213, line 2, *for* $\dfrac{m+n}{n} y$, *read* $\dfrac{m+n}{m} y$.

,, 228, second line from bottom, *for* Art. 180, *read* Art. 176.

,, 301, fourth line from bottom, *for* Art. 185, *read* Art. 183.

DIFFERENTIAL CALCULUS.

CHAPTER I.

FIRST PRINCIPLES—DIFFERENTIATION.

1. **Functions.**—The student, from his previous acquaintance with Algebra and Trigonometry, is supposed to understand what is meant when one quantity is said to be a function of another. Thus, in trigonometry, the sine, cosine, tangent, &c., of an angle are said to be functions of the angle, having each a single value if the angle is given, and varying when the angle varies. In like manner any algebraic expression in x is said to be a function of x. Geometry also furnishes us with simple illustrations. For instance, the area of a square, or of any regular polygon of a given number of sides, is a function of its side; and the volume of a sphere, of its radius.

In general, whenever two quantities are so related, that *any change made in the one produces a corresponding variation in the other*, then the latter is said to be a function of the former.

This relation between two quantities is usually represented by the letters F, f, ϕ, &c.

Thus the equations

$$u = F(x), \quad v = f(x), \quad w = \phi(x),$$

denote that u, v, w, are regarded as functions of x, whose values are determined for any particular value of x, when the form of the function is known.

2. **Dependent and Independent Variables, Constants.**—In each of the preceding expressions, x is said to be

the *independent* variable, to which any value may be assigned at pleasure; and u, v, w, are called *dependent* variables, as their values depend on that of x, and are determined when it is known.

Thus, in the equations

$$y = 10^x, \quad y = x^3, \quad y = \sin x,$$

the value of y depends on that of x, and is in each case determined when the value of x is given.

If we suppose any series of values, positive or negative, assigned to the independent variable x, then every function of x will assume a corresponding series of values. If a quantity retain the same value, whatever change be given to x, it is said to be a *constant* with respect to x. We usually denote constants by a, b, c, &c., the first letters of the alphabet; variables by the last, viz., u, v, w, x, y, z.

3. Algebraic and Transcendental Functions.—Functions which consist of a finite number of terms, involving integral and fractional powers of x, together with constants solely, are called algebraic functions—thus

$$(x - a)^n, \quad \frac{\sqrt{x^2 - a^2}}{(x^2 + a^2)^{\frac{3}{2}}}, \quad (a + x)(b - x)^{\frac{1}{2}}, \text{ &c.,}$$

are algebraic expressions.

Functions which do not admit of being represented as ordinary algebraic expressions in a *finite number of terms* are called transcendental: thus, $\sin x$, $\cos x$, $\tan x$, e^x, $\log x$, &c., are transcendental functions; for they cannot be expressed in terms of x except by a series containing an *infinite* number of terms.

Algebraic functions are ultimately reducible to the following elementary forms: (1). Sum, or difference $(u + v, u - v)$. (2). Product, and its inverse, quotient $\left(uv, \dfrac{u}{v}\right)$. Powers, and their inverse, roots $(u^m, u^{\frac{1}{m}})$.

The elementary transcendental functions are also ultimately reducible to: (1). The sine, and its inverse, $(\sin u, \sin^{-1} u)$. (2). The exponential, and its inverse, logarithm $(e^u, \log u)$.

4. Continuous Functions.—A function $\phi(x)$ is said to be a *continuous* function of x, between the limits a and b, when, to each value of x, between these limits, corresponds a finite value of the function, and when an infinitely small change in the value of x produces only an infinitely small change in the function. If these conditions be not fulfilled the function is discontinuous. It is easily seen that all algebraic expressions, such as

$$a_0 x^n + a_1 x^{n-1} + \ldots a_n,$$

and all circular expressions, $\sin x$, $\tan x$, &c., are, *in general*, continuous functions, as also e^x, $\log x$, &c. In such cases, accordingly, it follows that if x receive a very small change, the corresponding change in the function of x is also very small.

5. Increments and Differentials.—In the Differential Calculus we investigate the changes which any function undergoes when the variable on which it depends is made to pass through a series of different stages of magnitude.

If the variable x be supposed to receive any change, such change is called an *increment*; this increment of x is usually represented by the notation Δx.

When the increment, or *difference*, is supposed *infinitely small* it is called a *differential*, and represented by dx, i.e. *an infinitely small difference is called a differential*.

In like manner, if u be a function of x, and x becomes $x + \Delta x$, the corresponding value of u is represented by $u + \Delta u$; i. e. the increment of u is denoted by Δu.

6. Limiting Ratios, Derived Functions.—If u be a function of x, then for finite increments, it is obvious that the ratio of the increment of u to the corresponding increment of x has, in general, a finite value. Also when the increment of x is regarded as being infinitely small, we assume that the ratio *above mentioned* has still a definite *limiting* value. In the Differential Calculus we investigate the values of these *limiting ratios* for different forms of functions.

The ratio of the increment of u to that of x in the limit, when both are *infinitely small*, is denoted by $\dfrac{du}{dx}$. When

$u = f(x)$, this limiting ratio is denoted by $f'(x)$, and is called the *first derived function** of $f(x)$.

Thus; let x become $x + h$, where $h = \Delta x$, then u becomes

$$f(x + h), \text{ i. e. } u + \Delta u = f(x + h),$$

$$\therefore \Delta u = f(x + h) - f(x),$$

$$\frac{\Delta u}{\Delta x} = \frac{f(x + h) - f(x)}{h}.$$

The limiting value of this expression when h is infinitely small is called the first derived function of $f(x)$, and represented by $f'(x)$.

Again, since the ratio $\dfrac{\Delta u}{\Delta x}$ has $f'(x)$ for its limiting value, if we assume

$$\frac{\Delta u}{\Delta x} = f'(x) + \varepsilon,$$

ε must become evanescent along with Δx; also $\dfrac{\Delta u}{\Delta x}$ becomes $\dfrac{du}{dx}$ at the same time; hence we have

$$\frac{du}{dx} = f'(x). \qquad (1)$$

This result may be stated otherwise, thus :—If u_1 denote the value of u when x becomes x_1, then the value of the ratio $\dfrac{u_1 - u}{x_1 - x}$, when $x_1 - x$ is evanescent, is called *the first derived function* of u, and denoted by $\dfrac{du}{dx}$.

---

* The method of derived functions was introduced by Lagrange, and the different derived functions of $f(x)$ were defined by him to be, the coefficients of the powers of h in the expansion of $f(x + h)$: that this definition of the first derived function agrees with that given in the text will be seen subsequently.

This agreement was also pointed out by Lagrange. See "Théorie des Fonctions Analytiques," Nos. 3, 9.

If x_1 be greater than x, then u_1 is also greater than u, provided $\dfrac{u_1 - u}{x_1 - x}$ is positive; and hence, in the limit, when $x_1 - x$ is evanescent, u_1 is greater or less than u according as $\dfrac{du}{dx}$ is positive or negative. Hence, if we suppose x to increase, then any function of x increases or diminishes at the same time, according as its derived function, taken with respect to x, is positive or negative. This principle is of great importance in tracing the different stages of a function of x, corresponding to a series of values of x.

7. **Differential, and Differential Coefficient, of** $f(x)$.

Let $u = f(x)$; then since
$$\frac{du}{dx} = f'(x),$$
we have $\quad du = d\left(f(x)\right) = f'(x)\,dx,$

where dx is regarded as being infinitely small. In this case dx is, as already stated, the *differential* of x, and du or $f'(x)\,dx$, is called the corresponding *differential* of u. Also $f'(x)$ is called the *differential coefficient* of $f(x)$, being the coefficient of dx in the differential of $f(x)$.

8. **Algebraic Illustration.**—That a fraction whose numerator and denominator are both evanescent, or infinitely small, may have a finite determinate value, is evident from algebra. For example, we have $\dfrac{a}{b} = \dfrac{na}{nb}$ whatever n may be. If n be regarded as an *infinitely small* number, the numerator and denominator of the fraction both become infinitely small magnitudes, while their ratio remains unaltered and equal to $\dfrac{a}{b}$.

It will be observed that this agrees with our ordinary idea of a ratio; for the value of a ratio depends on the *relative*, and not on the *absolute* magnitude of the terms which compose it.

Again, if
$$u = \frac{na + n^2 a'}{nb + n^2 b'},$$

in which n is regarded as infinitely small, and a, b, a' and b'

represent finite magnitudes, the terms of the fraction are both infinitely small,

but their ratio is $\dfrac{a + na'}{b + nb'}$,

the *limiting value* of which, as n is diminished indefinitely, is $\dfrac{a}{b}$. Again, if we suppose n *indefinitely increased*, the limiting value of the fraction is $\dfrac{a'}{b'}$. For

$$\frac{a + a'n}{b + b'n} = \frac{a'}{b'} + \frac{ab' - ba'}{b'(b + b'n)};$$

but the fraction $\dfrac{ab' - ba'}{b'(b + b'n)}$ diminishes indefinitely as n increases indefinitely, and may be made less than any assignable magnitude, however small. Accordingly the *limiting value* of the fraction in this case is $\dfrac{a'}{b'}$.

9. **Trigonometrical Illustration.**—To find the values of $\dfrac{\sin \theta}{\tan \theta}$, and $\dfrac{\sin \theta}{\theta}$, when θ is regarded as infinitely small.

Here $\dfrac{\sin \theta}{\tan \theta} = \cos \theta$, and when $\theta = 0$, $\cos \theta = 1$.

Hence, in the limit, when $\theta = 0$,* we have $\dfrac{\sin \theta}{\tan \theta} = 1$, and, $\therefore \dfrac{\tan \theta}{\sin \theta} = 1$, at the same time.

Again, to find the value of $\dfrac{\theta}{\sin \theta}$, when θ is infinitely small. From geometrical considerations it is evident that if θ be the circular measure of an angle, we have

$$\tan \theta > \theta > \sin \theta,$$

or $\dfrac{\tan \theta}{\sin \theta} > \dfrac{\theta}{\sin \theta} > 1;$

---

* *If a variable quantity be supposed to diminish gradually, till it be less than anything finite which can be assigned, it is said in that state to be indefinitely small or evanescent; for abbreviation, such a quantity is often denoted by cypher.*

A discussion of infinitesimals, or infinitely small quantities of different orders, will be found in the next Chapter.

but in the limit, i.e. when θ is infinitely small,

$$\frac{\tan \theta}{\sin \theta} = 1,$$

and therefore, at the same time, we have

$$\frac{\theta}{\sin \theta} = 1.$$

This shows that in a circle the ultimate ratio of an arc to its chord is unity, when they are both regarded as evanescent.

10. **Geometrical Illustration.**—Assuming that the relation $y = f(x)$ may in all cases be represented by a curve, where

$$y = f(x)$$

expresses the equation connecting the co-ordinates (x, y) of each of its points; then, if the axes be rectangular, and two points (x, y), (x_1, y_1) be taken on the curve, it is obvious that $\frac{y_1 - y}{x_1 - x}$ represents the tangent of the angle which the chord joining the points (x, y), (x_1, y_1) makes with the axis of x.

If, now, we suppose the points taken *infinitely near* to each other, so that $x_1 - x$ becomes evanescent, then the chord becomes the *tangent* at the point (x, y), but

$$\frac{y_1 - y}{x_1 - x} \text{ becomes } \frac{dy}{dx} \text{ or } f'(x) \text{ in this case.}$$

Hence, $f'(x)$ represents the *trigonometrical tangent of the angle which the line touching the curve at the point (x, y) makes with the axis of x*. We see, accordingly, that to draw the tangent at any point to the curve

$$y = f(x)$$

is the same as to find the *derived function* $f'(x)$ of y with respect to x. Hence, also, the equation of the tangent to the curve at a point (x, y) is evidently

$$y - Y = f'(x)(x - X), \qquad (2)$$

where X, Y are the current co-ordinates of any point on the tangent. At the points for which the tangent is parallel to the axis of x, we have $f'(x) = 0$; at the points where the tangent is perpendicular to the axis, $f'(x) = \infty$. For all other points $f'(x)$ has a *determinate finite* real value in general. This conclusion verifies the statement, that the ratio of the increment of the dependent variable to that of the independent variable has, in general, a *finite determinate* magnitude, when the increment becomes infinitely small.

This has been so admirably expressed, and its connexion with the fundamental principles of the Differential Calculus so well explained, by M. Navier, that I cannot forbear introducing the following extract from his "Leçons d'Analyse":—

"Among the properties which the function $y = f(x)$, or the line which represents it, possesses, the most remarkable—in fact that which is the principal object of the Differential Calculus, and which is constantly introduced in all practical applications of the Calculus—is the degree of rapidity with which the function $f(x)$ varies when the independent variable x is made to vary from any assigned value. This degree of rapidity of the increment of the function, when x is altered, may differ, not only from one function to another, but also in the same function, according to the value attributed to the variable. In order to form a precise notion on this point, let us attribute to x a determined value represented by ON, to which will correspond an equally determined value of y, represented by PN. Let us now suppose, starting from this value, that x increases by any quantity denoted by Δx, and represented by NM, the function y will vary in consequence by a certain quantity, denoted by Δy, and we shall have

$$y + \Delta y = f(x + \Delta x), \quad \text{or} \quad \Delta y = f(x + \Delta x) - f(x).$$

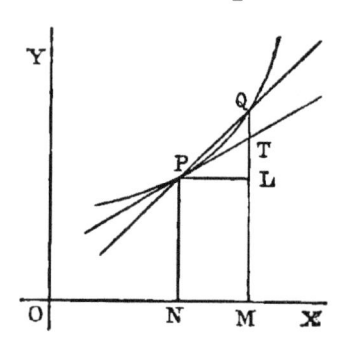

Fig. 1.

The new value of y is represented in the figure by QM, and QL represents Δy, or the variation of the function.

The ratio $\frac{\Delta y}{\Delta x}$ of the increment of the function to that of the independent variable, of which the expression is

$$\frac{f(x + \Delta x) - f(x)}{\Delta x},$$

is represented by the trigonometrical tangent of the angle QPL made by the secant PQ with the axis of x.

"It is plain that this ratio $\frac{\Delta y}{\Delta x}$ is the natural expression of the property referred to, that is, of the degree of rapidity with which the function y increases when we increase the independent variable x; for the greater the value of this ratio, the greater will be the increment Δy when x is increased by a given quantity Δx. But it is very important to remark, that the value of $\frac{\Delta y}{\Delta x}$ (except in the case when the line PQ becomes a right line) depends not only on the value attributed to x, that is to say, on the position of P on the curve, but also on the *absolute value of the increment* Δx. If we were to leave this increment arbitrary, it would be impossible to assign to the ratio $\frac{\Delta y}{\Delta x}$ any precise value, and it is accordingly necessary to adopt a convention which shall remove all uncertainty in this respect.

"Suppose that after having given to Δx any value, to which will correspond a certain value Δy and a certain direction of the secant PQ, we diminish progressively the value of Δx, so that the increment ends by becoming evanescent; the corresponding increment Δy will vary in consequence, and will equally tend to become evanescent. The point Q will tend to coincide with the point P, and the secant PQ with the tangent PT drawn to the curve at the point P. The ratio $\frac{\Delta y}{\Delta x}$ of the increments will equally approach to a certain limit, represented by the trigonometrical tangent of the angle TPL made by the tangent with the axis of x.

"We accordingly observe that when the increment Δx,

and consequently Δy, diminish progressively and tend to vanish, the ratio $\frac{\Delta y}{\Delta x}$ of these increments approaches in general to a limit whose value is finite and determinate. Hence the value of $\frac{\Delta y}{\Delta x}$ corresponding to this limit must be considered as giving the true and precise measure of the *rapidity with which the function $f(x)$ varies when the independent variable x is made to vary* from an assigned value; for there does not remain anything arbitrary in the expression of this value, as it no longer depends on the absolute values of the increments Δx and Δy, nor on the figure of the curve at any finite distance at either side of the point P. It depends solely on the *direction* of the curve at this point, that is, on the inclination of the tangent to the axis of x. The ratio just determined expresses what Newton called the *fluxion* of the ordinate. As to the mode of finding its value in each particular case, it is sufficient to consider the general expression

$$\frac{\Delta y}{\Delta x} = \frac{f(x + \Delta x) - f(x)}{\Delta x},$$

and to see what is the limit to which this expression tends, as Δx takes smaller and smaller values and tends to vanish. This limit will be a certain function of the independent variable x, whose form depends on that of the given function $f(x)$..... We shall add one other remark; which is, that the differentials represented by dx and dy denote always quantities of the same nature as those denoted by the variables x and y. Thus in geometry, when x represents a line, an area, or a volume, the differential dx also represents a line, an area, or a volume. These differentials are always supposed to be less than any assigned magnitude, however small; but this hypothesis does not alter the nature of these quantities: dx and dy are always homogeneous with x and y, that is to say, present always the same number of dimensions of the unit by means of which the values of these variables are expressed."

10*a*. **Limit of a Variable Magnitude.**—As the conception of a limit is fundamental in the Calculus, it may be well to add a few remarks in further elucidation of its meaning:—

In general, *when a variable magnitude tends continually to equality with a certain fixed magnitude, and approaches nearer to it than any assignable difference, however small, this fixed magnitude is called the limit of the variable magnitude.* For example, if we inscribe, or circumscribe, a polygon to any closed curve, and afterwards conceive each side indefinitely diminished, and consequently their number indefinitely increased, then the closed curve is said to be the *limit of either polygon*. By this means the total length of the curve is the limit of the perimeter either of the inscribed or circumscribed polygon. In like manner, the area of the curve is the limit to the area of either polygon. For instance, since the area of any polygon circumscribed to a circle is obviously equal to the rectangle under the radius of the circle and the semi-perimeter of the polygon, it follows that the area of a circle is represented by the product of its radius and its semi-circumference. Again, since the length of the side of a regular polygon inscribed in a circle bears to that of the corresponding arc the same ratio as the perimeter of the polygon to the circumference of the circle, it follows that the ultimate ratio of the chord to the arc is one of equality, as shown in Art. 9. The like result follows immediately for any curve.

The following principles concerning limits are of frequent application:—(1) *The limit of the product of two quantities, which vary together, is the product of their limits;* (2) *The limit of the quotient of the quantities is the quotient of their limits.*

For, let P and Q represent the two quantities, and p and q their respective limits; then if

$$P = p + a, \qquad Q = q + \beta,$$

a and β denote quantities which diminish indefinitely as P and Q approach their limits, and which become evanescent in the limit.

Again, we have

$$PQ = pq + p\beta + qa + a\beta.$$

Accordingly, in the limit, we have

$$PQ = pq.$$

Again, $\dfrac{P}{Q} = \dfrac{p+a}{q+\beta} = \dfrac{p}{q} + \dfrac{qa - p\beta}{q(q+\beta)}.$

The numerator of the last fraction becomes evanescent in the limit, while the denominator becomes q^2, and consequently the limit of $\dfrac{P}{Q}$ is $\dfrac{p}{q}$.

11. Differentiation.—The process of finding the derived function, or the differential coefficient of any expression, is called *differentiating* the expression.

We proceed to explain this process by applying it to a few elementary examples.

EXAMPLES.

1. $\qquad y = x^2.$

Substitute $x + h$ for x, and denote the new value of y by y_1, then

$$y_1 = (x+h)^2 = x^2 + 2xh + h^2;$$

$$\therefore \frac{y_1 - y}{h} \text{ or } \frac{\Delta y}{\Delta x} = 2x + h.$$

If h be taken an infinitely small quantity, we get in the limit

$$\frac{dy}{dx} = 2x;$$

or if $\qquad f(x) = x^2$, we have $f'(x) = 2x.$

2. $\qquad y = \dfrac{1}{x}.$

Here $\qquad y_1 = \dfrac{1}{x+h}.$

$$y_1 - y = \frac{1}{x+h} - \frac{1}{x} = -\frac{h}{x(x+h)};$$

$$\therefore \frac{y_1 - y}{h}, \text{ or } \frac{\Delta y}{\Delta x} = -\frac{1}{x(x+h)},$$

which equation, when h is evanescent, becomes

$$\frac{dy}{dx} = -\frac{1}{x^2}, \text{ or } \frac{d\left(\frac{1}{x}\right)}{dx} = -\frac{1}{x^2}.$$

12. Differentiation of the Algebraic Sum of a Finite Number of Functions.—Let

$$y = u + v - w + \&c.;$$

then, if $x_1 = x + h$, we get

$$y_1 = u_1 + v_1 - w_1 + \ldots;$$

$$\therefore \frac{y_1 - y}{h} = \frac{u_1 - u}{h} + \frac{v_1 - v}{h} - \frac{w_1 - w}{h} + \ldots,$$

which becomes in the limit, when h is infinitely small,

$$\frac{dy}{dx} = \frac{du}{dx} + \frac{dv}{dx} - \frac{dw}{dx} + \ldots$$

Hence, if a function consist of several terms, *its derived function is the sum of the derived functions of its several parts, taken with their proper signs.*

It is evident that the differential of a constant is zero.

13. Differentiation of the Product of Two Functions.—Let $y = uv$, where u, v, are both functions of x; and suppose Δy, Δu, Δv, to be the increments of y, u, v, corresponding to the increment Δx in x. Then

$$\Delta y = (u + \Delta u)(v + \Delta v) - uv$$

$$= u\Delta v + v\Delta u + \Delta u \, \Delta v,$$

or

$$\frac{\Delta y}{\Delta x} = u \frac{\Delta v}{\Delta x} + (v + \Delta v) \frac{\Delta u}{\Delta x}.$$

Now suppose Δx to be infinitely small, then

$$\frac{\Delta y}{\Delta x}, \frac{\Delta v}{\Delta x}, \frac{\Delta u}{\Delta x},$$

become in the limit

$$\frac{dy}{dx}, \frac{dv}{dx}, \text{ and } \frac{du}{dx};$$

also, since Δv vanishes at the same time, the last term disappears from the equation, and thus we arrive at the result

$$\frac{dy}{dx} = u \frac{dv}{dx} + v \frac{du}{dx}. \qquad (3)$$

Hence, to differentiate the product of two functions, *multiply each of the factors by the differential coefficient of the other, and add the products thus found.*

Otherwise thus: let $f(x)$, $\phi(x)$, denote the functions, and h the increment of x, then

$$y_1 = f(x+h)\,\phi(x+h);$$

$$\therefore \frac{y_1 - y}{h} = \frac{f(x+h)\,\phi(x+h) - f(x)\,\phi(x)}{h}$$

$$= \frac{f(x+h) - f(x)}{h}\,\phi(x+h) + f(x)\,\frac{\phi(x+h) - \phi(x)}{h}.$$

Now, in the limit,

$$\frac{f(x+h) - f(x)}{h} = f'(x), \quad \phi(x+h) = \phi(x),$$

and

$$\frac{\phi(x+h) - \phi(x)}{h} = \phi'(x),$$

and, accordingly,

$$\frac{dy}{dx} = f(x)\,\phi'(x) + \phi(x)\,f'(x),$$

which agrees with the preceding result.

When $y = au$, where a is a constant with respect to x, we have evidently

$$\frac{dy}{dx} = a\frac{du}{dx}.$$

14. Differentiation of the Product of any Number of Functions.—First let

$$y = uvw;$$

suppose

$$vw = z,$$

then

$$y = uz,$$

and, by Art. 13, we have

$$\frac{dy}{dx} = u\frac{dz}{dx} + z\frac{du}{dx};$$

but, by the same Article,
$$\frac{dz}{dx} = w\frac{dv}{dx} + v\frac{dw}{dx};$$
hence
$$\frac{dy}{dx} = vw\frac{du}{dx} + wu\frac{dv}{dx} + uv\frac{dw}{dx}.$$

This process of reasoning can be easily extended to any number of functions.

The preceding result admits of being written in the form
$$\frac{1}{y}\frac{dy}{dx} = \frac{1}{u}\frac{du}{dx} + \frac{1}{v}\frac{dv}{dx} + \frac{1}{w}\frac{dw}{dx},$$
and in general, if $y = y_1 \cdot y_2 \cdot y_3 \ldots y_n$, it can be easily proved in like manner that
$$\frac{1}{y}\frac{dy}{dx} = \frac{1}{y_1}\frac{dy_1}{dx} + \frac{1}{y_2}\frac{dy_2}{dx} \ldots + \frac{1}{y_n}\frac{dy_n}{dx}. \qquad (4)$$

15. Differentiation of a Quotient—Let
$$y = \frac{u}{v}, \text{ then } u = yv;$$
therefore, by Art. 13,
$$\frac{du}{dx} = y\frac{dv}{dx} + v\frac{dy}{dx},$$
or
$$v\frac{dy}{dx} = \frac{du}{dx} - y\frac{dv}{dx} = \frac{du}{dx} - \frac{u}{v}\frac{dv}{dx}$$
$$= \frac{v\frac{du}{dx} - u\frac{dv}{dx}}{v};$$
$$\therefore \frac{dy}{dx} = \frac{v\frac{du}{dx} - u\frac{dv}{dx}}{v^2}. \qquad (5)$$

This may be written in the following form, which is often useful:
$$\frac{d}{dx}\left(\frac{u}{v}\right) = \frac{1}{v}\frac{du}{dx} - \frac{u}{v^2}\frac{dv}{dx}.$$

Hence, to differentiate a fraction, *multiply the denominator into the derived function of the numerator, and the numerator into the derived function of the denominator; take the latter product from the former, and divide by the square of the denominator.*

In the particular case where u is a constant with respect to x (a suppose), we obviously have

$$\frac{d}{dx}\left(\frac{a}{v}\right) = -\frac{a}{v^2}\frac{dv}{dx}. \qquad (6)$$

EXAMPLES.

1. $u = \dfrac{a-x}{a+x}.$ *Ans.* $\dfrac{du}{dx} = -\dfrac{2a}{(a+x)^2}.$

2. $u = (a+x)(b+x).$ $\dfrac{du}{dx} = a+b+2x.$

16. Differentiation of an Integral Power.—Let $y = x^n$, where n is a *positive* integer.

Suppose y_1 to be the value of y, when x becomes x_1, then

$$\frac{y_1 - y}{x_1 - x} = \frac{x_1^n - x^n}{x_1 - x} = x_1^{n-1} + xx_1^{n-2} + \ldots + x^{n-1}.$$

Now, suppose $x_1 - x$ to be evanescent. In this case we may write x for x_1 in the right-hand side of the preceding equation, when it becomes nx^{n-1}; but the left-hand side, in the limit, is represented by $\dfrac{dy}{dx}$.

Hence
$$\frac{dy}{dx} = nx^{n-1},$$

or
$$\frac{d(x^n)}{dx} = nx^{n-1}.$$

This result follows also from Art. 14; for, making

$$y_1 = y_2 = y_3 = \ldots = y_n = u,$$

we evidently get from (4),

$$\frac{d(u^n)}{dx} = nu^{n-1}\frac{du}{dx}. \qquad (7)$$

This reduces to the preceding on making $u = x$.

Differentiation of a Function of a Function.

17. Differentiation of a Fractional Power.—Let $y = u^{\frac{m}{n}}$,

then $\quad y^n = u^m$, and $\dfrac{d(y^n)}{dx} = \dfrac{d(u^m)}{dx}$;

hence, by (7),

$$ny^{n-1}\frac{dy}{dx} = mu^{m-1}\frac{du}{dx};$$

$$\therefore \frac{d(u^{\frac{m}{n}})}{dx} = \frac{dy}{dx} = \frac{m}{n}\frac{u^{m-1}}{y^{n-1}}\frac{du}{dx} = \frac{m}{n}u^{\frac{m}{n}-1}\frac{du}{dx}. \qquad (8)$$

18. Differentiation of a Negative Power.—Let $y = u^{-m}$, then $y = \dfrac{1}{u^m}$, and by (6) we get

$$\frac{d}{dx}(u^{-m}) = -\frac{mu^{m-1}\frac{du}{dx}}{u^{2m}} = -mu^{-m-1}\frac{du}{dx}. \qquad (9)$$

Combining the results established in (7), (8), and (9), we find that

$$\frac{d(u^m)}{dx} = mu^{m-1}\frac{du}{dx}$$

for all values of m, positive, negative, or fractional. When applied to the differentiation of any power of x we get the following rule:—*Diminish the index by unity, and multiply the power of x thus obtained by the original index;* the result is the required differential coefficient, with respect to x.

19. Differentiation of a Function of a Function.—Let $y = f(x)$ and $u = \phi(y)$, to find $\dfrac{du}{dx}$. Suppose y_1, u_1, to be the values of y and u corresponding to the value x_1 for x; then if Δy, Δu, Δx, denote the corresponding increments, we have evidently

$$\frac{u_1 - u}{x_1 - x} = \frac{u_1 - u}{y_1 - y} \cdot \frac{y_1 - y}{x_1 - x},$$

or

$$\frac{\Delta u}{\Delta x} = \frac{\Delta u}{\Delta y}\frac{\Delta y}{\Delta x}.$$

As this relation holds for all corresponding increments, however small, it must hold in the limit,* when Δx is evanescent; in which case it becomes

$$\frac{du}{dx} = \frac{du}{dy}\frac{dy}{dx}. \qquad (10)$$

Hence *the derived function with respect to x of u is the product of its derived with respect to y; and the derived of y with respect to x.*

20. Differentiation of an Inverse Function.—To prove that

$$\frac{dx}{dy} = \frac{1}{\frac{dy}{dx}}.$$

Suppose that from the equation

$$y = f(x) \qquad (a)$$

the equation

$$x = \phi(y) \qquad (b)$$

is deduced, and let x_1, y_1, be corresponding values of x, y, which satisfy the equation (a), it is evident that they will also satisfy the equation (b). But

$$\frac{y_1 - y}{x_1 - x} \times \frac{x_1 - x}{y_1 - y} = 1.$$

As this equation holds for all finite increments, it must hold when $x_1 - x$ and $y_1 - y$ are infinitely small; therefore we have in the limit

$$\frac{dy}{dx}\cdot\frac{dx}{dy} = 1. \qquad (11)$$

The same result may also be arrived at from Art. 19, as follows:—

When $y = f(x)$, and $u = \phi(y)$,

---

* The Student will observe that this is a case of the principle (Art. 10a) that the limit of the product of two quantities is equal to the product of their limits.

we have, in all cases,
$$\frac{du}{dx} = \frac{du}{dy}\frac{dy}{dx}.$$

This result must still hold in the particular case when $u = x$, in which case it becomes
$$1 = \frac{dx}{dy}\frac{dy}{dx}.$$

EXAMPLES.

1. $u = (a^2 - x^2)^5$.

Let $a^2 - x^2 = y$, then $u = y^5$,

$$\frac{du}{dy} = 5y^4, \text{ and } \frac{dy}{dx} = -2x.$$

Hence $\frac{du}{dx} = -10x(a^2 - x^2)^4$.

2. $u = (a + bx^3)^4$.   Ans. $\frac{du}{dx} = 12bx^2(a + bx^3)^3$.

3. $u = (1 + x^2)^{\frac{1}{2}}$. $\frac{du}{dx} = \frac{x}{(1 + x^2)^{\frac{1}{2}}}$.

4. $u = (1 + x^n)^m$. $\frac{du}{dx} = mnx^{n-1}(1 + x^n)^{m-1}$.

We next proceed to determine the derived functions of the elementary trigonometrical and circular functions.

21. **Differentiation of** $\sin x$.—Let

$$y = \sin x, \quad y_1 = \sin(x + h),$$

$$\frac{y_1 - y}{h} = \frac{\sin(x + h) - \sin x}{h} = \frac{2\sin\frac{h}{2}\cos\left(x + \frac{h}{2}\right)}{h}.$$

But by Art. 9, the limit of $\dfrac{\sin\frac{h}{2}}{\frac{h}{2}} = 1$; moreover, the limit of $\cos\left(x + \dfrac{h}{2}\right)$ is $\cos x$.

Hence
$$\frac{d(\sin x)}{dx} = \cos x. \qquad (12)$$

22. Differentiation of $\cos x$.

$$y = \cos x, \quad y_1 = \cos(x+h),$$

$$\frac{y_1 - y}{h} = \frac{\cos(x+h) - \cos x}{h} = -\frac{2\sin\dfrac{h}{2}\sin\left(x+\dfrac{h}{2}\right)}{h}.$$

Hence, in the limit,

$$\frac{d\cos x}{dx} = -\sin x. \qquad (13)$$

This result might be deduced from the preceding, by substituting $\dfrac{\pi}{2} - z$ for x, and applying the principle of Art. 19.

It may be noted that (12) and (13) admit also of being written in the following symmetrical form:—

$$\frac{d\sin x}{dx} = \sin\left(x + \frac{\pi}{2}\right),$$

$$\frac{d\cos x}{dx} = \cos\left(x + \frac{\pi}{2}\right).$$

23. Differentiation of $\tan x$.

$$y = \tan x, \quad y_1 = \tan(x+h),$$

$$\frac{y_1 - y}{h} = \frac{\tan(x+h) - \tan x}{h} = \frac{\dfrac{\sin(x+h)}{\cos(x+h)} - \dfrac{\sin x}{\cos x}}{h}$$

$$= \frac{\sin h}{h \cos x \cos(x+h)},$$

which becomes $\dfrac{1}{\cos^2 x}$ in the limit.

Hence $\dfrac{d(\tan x)}{dx} = \dfrac{1}{\cos^2 x} = \sec^2 x.$ \hfill (14)

Otherwise thus,

$$\frac{d(\tan x)}{dx} = \frac{d \cdot \dfrac{\sin x}{\cos x}}{dx} = \frac{\cos x \dfrac{d \sin x}{dx} - \sin x \dfrac{d \cos x}{dx}}{\cos^2 x}$$

$$= \frac{\cos^2 x + \sin^2 x}{\cos^2 x} = \frac{1}{\cos^2 x}.$$

24. Differentiation of $\cot x$.—Proceed as in the last, and we get $\dfrac{d(\cot x)}{dx} = -\dfrac{1}{\sin^2 x} = -\operatorname{cosec}^2 x.$ \hfill (15)

This result can also be derived from the preceding, by putting $\dfrac{\pi}{2} - z$ for x, as in Art. 22.

25. Differentiation of $\sec x$.

$$y = \sec x = \frac{1}{\cos x};$$

$$\therefore \frac{dy}{dx} = \frac{\sin x}{\cos^2 x} = \tan x \sec x. \hfill (16)$$

Similarly $\dfrac{d \operatorname{cosec} x}{dx} = - \cot x \operatorname{cosec} x.$

26. Differentiation of $y = \sin^{-1} x$.

Here $x = \sin y,\ \therefore \dfrac{dx}{dy} = \cos y.$

Hence, by Art. 20, we get

$$\frac{dy}{dx} = \frac{1}{\cos y} = \pm \frac{1}{\sqrt{1-x^2}}.$$

The ambiguity of the sign in this case arises from the ambiguity of the expression $y = \sin^{-1} x$; for if y satisfy this equation for a particular value of x, so also does $\pi - y$; as also $2\pi + y$, &c. If, however, we assign always to y its *least value*, i.e. the *acute* angle whose sine is represented by x, then the sign of the differential coefficient is determinate, and is evidently positive; since an angle increases with its sine, so long as it is *acute*. Accordingly, with the preceding limitation,

$$\frac{d . \sin^{-1} x}{dx} = \frac{1}{\sqrt{1-x^2}}. \qquad (17)$$

In like manner we find

$$\frac{d . \cos^{-1} x}{dx} = -\frac{1}{\sqrt{1-x^2}}, \qquad (18)$$

with the same limitation.

This latter result can be at once deduced from the preceding by aid of the elementary equation

$$\sin^{-1} x + \cos^{-1} x = \frac{\pi}{2}.$$

27. Differentiation of $\tan^{-1} x$.

$$y = \tan^{-1} x, \therefore x = \tan y;$$

hence
$$\frac{dx}{dy} = \frac{1}{\cos^2 y};$$

$$\therefore \frac{d . \tan^{-1} x}{dx} = \frac{dy}{dx} = \cos^2 y = \frac{1}{1+x^2}. \qquad (19)$$

Similarly,
$$\frac{d . \cot^{-1} x}{dx} = -\frac{1}{1+x^2}.$$

28. Geometrical Demonstration.—The results arrived at in the preceding Articles admit also of easy demonstration by geometrical construction. We shall illustrate this method by applying it to the case of $\sin\theta$.

Suppose $XPQY$ to be a quadrant of a circle having O as its centre, and construct as in figure. Let θ denote the angle XOP expressed in circular measure; then

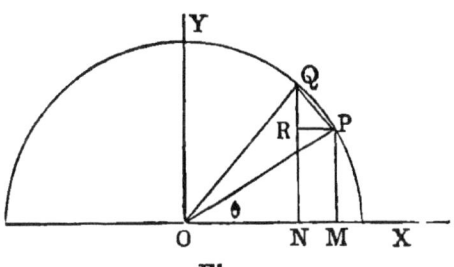

Fig. 2.

$$\theta = \frac{\text{arc } PX}{OP}, \text{ and } h = \Delta\theta = \frac{\text{arc } PQ}{OP}.$$

Accordingly,

$$\sin(\theta + h) - \sin\theta = \frac{QR}{OP} = \frac{QR}{PQ} \cdot \frac{PQ}{OP} = \cos PQR \cdot \frac{PQ}{OP};$$

$$\therefore \frac{\sin(\theta + h) - \sin\theta}{h} = \cos PQR \cdot \frac{PQ}{\text{arc } PQ}.$$

But we have seen, in Art. 9, that the limiting value of $\dfrac{PQ}{\text{arc } PQ}$ $= 1$; also $PQR = \theta$, at the same time; hence $\dfrac{d \sin\theta}{d\theta} = \cos\theta$, as before.

The student will find no difficulty in applying the preceding construction to the differentiation of $\cos\theta$, $\sin^{-1}\theta$, and $\cos^{-1}\theta$. The differential coefficients of $\tan\theta$, $\tan^{-1}\theta$, &c., can, in like manner, be easily obtained by geometrical construction.

EXAMPLES.

1. $y = \sin(nx + a)$. $\quad \dfrac{dy}{dx} = n\cos(nx + a)$.

2. $y = \cos mx \cos nx$. $\quad \dfrac{dy}{dx} = -(m\cos nx \sin mx + n\cos mx \sin nx)$.

3. $y = \sin^n x$. $\quad \dfrac{dy}{dx} = n\sin^{n-1} x \cos x$.

4. $y = \sin(1 + x^2)$. $\quad \dfrac{dy}{dx} = 2x \cos(1 + x^2)$.

5. Show that $\sin^2 x \dfrac{d}{dx}(\sin^m x \sin mx) = m \sin^{m+1} x \sin(m+1)x$.

Here $\dfrac{d}{dx}(\sin^m x \sin mx) = m \sin^{m-1} x (\cos x \sin mx + \sin x \cos mx)$
$\qquad\qquad\qquad = m \sin^{m-1} x \sin(m+1)x$; ∴ &c.

6. $y = (a \sin^2 x + b \cos^2 x)^n$. $\quad \dfrac{dy}{dx} = n(a-b) \sin 2x\, (a \sin^2 x + b \cos^2 x)^{n-1}$.

7. $y = \sin(\sin x)$.

Or $y = \sin u$, where $u = \sin x$. $\quad \dfrac{dy}{dx} = \cos x \cos(\sin x)$.

8. $y = \sin^{-1}(x^n)$. $\quad \dfrac{dy}{dx} = \dfrac{nx^{n-1}}{(1 - x^{2n})^{\frac{1}{2}}}$.

9. $y = \sin^{-1}(1 - x^2)^{\frac{1}{2}}$.

Here $(1 - x^2)^{\frac{1}{2}} = \sin y$; ∴ $x = \cos y$.

$1 = -\sin y \dfrac{dy}{dx}$; $\quad ∴ \dfrac{dy}{dx} = -\dfrac{1}{\sqrt{1-x^2}}$.

10. $y = \cos^{-1} \dfrac{b + a \cos x}{a + b \cos x}$. $\quad \dfrac{dy}{dx} = \dfrac{\sqrt{a^2 - b^2}}{a + b \cos x}$.

11. $y = \sec^n x$. $\quad \dfrac{dy}{dx} = n \sec^n x \tan x$.

12. $y = \sec^{-1}(x^2)$. $\quad \dfrac{dy}{dx} = \dfrac{2}{x \sqrt{x^4 - 1}}$.

29. **Differentiation of $\log_a x$.**

Let $\qquad y = \log_a x, \quad y_1 = \log_a(x + h)$,

$$\dfrac{y_1 - y}{h} = \dfrac{\log_a(x+h) - \log_a x}{h} = \dfrac{\log_a\left(1 + \dfrac{h}{x}\right)}{h}.$$

Hence $\dfrac{dy}{dx}$ is equal to the limiting value of

$$\dfrac{1}{h} \log_a\left(1 + \dfrac{h}{x}\right),$$

when h is infinitely small.

Again, let $h = xu$, then

$$\dfrac{1}{h}\log_a\left(1 + \dfrac{h}{x}\right) = \dfrac{1}{x}\dfrac{\log_a(1+u)}{u} = \dfrac{1}{x}\log_a(1+u)^{\frac{1}{u}},$$

∴ $\dfrac{dy}{dx} = \dfrac{1}{x}$ multiplied by the value of $\log_a (1 + u)^{\frac{1}{u}}$ when u is infinitely small.

To find the value of the latter expression, let $\dfrac{1}{u} = z$, then $(1 + u)^{\frac{1}{u}}$ becomes $\left(1 + \dfrac{1}{z}\right)^z$, in which z is regarded as infinitely great. Suppose the *limiting value of this expression to be represented by the letter* e, according to the usual notation. We can then find the value of e as follows by the Binomial Theorem :—

$$\left(1 + \dfrac{1}{z}\right)^z = 1 + \dfrac{z}{1} \cdot \dfrac{1}{z} + \dfrac{z(z-1)}{1 \cdot 2} \cdot \dfrac{1}{z^2} + \ldots$$

$$= 1 + \dfrac{1}{1} + \dfrac{\left(1 - \dfrac{1}{z}\right)}{1 \cdot 2} + \dfrac{\left(1 - \dfrac{1}{z}\right)\left(1 - \dfrac{2}{z}\right)}{1 \cdot 2 \cdot 3} + \&c.$$

The limiting* value of which, when $z = \infty$, is evidently

$$1 + \dfrac{1}{1} + \dfrac{1}{1 \cdot 2} + \dfrac{1}{1 \cdot 2 \cdot 3} + \dfrac{1}{1 \cdot 2 \cdot 3 \cdot 4} + \&c.$$

By taking a sufficient number of terms of this series, we can approximate to the value of e as nearly as we please. The ultimate value can be shown to be an incommensurable quantity, and is the base of the natural or Napierian system of logarithms. When taken to nine decimal places, its value is 2.718281828.

Again, since $(1 + u)^{\frac{1}{u}} = e$ when $u = 0$, we get

$$\dfrac{d \cdot \log_a x}{dx} = \dfrac{\log_a e}{x}. \qquad (20)$$

Also, since the calculation of logarithms to any other base starts from the logarithms of some numbers to the base e;

---

* It will be shown in Chapter 3, without assuming the Binomial expansion, that e is the limit of the sum of the series

$$1 + \dfrac{1}{1} + \dfrac{1}{1 \cdot 2} + \dfrac{1}{1 \cdot 2 \cdot 3} + \&c., \textit{ad infinitum}.$$

and moreover, since the logarithms of all numbers are expressed by their logarithms to the base e multiplied by the modulus of transformation, the system whose base is e is fundamental in analysis, and we shall denote it by the symbol log without a suffix. In this case, since $\log e = 1$, we have

$$\frac{d}{dx}(\log x) = \frac{1}{x}. \qquad (21)$$

Again,

$$\frac{d}{dx}(\log_{10} x) = \frac{\log_{10} e}{x} = \frac{M}{x}, \qquad (22)$$

where M or $\log_{10} e$ is the modulus of Briggs' or the ordinary tabulated system of logarithms. The value of this modulus, when calculated to ten decimal places, is

$$0.4342944819.$$

On the method of its determination see Galbraith's "Algebra," p. 379.

If x be a large number, it is evident, from the preceding, that the *tabular difference* (as given in Logarithmic Tables), i.e. the difference between $\log_{10}(x+1)$ and $\log_{10} x$, is $\frac{M}{x}$, approximately. The student can readily verify this result by reference to the Tables.

30. **Differentiation of a^x.**

Let $y = a^x$, then $\log y = x \log a$;

$$\therefore \frac{d(\log y)}{dx} = \log a;$$

but

$$\frac{d(\log y)}{dx} = \frac{d(\log y)}{dy}\frac{dy}{dx} = \frac{1}{y}\frac{dy}{dx};$$

$$\therefore \frac{d \cdot a^x}{dx} = \frac{dy}{dx} = y \log a = a^x \log a. \qquad (23)$$

Also, since $\log e = 1$, we have

$$\frac{d \cdot e^x}{dx} = e^x. \qquad (24)$$

Logarithmic Differentiation.

EXAMPLES.

1. $y = \log(\sin x)$.

Let $\sin x = z$, then $y = \log z$.

And since
$$\frac{dy}{dx} = \frac{dy}{dz} \cdot \frac{dz}{dx},$$

we get
$$\frac{dy}{dx} = \frac{\cos x}{\sin x} = \cot x.$$

2. $y = \log \sqrt{a^2 - x^2} = \frac{1}{2} \log(a^2 - x^2)$; $\dfrac{dy}{dx} = -\dfrac{x}{a^2 - x^2}$.

3. $y = c^{nx}$. Ans. $\dfrac{dy}{dx} = n e^{nx}$.

4. $y = \log \sqrt{\dfrac{1 - \cos x}{1 + \cos x}}$:

$$\sqrt{\frac{1 - \cos x}{1 + \cos x}} = \sqrt{\frac{2 \sin^2 \frac{x}{2}}{2 \cos^2 \frac{x}{2}}} = \tan \frac{x}{2};$$

$\therefore y = \log \tan \dfrac{x}{2}$. Hence $\dfrac{dy}{dx} = \dfrac{1}{\sin x}$.

31. Logarithmic Differentiation.—When the function to be differentiated consists of products and quotients of functions, it is in general useful to take the logarithm of the function, and to differentiate it. This process is called logarithmic differentiation.

EXAMPLES.

1. $y = y_1 \cdot y_2 \cdot y_3 \ldots y_n$, $\log y = \log y_1 + \log y_2 + \ldots + \log y_n$.

Hence
$$\frac{1}{y}\frac{dy}{dx} = \frac{1}{y_1}\frac{dy_1}{dx} + \frac{1}{y_2}\frac{dy_2}{dx} + \ldots + \frac{1}{y_n}\frac{dy_n}{dx}.$$

This furnishes another proof of formula (4), p. 15.

2. $y = \dfrac{\sin^m x}{\cos^n x}$. Here, $\log y = m \log \sin x - n \log \cos x$;

$\therefore \dfrac{1}{y}\dfrac{dy}{dx} = m \dfrac{\cos x}{\sin x} + n \dfrac{\sin x}{\cos x}$; $\therefore \dfrac{dy}{dx} = \dfrac{\sin^{m-1} x}{\cos^{n+1} x}(m \cos^2 x + n \sin^2 x).$

First Principles—Differentiation.

3. $$y = \frac{(x-1)^{\frac{5}{2}}}{(x-2)^{\frac{3}{4}}(x-3)^{\frac{7}{3}}}.$$

Here $\log y = \frac{5}{2} \log (x-1) - \frac{3}{4} \log (x-2) - \frac{7}{3} \log (x-3)$;

hence $\dfrac{1}{y} \dfrac{dy}{dx} = \dfrac{5}{2} \dfrac{1}{x-1} - \dfrac{3}{4} \dfrac{1}{x-2} - \dfrac{7}{3} \dfrac{1}{x-3} = - \dfrac{7x^2 + 30x - 97}{12 \cdot (x-1)(x-2)(x-3)}$;

$$\therefore \frac{dy}{dx} = - \frac{(x-1)^{\frac{3}{2}}(7x^2 + 30x - 97)}{12 \cdot (x-2)^{\frac{7}{4}}(x-3)^{\frac{10}{3}}}.$$

4. $y = x(a^2 + x^2)\sqrt{a^2 - x^2}$. $\dfrac{dy}{dx} = \dfrac{a^4 + a^2 x^2 - 4x^4}{\sqrt{a^2 - x^2}}.$ ✓

5. $y = x^x$. Here $\log y = x \log x$.

Hence $\dfrac{1}{y} \dfrac{dy}{dx} = (\log x + 1)$; $\quad \therefore \dfrac{d \cdot x^x}{dx} = x^x (1 + \log x)$.

6. $y = e^{x^x}$. Here $\log y = x^x$,

$$\frac{1}{y} \frac{dy}{dx} = \frac{d \cdot x^x}{dx} = x^x (1 + \log x);$$

$$\therefore \frac{dy}{dx} = e^{x^x} x^x (1 + \log x).$$

7. $y = u^v$, where u and v are both functions of x.

Here $\qquad \log y = v \log u$,

$$\therefore \frac{1}{y} \frac{dy}{dx} = \log u \frac{dv}{dx} + \frac{v}{u} \frac{du}{dx};$$

$$\therefore \frac{dy}{dx} = u^v \left(\log u \frac{dv}{dx} + \frac{v}{u} \frac{du}{dx} \right) = u^v \log u \frac{dv}{dx} + vu^{v-1} \frac{du}{dx}.$$

32. The expression to be differentiated frequently admits of being transformed to a simpler shape. In such cases the student will find it an advantage to reduce the expression to its simplest form before proceeding to its differentiation.

Examples.

1. $y = \sin^{-1} \dfrac{x}{\sqrt{1 + x^2}}.$

Here $\dfrac{x}{\sqrt{1 + x^2}} = \sin y$, or $\dfrac{x^2}{1 + x^2} = \sin^2 y$; hence $x = \tan y$,

and we get $\dfrac{dy}{dx} = \cos^2 y = \dfrac{1}{1 + x^2}.$

Logarithmic Differentiation.

2.
$$y = \tan^{-1}\frac{\sqrt{1+x^2}+\sqrt{1-x^2}}{\sqrt{1+x^2}-\sqrt{1-x^2}}.$$

Here
$$\tan y = \frac{\sqrt{1+x^2}+\sqrt{1-x^2}}{\sqrt{1+x^2}-\sqrt{1-x^2}}.$$

$$\frac{\sqrt{1+x^2}}{\sqrt{1-x^2}} = \frac{\tan y + 1}{\tan y - 1};$$

$$\therefore x^2 = \frac{(1+\tan y)^2-(1-\tan y)^2}{(1+\tan y)^2+(1-\tan y)^2} = \frac{2\tan y}{1+\tan^2 y} = \sin 2y.$$

Hence
$$\frac{dy}{dx}\cos 2y = x,$$

$$\therefore \frac{dy}{dx} = \frac{x}{\cos 2y} = \frac{x}{\sqrt{1-x^4}}.$$

3.
$$y = \log\sqrt{\frac{\sqrt{1+x}+\sqrt{1-x}}{\sqrt{1+x}-\sqrt{1-x}}} = \frac{1}{2}\log\frac{\sqrt{1+x}+\sqrt{1-x}}{\sqrt{1+x}-\sqrt{1-x}}.$$

$$= \frac{1}{2}\log\frac{1+\sqrt{1-x^2}}{x} = \frac{1}{2}\log(1+\sqrt{1-x^2}) - \frac{1}{2}\log x.$$

Hence
$$\frac{dy}{dx} = -\frac{1}{2x\sqrt{1-x^2}}.$$

4.
$$y = \tan^{-1}\frac{\sqrt{1+x^2}-1}{x} + \tan^{-1}\frac{2x}{1-x^2}.$$

Let $x = \tan z$, and the student can easily prove that

$$y = \frac{5}{2}z; \text{ hence } \frac{dy}{dx} = \frac{5}{2}\frac{1}{1+x^2}.$$

Examples.

1. $y = \sec^{-1} x$.
   Ans. $\dfrac{dy}{dx} = \dfrac{1}{x\sqrt{x^2-1}}$.

2. $y = x \log x$.
   $\dfrac{dy}{dx} = 1 + \log x$.

3. $y = \log \tan x$.
   $\dfrac{dy}{dx} = \dfrac{2}{\sin 2x}$.

4. $y = \log \tan^{-1} x$.
   $\dfrac{dy}{dx} = \dfrac{1}{(1+x^2)\tan^{-1} x}$.

5. $y = a\sqrt{x}$.
   $\dfrac{dy}{dx} = \dfrac{a}{2\sqrt{x}}$.

6. $y = \sin(\log x)$.
   $\dfrac{dy}{dx} = \dfrac{\cos(\log x)}{x}$.

7. $y = \tan^{-1} \dfrac{x}{\sqrt{1-x^2}}$.
   $\dfrac{dy}{dx} = \dfrac{1}{\sqrt{1-x^2}}$.

8. $y = \tan^{-1} \dfrac{\sqrt{x}+\sqrt{a}}{1-\sqrt{ax}}$.
   $\dfrac{dy}{dx} = \dfrac{1}{2\sqrt{x}(1+x)}$.

   Here $y = \tan^{-1}\sqrt{x} + \tan^{-1}\sqrt{a}$.

9. $y = \dfrac{x^{2n}}{(1+x^2)^n}$.
   $\dfrac{dy}{dx} = \dfrac{2nx^{2n-1}}{(1+x^2)^{n+1}}$.

10. $y = \log\left(\dfrac{1+x}{1-x}\right)^{\frac{1}{4}} - \frac{1}{2}\tan^{-1} x$.
    $\dfrac{dy}{dx} = \dfrac{x^2}{1-x^4}$.

11. $y = \log\sqrt{\dfrac{\sqrt{1+x^2}+x}{\sqrt{1+x^2}-x}}$.
    $\dfrac{dy}{dx} = \dfrac{1}{\sqrt{1+x^2}}$.

12. $y = \sin^{-1}\dfrac{3+2x}{\sqrt{13}}$.
    $\dfrac{dy}{dx} = \dfrac{1}{\sqrt{1-3x-x^2}}$.

13. $y = \log\dfrac{(1+x^2)^{\frac{1}{4}}}{(1+x)^{\frac{1}{2}}} + \frac{1}{2}\tan^{-1} x$.
    $\dfrac{dy}{dx} = \dfrac{x}{(1+x)(1+x^2)}$.

14. $y = \dfrac{1-x}{\sqrt{1+x^2}}$.
    $\dfrac{dy}{dx} = -\dfrac{(1+x)}{(1+x^2)^{\frac{3}{2}}}$.

Examples. 31

15. $y = \dfrac{(1-x^2)^{\frac{3}{2}} \sin^{-1} x}{x}$. Ans. $\dfrac{dy}{dx} = \dfrac{1-x^2}{x} - \dfrac{1+2x^2}{x^2}(1-x^2)^{\frac{1}{2}} \cdot \sin^{-1} x$.

16. $y = \dfrac{1 - \tan x}{\sec x}$. $\dfrac{dy}{dx} = -(\cos x + \sin x)$.

17. $y = \log \dfrac{\sqrt{1-x^2} + x\sqrt{2}}{\sqrt{1-x^2}}$. $\dfrac{dy}{dx} = \dfrac{\sqrt{2}}{(\sqrt{1-x^2} + x\sqrt{2})(1-x^2)}$.

18. $y = \dfrac{e^{a \tan^{-1} x}(ax - 1)}{(1+x^2)^{\frac{1}{2}}}$. $\dfrac{dy}{dx} = \dfrac{(1+a^2)\, x\, e^{a \tan^{-1} x}}{(1+x^2)^{\frac{3}{2}}}$.

19. $y = \log \dfrac{1+x}{1-x} + \tfrac{1}{2} \log \dfrac{1+x+x^2}{1-x+x^2} + \sqrt{3}\,\tan^{-1} \dfrac{x\sqrt{3}}{1-x^2}$. $\dfrac{dy}{dx} = \dfrac{6}{1-x^6}$.

20. $y = \log\{(2x-1) + 2\sqrt{x^2 - x - 1}\}$. $\dfrac{dy}{dx} = \dfrac{1}{(x^2 - x - 1)^{\frac{1}{2}}}$.

21. $y = \log \sqrt{\dfrac{1 + x\sqrt{2} + x^2}{1 - x\sqrt{2} + x^2}} + \tan^{-1} \dfrac{x\sqrt{2}}{1-x^2}$. $\dfrac{dy}{dx} = \dfrac{2\sqrt{2}}{1+x^4}$.

22. $y = e^{x^x} \tan^{-1} x$. $\dfrac{dy}{dx} = e^{x^x}\left(\dfrac{1}{1+x^2} + x^x \tan^{-1} x\,(1 + \log x)\right)$.

23. Being given that $y = x^3 (1-x^2)^{\frac{1}{2}}\left(1 - \dfrac{x^2}{2}\right)^{\frac{1}{2}}$; if

$$\dfrac{dy}{dx} = \dfrac{cx^2 + c'x^4 + c''x^6}{(1-x^2)^{\frac{1}{2}}\left(1 - \dfrac{x^2}{2}\right)^{\frac{1}{2}}},$$

determine the values of c, c', c''.   Ans. $c = 3$, $c' = -6$, $c'' = \tfrac{5}{2}$.

24. $y = \log(\log x)$. $\dfrac{dy}{dx} = \dfrac{1}{x \log x}$.

25. $y = \cos^{-1} \dfrac{3 + 5\cos x}{5 + 3\cos x}$. $\dfrac{dy}{dx} = \dfrac{4}{5 + 3\cos x}$.

26. $y = \sin^{-1} \dfrac{1-x^2}{1+x^2}$. $\dfrac{dy}{dx} = \dfrac{-2}{1+x^2}$.

27. $y = e^{ax} \sin^m rx$. $\dfrac{dy}{dx} = e^{ax} \sin^{m-1} rx\,(a \sin rx + mr \cos rx)$.

28. $y = e^{ax} \sin rx$. $\dfrac{dy}{dx} = e^{ax} \sqrt{a^2 + r^2}\,\sin(rx + \phi)$,

where $\tan \phi = \dfrac{r}{a}$.

29. $y = \log(\sqrt{x-a} + \sqrt{x-b})$. Ans. $\dfrac{dy}{dx} = \dfrac{1}{2\sqrt{(x-a)(x-b)}}$.

30. $y = 2\tan^{-1}\left(\dfrac{1-x}{1+x}\right)^{\frac{1}{2}}$.

Here $\dfrac{1-x}{1+x} = \tan^2\dfrac{y}{2}$; $\therefore x = \cos y$; $\therefore \dfrac{dy}{dx} = -\dfrac{1}{(1-x^2)^{\frac{1}{2}}}$.

31. $y = x^{x^n}$. $\dfrac{dy}{dx} = x x^{n+n-1}(n\log x + 1)$.

32. $y = (1+x^2)^{\frac{m}{2}}\sin(m\tan^{-1}x)$. $\dfrac{dy}{dx} = m(1+x^2)^{\frac{m-1}{2}}\cos\{(m-1)\tan^{-1}x\}$.

33. $y = \log\sqrt{\dfrac{a\cos x - b\sin x}{a\cos x + b\sin x}}$. $\dfrac{dy}{dx} = \dfrac{-ab}{a^2\cos^2 x - b^2\sin^2 x}$.

34. Define the differential coefficient of a function of a variable quantity, with respect to that quantity, and shew that it measures the rate of increase of the function as compared with the rate of increase of the variable.

35. If $y = \dfrac{1}{x}$, prove the relation

$$\dfrac{dy}{\sqrt{1+y^4}} + \dfrac{dx}{\sqrt{1+x^4}} = 0.$$

36. If $u = \log\dfrac{x^2+ax+\sqrt{(x^2+ax)^2-bx}}{x^2+ax-\sqrt{(x^2+ax)^2-bx}}$, prove that $\dfrac{du}{dx}$ is of the form

$\dfrac{Ax+B}{\sqrt{(x^2+ax)^2-bx}}$, and determine the values of A and B. Ans. $A = 3$, $B = a$.

37. Prove that $\dfrac{d}{d\theta}\left(\sin\theta\cos\theta\sqrt{1-c^2\sin^2\theta}\right) = \dfrac{A\sin^4\theta + B\sin^2\theta + C}{\sqrt{1-c^2\sin^2\theta}}$,

and determine the values of A, B, C. Ans. $A = 3c^2$, $B = -2(1+c^2)$, $C = 1$.

38. If $u = x + \dfrac{1}{2}\dfrac{x^3}{3} + \dfrac{1\cdot 3}{2\cdot 4}\dfrac{x^5}{5} + \dfrac{1\cdot 3\cdot 5}{2\cdot 4\cdot 6}\dfrac{x^7}{7} + \ldots$ ad inf.; find the sum

of the series represented by $\dfrac{du}{dx}$. Ans. $(1-x^2)^{-\frac{1}{2}}$.

39. Reduce to its simplest form the expression

$$\dfrac{3a^2}{(x^2+a)^{\frac{1}{2}}(x^2+2a)^{\frac{1}{2}}} - \dfrac{d}{dx}\cdot\dfrac{x(x^2+2a)^{\frac{1}{2}}}{(x^2+a)^{\frac{3}{2}}}. \quad \text{Ans.}\ \dfrac{1}{(x^2+a)^{\frac{1}{2}}(x^2+2a)^{\frac{1}{2}}}.$$

40. If $\sin y = x\sin(a+y)$, prove that $\dfrac{dy}{dx} = \dfrac{\sin^2(a+y)}{\sin a}$.

Examples. 33

41. If $x(1+y)^{\frac{1}{2}} + y(1+x)^{\frac{1}{2}} = 0$, find $\dfrac{dy}{dx}$.

In this case $\quad x^2(1+y) = y^2(1+x)$;

$$\therefore x^2 - y^2 = yx(y-x),$$

or $\quad x+y+xy = 0$; $\therefore y = -\dfrac{x}{1+x}$; $\therefore \dfrac{dy}{dx} = -\dfrac{1}{(1+x)^2}$.

42. $y = \log(x + \sqrt{x^2 - a^2}) + \sec^{-1}\dfrac{x}{a}$. $\quad \dfrac{dy}{dx} = \dfrac{1}{x}\sqrt{\dfrac{x+a}{x-a}}$.

43. If x and y are given as functions of t by the equations

$$x = f(t); \quad y = F(t);$$

find the value of $\dfrac{dy}{dx}$ in terms of t. $\quad \dfrac{dy}{dx} = \dfrac{F'(t)}{f'(t)}$.

44. $\quad y = \cfrac{x^2}{1 + \cfrac{x^2}{1 + \cfrac{x^2}{1 + \&c., \text{ ad infinitum.}}}}$

Hence $y = \dfrac{x^2}{1+y}$. $\quad \dfrac{dy}{dx} = \dfrac{x}{\sqrt{x^2 + \frac{1}{4}}}$.

45. $x = e^{\frac{x-y}{y}}$.

Hence $y = \dfrac{x}{1 + \log x}$. $\quad \dfrac{dy}{dx} = \dfrac{\log x}{(1 + \log x)^2}$.

D

CHAPTER II.

SUCCESSIVE DIFFERENTIATION.

33. Successive Derived Functions.—In the preceding chapter we have considered the process of finding the derived functions of different forms of functions of a single variable.

If the primitive function be represented by $f(x)$, then, as already stated, its *first* derived function is denoted by $f'(x)$. If this new function, $f'(x)$, be treated in the same manner, its derived function is called the *second* derived of the original function $f(x)$, and is denoted by $f''(x)$.

In like manner the derived function of $f''(x)$ is the *third* derived of $f(x)$, and represented by $f'''(x)$, &c.

In accordance with this notation, the successive derived functions of $f(x)$ are represented by

$$f'(x), \quad f''(x), \quad f'''(x), \ldots f^{(n)}(x),$$

each of which is the derived function of the preceding.

34. Successive Differential Coefficients.

If $\quad y = f(x)$ we have $\dfrac{dy}{dx} = f'(x)$.

Hence, differentiating both sides with regard to x, we get

$$\frac{d}{dx}\left(\frac{dy}{dx}\right) = \frac{d}{dx}f'(x) = f''(x).$$

Let $\quad \dfrac{d}{dx}\left(\dfrac{dy}{dx}\right)$ be represented by $\dfrac{d^2y}{dx^2}$,

then $\quad \dfrac{d^2y}{dx^2} = f''(x).$

In like manner $\dfrac{d}{dx}\left(\dfrac{d^2y}{dx^2}\right)$ is represented by $\dfrac{d^3y}{dx^3}$, and so on;

hence $\quad\dfrac{d^3y}{dx^3} = f'''(x),$ &c. . . . $\dfrac{d^ny}{dx^n} = f^{(n)}(x).\quad$ (1)

The expressions

$$\frac{dy}{dx},\ \frac{d^2y}{dx^2},\ \frac{d^3y}{dx^3},\ \ldots\ \frac{d^ny}{dx^n}$$

are called the *first, second, third*, . . . n^{th} differential coefficients of y regarded as a function of x.

These functions are sometimes represented by

$$y',\ y'',\ y''',\ \ldots\ y^{(n)},$$

a notation which will often be found convenient in abbreviating the labour of forming the successive differential coefficients of a given expression. From the mode of arriving at them, the successive differential coefficients of a function are evidently the same as its successive derived functions considered in the preceding Article.

35. **Successive Differentials.**—The preceding result admits of being considered also in connexion with differentials; for, since x is the independent variable, its increment, dx, may be always taken of the *same infinitely small value*. Hence, in the equation $dy = f'(x)\,dx$ (Art. 7), we may regard dx as constant, and we shall have, on proceeding to the next differentiation,

$$d(dy) = dx\,d\left[f'(x)\right] = (dx)^2 f''(x),$$

since $\quad d\left[f'(x)\right] = f''(x)\,dx.$

Again, representing $\quad d(dy)$ by $d^2y,$

we have $\quad d^2y = f''(x)(dx)^2;$

if we differentiate again, we get

$$d^3y = f'''(x)(dx^3);$$

and in general

$$d^ny = f^{(n)}(x)(dx)^n.$$

From this point of view we see the reason why $f^{(n)}(x)$ is called the nth *differential coefficient* of $f(x)$.

In the preceding results it may be observed that if dx be regarded as an *infinitely small quantity*, or an *infinitesimal* of the first order, $(dx)^2$, being *infinitely small in comparison with dx*, may be called an infinitely small quantity or an infinitesimal of the second order; as also d^2y, if $f''(x)$ be finite. In general, $d^n y$, being of the same order as $(dx)^n$, is called an *infinitesimal* of the nth order.

36. **Infinitesimals.**—We may premise that the expressions great and small, as well as *infinitely* great and infinitely small, are to be understood as *relative* terms. Thus, a magnitude which is regarded as being infinitely great in comparison with a *finite* magnitude is said to be infinitely great. Similarly, a magnitude which is infinitely small in comparison with a finite magnitude is said to be *infinitely small*. If any finite magnitude be conceived to be divided into an infinitely great number of *equal* parts, each part will be infinitely small with regard to the finite magnitude; and may be called an *infinitesimal* of the first order. Again, if one of these infinitesimals be conceived to be divided into an infinite number of equal parts, each of these parts is infinitely small in comparison with the former infinitesimal, and may be regarded as an *infinitesimal* of the *second* order, and so on.

Since, in general, the number by which any measurable quantity is represented depends upon the *unit* with which the quantity is compared, it follows that a finite magnitude may be represented by a very great, or by a very small number, according to the unit to which it is referred. For example, the diameter of the earth is very great in comparison with the length of one foot, but very small in comparison with the distance of the earth from the nearest fixed star, and it would, accordingly, be represented by a very large, or a very small number, according to which of these distances is assumed as the unit of comparison. Again, with respect to the latter distance taken as the unit, the diameter of the earth may be regarded as a very small magnitude of the first order, and the length of a foot as one of a higher order of smallness in comparison. Similar remarks apply to other magnitudes.

Again, in the comparison of numbers, if the fraction (one million)th or $\dfrac{1}{10^6}$, which is very small in comparison with unity, be regarded as a small quantity of the first order, the fraction $\frac{1}{10^{12}}$, being the same fractional part of $\frac{1}{10^6}$ that this is of 1, must be regarded as a small quantity of the second order, and so on.

If now, instead of the series $\frac{1}{10^6}$, $\left(\frac{1}{10^6}\right)^2$, $\left(\frac{1}{10^6}\right)^3$, ... we consider the series $\frac{1}{n}$, $\frac{1}{n^2}$, $\frac{1}{n^3}$, ... in which n is supposed to be increased without limit, then each term in the series is infinitely small in comparison with the preceding one, being derived from it by multiplying by the infinitely small quantity $\frac{1}{n}$. Hence, if $\frac{1}{n}$ be regarded as an infinitesimal of the first order, $\frac{1}{n^2}$, $\frac{1}{n^3}$, ... $\frac{1}{n^r}$, may be regarded as infinitesimals of the *second*, *third*, ... r^{th} orders.

37. Geometrical Illustration of Infinitesimals.—The following geometrical results will help to illustrate the theory of infinitesimals, and also will be found of importance in the application of the Differential Calculus to the theory of curves.

Suppose two points, A, B, taken on the circumference of a circle; join B to E, the other extremity of the diameter AE, and produce EB to meet the tangent at A in D. Then since the triangles ADB and EAB are equiangular, we have

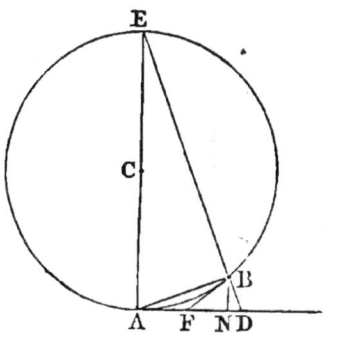

Fig. 3.

$$\frac{AB}{AD} = \frac{BE}{AE}, \text{ and } \frac{BD}{AD} = \frac{AB}{AE}.$$

Now suppose the point B to approach the point A and to become indefinitely near to it, then BE becomes ultimately equal to AE, and, therefore, at the same time, $\frac{AB}{AD} = 1$.

Again, $\dfrac{BD}{AD}$ becomes infinitely small along with $\dfrac{AB}{AE}$, i.e. BD becomes infinitely small in comparison with AD or AB. Hence BD is an *infinitesimal of the second order* when AB is taken as one of the first order.

Moreover, since $DE - AE < BD$, it follows that, *when one side of a right-angled triangle is regarded as an infinitely small quantity of the first order, the difference between the hypothenuse and the remaining side is an infinitely small quantity of the second order.*

Next, draw BN perpendicular to AD, and BF a tangent at B; then, since $AB > AN$, we get $AD - AB < AD - AN < DN$;

$$\therefore \frac{AD - AB}{BD} < \frac{DN}{BD} < \frac{AD}{DE}.$$

Consequently, $\dfrac{AD - AB}{BD}$ becomes infinitely small along with AD; $\therefore AD - AB$ is *an infinitesimal of the third order*. Moreover, as $BF = FD$, we have $AD = AF + BF$; $\therefore AF + BF - AB$ is an infinitely small quantity of the third order; but $AF + FB$ is $>$ arc AB, hence we infer that *the difference between the length of the arc AB and its chord is an infinitely small quantity of the third order, when the arc is an infinitely small quantity of the first.* In like manner it can be seen that $BD - BN$ is an infinitesimal of the *fourth* order, and so on.

Again, if AB represent an elementary portion of any continuous* curve, to which AF and BF are tangents, since the length of the arc AB is less than the sum of the tangents AF and BF, we may extend the result just arrived at to all such curves.

---

* In this extension of the foregoing proof it is assumed that the ultimate ratio of the tangents drawn to a continuous curve at two indefinitely near points is, in general, a ratio of equality. This is easily shown in the case of an ellipse, since the ratio of the tangents is the same as that of the parallel diameters. Again, it can be seen without difficulty that an indefinite number of ellipses can be drawn touching a curve at two points arbitrarily assumed on the curve; if now we suppose the points to approach one another indefinitely along the curve, the property in question follows immediately for any continuous curve.

Hence, the *difference between the length of an infinitely small portion of any continuous curve and its chord is an infinitely small quantity of the third order*, i.e. the difference between them is ultimately an infinitely small quantity of the *second* order in comparison with the length of the chord.

The same results might have been established from the expansions for sin a and cos a, when a is considered as infinitely small.

If in the general case of any continuous curve we take two points A, B, on the curve, join them, and draw BE perpendicular to AB, meeting in E the *normal* drawn to the curve at the point A; then all the results established above for the circle still hold. When the point B is taken infinitely near to A, the line AE becomes the diameter of the *circle of curvature* belonging to the point A; for, it is evident that the circle which passes through A and B, and has the same tangent at A as the given curve, has *a contact of the second order* with it. See "Salmon's Conic Sections," Art. 239.

Examples.

1. In a triangle, if the vertical angle be very small in comparison with either of the base angles, prove that the difference between the sides is very small in comparison with either of them; and hence, that these sides may be regarded as ultimately equal.

2. In a triangle, if the external angle at the vertex be very small, show that the difference between the sum of the sides and the base is a very small quantity of the *second* order.

3. If the base of a triangle be an infinitesimal of the first order, as also its base angles, show that the difference between the sum of its sides and its base is an infinitesimal of the *third* order.

This furnishes an additional proof that the difference between the length of an arc of a continuous curve and that of its chord is ultimately an infinitely small quantity of the third order.

4. If a right line be displaced, through an infinitely small angle, prove that the projections on it of the displacements of its extremities are equal.

5. If the side of a regular polygon inscribed in a circle be a very small magnitude of the first order in comparison with the radius of the circle, show that the difference between the circumference of the circle and the perimeter of the polygon is a very small magnitude of the second order.

38. Fundamental Principle of the Infinitesimal Calculus.—We shall now proceed to enunciate the fundamental principle of the Infinitesimal Calculus as conceived by Leibnitz:* it may be stated as follows:—

If the difference between two quantities be infinitely small in comparison with either of them, then the ratio of the quantities becomes unity in the limit, and either of them can be in general replaced by the other in any expression.

For let a, β, represent the quantities, and suppose

$$a = \beta + i, \text{ or } \frac{a}{\beta} = 1 + \frac{i}{\beta}.$$

Now the ratio $\dfrac{i}{\beta}$ becomes evanescent whenever i is infinitely small in comparison with β. This may take place in three different ways: (1) when β is finite, and i infinitely small: (2) when i is finite, and β infinitely great; (3) when β is infinitely small, and i also infinitely small of a higher order: thus, if $i = k\beta^2$, then $\dfrac{i}{\beta} = k\beta$, which becomes evanescent along with β.

---

* This principle is stated for finite magnitudes by Leibnitz, as follows:—
"Cæterum æqualia esse puto, non tantum quorum differentia est omnino nulla, sed et quorum differentia est incomparabiliter parva." . . . "Scilicet eas tantum homogeneas quantitates comparabiles esse, cum Euc. Lib. 5, defin. 5, censeo, quarum una numero sed finito multiplicata, alteram superare potest; et quæ tali quantitate non differunt, æqualia esse statuo. quod etiam Archimedes sumsit, aliique post ipsum omnes." Leibnitii Opera, Tom. 3, p. 328.

The foregoing can be identified with the fundamental principle of Newton, as laid down in his Prime and Ultimate Ratios, Lemma I.: "Quantitates, ut et quantitatum rationes, quæ ad æqualitatem tempore quovis finito constanter tendunt, et ante finem temporis illius proprius ad invicem accedunt quam pro datâ quavis differentiâ, fiunt ultimo æquales."

All applications of the infinitesimal method depend ultimately either on the limiting ratios of infinitely small quantities, or on the limiting value of the sum of an infinitely great number of infinitely small quantities; and it may be observed that the difference between the method of infinitesimals and that of limits (when exclusively adopted) is, that in the latter method it is usual to retain evanescent quantities of higher orders until the *end* of the calculation, and then to neglect them, on proceeding to the limit; while in the infinitesimal method such quantities are neglected from the commencement, from the knowledge that they cannot affect the *final result*, as they necessarily disappear in the limit.

Accordingly, in any of the preceding cases, the fraction $\dfrac{a}{\beta}$ becomes unity in the limit, and we can, in general, substitute a instead of β in any function containing them. Thus, an infinitely small quantity is neglected in comparison with a finite one, as their ratio is evanescent; and similarly an infinitesimal of any order may be neglected in comparison with one of a lower order.

Again, two infinitesimals a, β, are said to be of the same order if the fraction $\dfrac{\beta}{a}$ tends to a finite limit. If $\dfrac{\beta}{a^n}$ tends to a finite limit, β is called an infinitesimal of the nth order in comparison with a.

As an example of this method, let it be proposed to determine the direction of the tangent at a point (x, y) on a curve whose equation is given in rectangular co-ordinates.

Let $x + a$, $y + \beta$, be the co-ordinates of a near point on the curve, and, by Art. 10, the direction of the tangent depends on the limiting value of $\dfrac{\beta}{a}$. To find this, we substitute $x + a$ for x, and $y + \beta$ for y in the equation, and neglecting all powers of a and β beyond the first, we solve for $\dfrac{\beta}{a}$, and thus obtain the required solution.

For example, let the equation of the curve be $x^3 + y^3 = 3axy$: then, substituting as above, we get

$$x^3 + 3x^2 a + y^3 + 3y^2 \beta = 3axy + 3ax\beta + 3aya:$$

hence, on subtracting the given equation, we get the

$$\text{limit of } \frac{\beta}{a} = \frac{x^2 - ay}{ax - y^2}.$$

39. Subsidiary Principle.—If $a_1 + a_2 + a_3 + \ldots + a_n$ represent the sum of a number of infinitely small quantities, which approaches to a finite limit when n is increased indefinitely, and if $\beta_1, \beta_2, \ldots \beta_n$ be another system of infinitely small quantities, such that

$$\frac{\beta_1}{a_1} = 1 + \epsilon_1, \quad \frac{\beta_2}{a_2} = 1 + \epsilon_2, \quad \ldots \quad \frac{\beta_n}{a_n} = 1 + \epsilon_n,$$

where $\epsilon_1, \epsilon_2, \ldots \epsilon_n$, are infinitely small quantities, then the limit of the sum of $\beta_1, \beta_2, \ldots \beta_n$ is ultimately the same as that of $a_1, a_2, \ldots a_n$.

For, from the preceding equations we have

$$\beta_1 + \beta_2 + \ldots + \beta_n = a_1 + a_2 + \ldots + a_n + a_1\epsilon_1 + a_2\epsilon_2 + \ldots + a_n\epsilon_n.$$

Now, if η be the greatest of the infinitely small quantities, $\epsilon_1, \epsilon_2, \ldots \epsilon_n$, we have

$$\beta_1 + \beta_2 + \ldots + \beta_n - (a_1 + a_2 + \ldots + a_n) < \eta (a_1 + a_2 \ldots + a_n);$$

but the factor $a_1 + a_2 + \ldots + a_n$ has a finite limit, by hypothesis, and as η is infinitely small, it follows that the limit of $\beta_1 + \beta_2 + \ldots + \beta_n$ is the same as that of $a_1 + a_2 + \ldots + a_n$.

This result can also be established otherwise as follows:—

The ratio $\dfrac{\beta_1 + \beta_2 + \ldots + \beta_n}{a_1 + a_2 + \ldots + a_n}$,

by an elementary algebraic principle, lies between the greatest and the least values of the fractions

$$\frac{\beta_1}{a_1}, \frac{\beta_2}{a_2}, \ldots \frac{\beta_n}{a_n};$$

it accordingly has unity for its limit under the supposed conditions: and hence the limiting value of $\beta_1 + \beta_2 + \ldots + \beta_n$ is the same as that of $a_1 + a_2 + \ldots + a_n$.

40. **Approximations.**—The principles of the Infinitesimal Calculus above established lead to rigid and accurate results in the limit, and may be regarded as the fundamental principles of the Calculus, the former of the Differential, and the latter of the Integral. These principles are also of great importance in practical calculations, in which approximate results only are required. For instance, in calculating a result to seven decimal places, if $\dfrac{1}{10^4}$ be regarded as a small quantity a, then a^2, a^3, &c., may in general be neglected.

Thus, for example, to find sin 30′ and cos 30′ to seven decimal places. The circular measure of 30′ is $\dfrac{\pi}{360}$, or .0087266;

denoting this by a, and employing the formulæ,
$$\sin a = a - \frac{a^3}{6}, \quad \cos a = 1 - \frac{a^2}{2},$$
it is easily seen that to seven decimal places we have
$$\frac{a^2}{2} = .0000381, \quad \frac{a^3}{6} = .0000001.$$

Hence $\quad \sin 30' = .0087265 \, ; \, \cos 30' = .9999619.$

In this manner the sine and the cosine of any small angle can be readily calculated.

Again, to find the error in the calculated value of the sine of an angle arising from a small error in the observed value of the angle. Denoting the angle by a, and the small error by a, we have
$$\sin(a + a) = \sin a \cos a + \cos a \sin a = \sin a + a \cos a,$$
neglecting higher powers of a. Hence the error is represented by $a \cos a$, approximately.

In like manner we get to the same degree of approximation
$$\tan(a + a) - \tan a = \frac{a}{\cos^2 a}.$$

Again, to the same degree of approximation we have
$$\frac{a + a}{b + \beta} = \frac{a}{b} + \frac{ba - a\beta}{b^2},$$
where a, β are supposed very small in comparison with a and b.

As another example, the method leads to an easy mode of approximating to the roots of nearly square numbers; thus
$$\sqrt{a^2 + a} = a + \frac{a}{2a}; \quad \sqrt{a^2 + a^2} = a + \frac{a^2}{2a} = a, \text{ whenever } a^2 \text{ may}$$
be neglected.

Likewise, $\quad \sqrt[3]{a^3 + a} = a + \dfrac{a}{3a^2},$ &c.

If $b = a + a$, where a is very small in comparison with a,
we have $\quad \sqrt{ab} = \sqrt{a^2 + aa} = a + \dfrac{a}{2} = \dfrac{a + b}{2}.$

Again, in a plane triangle, we have the formula
$$c^2 = a^2 + b^2 - 2ab \cos C = (a + b)^2 \sin^2 \frac{C}{2} + (a - b)^2 \cos^2 \frac{C}{2}.$$
Now if we suppose a and b nearly equal, and neglect $(a - b)^2$ in comparison with $(a + b)^2$, we have
$$c = \sqrt{(a + b)^2 \sin^2 \frac{C}{2} + (a - b)^2 \cos^2 \frac{C}{2}} = (a + b) \sin \frac{C}{2}.$$
This furnishes a simple approximation for the length of the base of a triangle when its sides are very nearly of equal length.

Examples.

1. Find the value of $(1 + a)(1 - 2a^2)(1 + 3a^3)$, neglecting a^4 and higher powers of a. *Ans.* $1 + a - 2a^2 + a^3$.

2. Find the value of $\sin(a + \alpha) \sin(b + \beta)$, neglecting terms of 2nd order in α and β. *Ans.* $\sin a \sin b + \alpha \cos a \sin b + \beta \sin a \cos b$.

3. If $m = u - e \sin u$, e being very small, find the value of $\tan \frac{1}{2} u$.

Ans. $(1 + e) \tan \frac{m}{2}$.

Here $\frac{u}{2} = \frac{m}{2} + \frac{e}{2} \sin u$; $\tan \frac{u}{2} = \tan\left(\frac{m}{2} + a\right)$, where $a = \frac{e}{2} \sin u$; $\therefore$ &c.

4. In a right-angled spherical triangle we have the relation $\cos c = \cos a \cos b$; determine the corresponding formula in plane trigonometry.

The circular measure of a is $\frac{a}{R}$, R being the radius of the sphere; hence, substituting $1 - \frac{a^2}{R^2}$ for $\cos a$, &c., and afterwards making $R = \infty$, we get $c^2 = a^2 + b^2$.

5. If a parallelogram be slightly distorted, find the relation connecting the changes of its diagonals.

Ans. $d \Delta d + d' \Delta d' = 0$, where d, d' denote the diagonals, and Δd, $\Delta d'$ the changes in their lengths. In the case of a rectangle the increments are equal, and of opposite signs.

6. Find the limiting value of
$$\frac{Aa^m + Ba^{m+1} + Ca^{m+2} + \&c.}{aa^n + ba^{n+1} + ca^{n+2} + \&c.}$$
when a becomes evanescent.

In this case the true value is that of $\frac{Aa^m}{aa^n} = \frac{A}{a} a^{m-n}$.

Hence the required value is zero, $\frac{A}{a}$, or infinity, according as $m >$, $=$, or $< n$.

7. Find the value of
$$\frac{1 - \frac{x^2}{6} + \frac{x^4}{120}}{1 - \frac{x^2}{2} + \frac{x^4}{24}},$$
neglecting powers of x beyond the 4th. *Ans.* $1 + \frac{x^2}{3} + \frac{2x^4}{15}$.

8. Find the limiting values of $\frac{x}{y}$ when $y = 0$, x and y being connected by the equation $y^3 = 2xy - x^2$.

Here, dividing by y^2 we get
$$\frac{x^2}{y^2} - 2\frac{x}{y} = -y.$$

If we solve for $\frac{x}{y}$ we have
$$\frac{x}{y} = 1 \pm (1 - y)^{\frac{1}{2}}.$$

Hence, in the limit, when $y = 0$, we have $\frac{x}{y} = 2$, or $\frac{x}{y} = 0$.

9. In fig. 3, Art. 37, if AB be regarded as a side of a regular inscribed polygon of a very great number of sides, show that, neglecting small quantities of the 4th order, the difference between the perimeter of the inscribed polygon and that of the circumscribed polygon of the same number of sides is represented by $\frac{\pi}{2} BD$.

Let n be the number of sides, then the difference in question is $n(AD - AB)$;

but $\quad n = \dfrac{\pi AE}{\text{arc } AB}; \quad \therefore n(AD - AB) = \dfrac{\pi AE(AD - AB)}{AB}$

$\quad = \pi AE \dfrac{DE - AE}{AE} = \pi(DE - AE) = \dfrac{\pi}{2} BD$, q. p.

This result shows how rapidly the perimeters of the circumscribed and inscribed polygons approximate to equality, as the number of sides becomes very great.

10. Assuming the earth to be a sphere of 40,000,000 mètres circumference, show that the difference between its circumference and the perimeter of a regular inscribed polygon of 1,000,000 sides is less than $\frac{1}{15}$th of a millimètre.

11. If one side b of a spherical triangle be small, find an expression for the difference between the other sides, as far as terms of the second order in b.

Here $\qquad \cos c = \cos a \cos b + \sin a \sin b \cos C.$

Let z denote the difference in question; i. e. $c = a - z$;

then $\qquad \cos a \cos z + \sin a \sin z = \cos a \cos b + \sin a \sin b \cos C;$

$\qquad \therefore \sin z - \sin b \cos C = \cot a (\cos b - \cos z).$

Since z and b are both small, we get, to terms of the second order,
$$z - b\cos C = \frac{\cot a}{2}(z^2 - b^2).$$

The first approximation gives $z = b\cos C$. If this be substituted for z in the right-hand side, we get, for the second approximation,
$$z = b\cos C - \frac{b^2 \sin^2 C \cot a}{2}.$$

We now proceed to find the successive derived functions in some elementary examples.

41. Derived Functions of x^m.

Let $y = x^m$,

then $\dfrac{dy}{dx} = mx^{m-1}$, $\dfrac{d^2y}{dx^2} = m(m-1)x^{m-2}$,

and in general, $\dfrac{d^n y}{dx^n} = m(m-1)(m-2)\ldots(m-n+1)x^{m-n}$.

If m be a positive integer, we have
$$\frac{d^m(x^m)}{dx^m} = 1\cdot 2\ldots m.$$

and all the higher derived functions vanish.

If m be a fractional, or a negative index, then none of the successive derived functions can vanish.

EXAMPLES.

1. If $u = ax^n + bx^{n-1} + cx^{n-2} + \&c.$, prove that
$$\frac{d^2u}{dx^2} = n(n-1)ax^{n-2} + (n-1)(n-2)bx^{n-3} + \&c.,$$

also $\dfrac{d^n u}{dx^n} = 1\cdot 2\ldots n\cdot a$, and $\dfrac{d^{n+1}u}{dx^{n+1}} = 0.$

2. $y = \dfrac{a}{x^n}$,

prove that $\dfrac{dy}{dx} = -\dfrac{na}{x^{n+1}}$, $\dfrac{d^2y}{dx^2} = \dfrac{n(n+1)a}{x^{n+2}}$,

and $\dfrac{d^m y}{dx^m} = (-1)^m \dfrac{n(n+1)\ldots(n+m-1)a}{x^{n+m}}.$

Examples. 47

3. $y = 2a\sqrt{x}$;

prove that $\quad \dfrac{dy}{dx} = \dfrac{a}{\sqrt{x}}, \quad \dfrac{d^2y}{dx^2} = -\dfrac{a}{2x^{\frac{3}{2}}}, \quad \dfrac{d^3y}{dx^3} = \dfrac{3}{4}\dfrac{a}{x^{\frac{5}{2}}},$

$$\dfrac{d^{n+1}y}{dx^{n+1}} = (-1)^n \dfrac{3 \cdot 5 \cdot 7 \cdots (2n-1)\,a}{2^n \cdot x^{n+\frac{1}{2}}}.$$

42. **If** $y = x^3 \log x$, **to find** $\dfrac{d^4y}{dx^4}$.

Here $\quad \dfrac{dy}{dx} = 3x^2 \log x + x^2$;

also $\quad \dfrac{d^2y}{dx^2} = 6x \log x + 3x + 2x = 6x \log x + 5x$,

$\dfrac{d^3y}{dx^3} = 6 \log x + 6 + 5, \quad \dfrac{d^4y}{dx^4} = \dfrac{6}{x}.$

It might have been observed that in this case all the terms in the successive differentials which do not contain $\log x$ will disappear from the final result—thus, by the last Article, $\dfrac{d^3(x^2)}{dx^3} = 0$, accordingly, that term may be neglected; and similar reasoning applies to the other terms. The work can therefore be simplified by neglecting such terms as we proceed.

The student will find no difficulty in applying the same mode of reasoning to the determination of the value of

$\dfrac{d^n y}{dx^n}$, where $y = x^{n-1} \log x$.

For, as in the last, we may neglect as we proceed all terms which do not contain $\log x$ as a factor, and thus we get in this case,

$$\dfrac{d^n y}{dx^n} = \dfrac{(n-1)\cdots 2 \cdot 1}{x} = \dfrac{\lfloor n-1}{x}.$$

43. Derived Functions of sin mx.

Let
$$y = \sin mx,$$

then
$$\frac{dy}{dx} = m \cos mx,$$

$$\frac{d^2y}{dx^2} = -m^2 \sin mx,$$

and, in general,
$$\left. \begin{array}{l} \dfrac{d^{2n}y}{dx^{2n}} = (-1)^n m^{2n} \sin mx, \\[1em] \dfrac{d^{2n+1}y}{dx^{2n+1}} = (-1)^n m^{2n+1} \cos mx. \end{array} \right\} \qquad (1)$$

It is easily seen that these may be combined in the single equation (Art. 22),

$$\frac{d^r (\sin mx)}{dx^r} = m^r \sin\left(mx + r\frac{\pi}{2}\right). \qquad (2)$$

In like manner we have

$$\frac{d^r \cos mx}{dx^r} = m^r \cos\left(mx + r\frac{\pi}{2}\right).$$

44. Derived Functions of e^{ax}.

Let $y = e^{ax}$,

then
$$\frac{dy}{dx} = ae^{ax}, \quad \frac{d^2y}{dx^2} = a^2 e^{ax}, \quad \ldots \quad \frac{d^n y}{dx^n} = a^n e^{ax}. \qquad (3)$$

This result may be written in the form

$$\left(\frac{d}{dx}\right)^n \cdot e^{ax} = a^n e^{ax}, \qquad (4)$$

where the symbol $\left(\dfrac{d}{dx}\right)^n$ denotes that the *process of differentiation is applied n times in succession* to the function e^{ax}.

In general, adopting the same notation, we have

$$\left\{ A_0 \left(\frac{d}{dx}\right)^n + A_1 \left(\frac{d}{dx}\right)^{n-1} + A_2 \left(\frac{d}{dx}\right)^{n-2} + \&c. + A_n \right\} e^{ax}$$

$$= A_0 \left(\frac{d}{dx}\right)^n e^{ax} + A_1 \left(\frac{d}{dx}\right)^{n-1} e^{ax} + A_2 \left(\frac{d}{dx}\right)^{n-2} e^{ax} + \&c.$$

$$= A_0 a^n e^{ax} + A_1 a^{n-1} e^{ax} + A_2 a^{n-2} e^{ax} + \&c.$$

$$= \left[A_0 a^n + A_1 a^{n-1} + A_2 a^{n-2} + \&c. \ A_n \right] e^{ax}.$$

This result, if $\phi(x)$ denote the expression

$$A_0 x^n + A_1 x^{n-1} + \ldots A_n,$$

may be written in the form

$$\phi \left(\frac{d}{dx}\right) e^{ax} = \phi(a) e^{ax}; \qquad (5)$$

in which $\phi(a)$ is supposed to contain only *positive integral* powers of a.

45. To find the nth Derived Function of $e^{ax} \cos bx$.— Let y represent the proposed expression,

then
$$\frac{dy}{dx} = ae^{ax} \cos bx - be^{ax} \sin bx$$

$$= e^{ax} (a \cos bx - b \sin bx);$$

if $\tan \phi = \dfrac{b}{a}$, we have $b = \sqrt{a^2 + b^2} \sin \phi$, and $a = \sqrt{a^2 + b^2} \cos \phi$.

Hence we get

$$\frac{dy}{dx} = (a^2 + b^2)^{\frac{1}{2}} e^{ax} \cos (bx + \phi).$$

E

Again,
$$\frac{d^2y}{dx^2} = (a^2 + b^2)^{\frac{1}{2}} e^{ax} [a \cos(bx + \phi) - b \sin(bx + \phi)]$$
$$= (a^2 + b^2) e^{ax} \cos(bx + 2\phi).$$

By repeating this process it is easily seen that we have in general, when n is any positive integer,

$$\frac{d^n y}{dx^n} = (a^2 + b^2)^{\frac{n}{2}} e^{ax} \cos(bx + n\phi). \tag{6}$$

46. To find the Derived Functions of $\tan^{-1}\left(\dfrac{1}{x}\right)$, and $\tan^{-1} x$.

Let $y = \tan^{-1}\left(\dfrac{1}{x}\right)$, or $x = \cot y$:

then
$$\frac{dy}{dx} = \frac{-1}{1 + x^2} = -\sin^2 y.$$

$$\frac{d^2y}{dx^2} = \frac{d}{dx}\left(\frac{dy}{dx}\right) = -\frac{d}{dx}(\sin^2 y) = -\frac{dy}{dx}\frac{d}{dy}(\sin^2 y)$$

$$= \sin^2 y \frac{d}{dy}(\sin^2 y) = \sin^2 y \sin 2y.$$

Again, $\dfrac{d^3y}{dx^3} = \dfrac{d}{dx}(\sin^2 y \sin 2y) = \dfrac{dy}{dx}\dfrac{d}{dy}(\sin^2 y \sin 2y)$

$$= -\sin^2 y \frac{d}{dy}(\sin^2 y \sin 2y)$$

$$= -1 . 2 . \sin^3 y \sin 3y. \qquad (Ex.\ 5,\ Art.\ 28.)$$

Hence, also $\dfrac{d^4y}{dx^4} = 1 . 2 . 3 . \sin^4 y \sin 4y;$

and in general, $\dfrac{d^n y}{dx^n} = (-1)^n \underline{|n-1|} \sin^n y \sin ny.$

Again, since
$$\tan^{-1} x = \frac{\pi}{2} - \tan^{-1}\frac{1}{x},$$

we have
$$\frac{d^n (\tan^{-1} x)}{dx^n} = (-1)^{n-1} \lfloor n-1 \sin^n y \sin ny, \qquad (7)$$

where $y = \cot^{-1} x$, as before.

This result can also be written in the form

$$\frac{d^n (\tan^{-1} x)}{dx^n} = (-1)^{n-1} \lfloor n-1 \frac{\sin\left(n \tan^{-1}\frac{1}{x}\right)}{(1+x^2)^{\frac{n}{2}}}. \qquad (8)$$

47. **If** $y = \sin(m \sin^{-1} x)$, **to prove that**

$$(1-x^2)\frac{d^2y}{dx^2} - x\frac{dy}{dx} + m^2 y = 0. \qquad (9)$$

Here
$$\frac{dy}{dx} = \frac{m \cos(m \sin^{-1} x)}{\sqrt{1-x^2}};$$

$$\therefore (1-x^2)\left(\frac{dy}{dx}\right)^2 = m^2 \cos^2(m \sin^{-1} x) = m^2 (1-y^2).$$

Hence, differentiating a second time, and dividing by $2\frac{dy}{dx}$, we get the required result.

48. Theorem of Leibnitz.—To find the nth differential coefficient of the product of two functions of x. Let $y = uv$; then, adopting the notation of Art. 34, we write

$$y', u', v', \text{ for } \frac{dy}{dx}, \frac{du}{dx}, \text{ and } \frac{dv}{dx},$$

and similarly, y'', u'', v'', &c., for the second and higher derived functions—thus,

$$y^{(n)} = \frac{d^n y}{dx^n}, \quad u^{(n)} = \frac{d^n u}{dx^n}, \text{ &c.}$$

Successive Differentiation.

Now, if we differentiate the equation $y = uv$, we have

$$y' = uv' + vu', \text{ by Art. 13.}$$

The next differentiation gives

$$y'' = uv'' + u'v' + v'u' + vu'' = uv'' + 2u'v' + vu''.$$

The third differentiation gives

$$y''' = uv''' + u'v'' + 2u'v'' + 2u''v' + v'u'' + vu'''$$
$$= uv''' + 3u'v'' + 3u''v' + vu''',$$

in which the coefficients are the same as those in the expansion of $(a + b)^3$.

Suppose that the same law holds for the nth differential coefficient, and that

$$y^{(n)} = uv^{(n)} + nu'v^{(n-1)} + \frac{n(n-1)}{1 \cdot 2} u''v^{(n-2)} + \&c.,$$
$$+ nu^{(n-1)}v' + u^{(n)}v;$$

then, differentiating again, we get

$$y^{(n+1)} = uv^{(n+1)} + u'v^{(n)} + n\left(u'v^{(n)} + u''v^{(n-1)}\right)$$
$$+ \frac{n(n-1)}{2}\left(u''v^{(n-1)} + u'''v^{(n-2)}\right) + \&c. \ldots + u^{(n+1)}v$$
$$= uv^{(n+1)} + (n+1)u'v^{(n)} + \frac{(n+1)n}{1 \cdot 2} u''v^{(n-1)} + \&c. \ldots,$$

in which it can be easily seen that the coefficients follow the law of the Binomial Expansion.

Accordingly, if this law hold for any integer value of n, it holds for the next higher integer; but we have shown that it holds when $n = 3$; therefore it holds for $n = 4$, &c.

Hence it holds for all positive integer values of n.

In the ordinary notation the preceding result becomes

$$\frac{d^n(uv)}{dx^n} = u\frac{d^n v}{dx^n} + n\frac{du}{dx}\frac{d^{n-1}v}{dx^{n-1}} + \frac{n(n-1)}{1 \cdot 2}\frac{d^2u}{dx^2}\frac{d^{n-2}v}{dx^{n-2}} + \&c.$$
$$+ v\frac{d^n u}{dx^n}. \qquad (10)$$

Applications of Leibnitz's Theorem.

49. To prove that

$$\left(\frac{d}{dx}\right)^n (e^{ax} u) = e^{ax}\left(a + \frac{d}{dx}\right)^n u, \qquad (11)$$

where n is a positive integer.

Let $v = e^{ax}$ in the preceding theorem; then, since

$$\frac{dv}{dx} = ae^{ax},\quad \frac{d^2v}{dx^2} = a^2 e^{ax},\ \ldots\ \frac{d^n v}{dx^n} = a^n e^{ax},$$

we have

$$\left(\frac{d}{dx}\right)^n (e^{ax} u) = e^{ax}\left(a^n u + na^{n-1}\frac{du}{dx} + \frac{n(n-1)}{1 \cdot 2} a^{n-2}\frac{d^2 u}{dx^2} + \&\text{c.} + \frac{d^n u}{dx^n}\right);$$

which may be written in the form

$$\left(\frac{d}{dx}\right)^n (e^{ax} u) = e^{ax}\left\{a^n + na^{n-1}\frac{d}{dx} + \frac{n(n-1)}{1 \cdot 2} a^{n-2}\left(\frac{d}{dx}\right)^2 + \&\text{c.} + \left(\frac{d}{dx}\right)^n\right\} u,$$

or

$$\left(\frac{d}{dx}\right)^n (e^{ax} u) = e^{ax}\left(a + \frac{d}{dx}\right)^n u;$$

where the symbolic expression $\left(a + \dfrac{d}{dx}\right)^n$ is supposed to be developed by the Binomial Theorem, and $\dfrac{du}{dx},\ \dfrac{d^2u}{dx^2},\ \ldots\ \dfrac{d^r u}{dx^r}$ substituted for $\left(\dfrac{d}{dx}\right) u,\ \left(\dfrac{d}{dx}\right)^2 u,\ \left(\dfrac{d}{dx}\right)^r u$, in the resulting expansion.

50. In general, if $\phi(a)$ represent any expression involving only *positive integral* powers of a, we shall have

$$\phi\left(\frac{d}{dx}\right) e^{ax} u = e^{ax} \phi\left(a + \frac{d}{dx}\right) u. \qquad (12)$$

For, let $\phi\left(\dfrac{d}{dx}\right)$, when expanded, be of the form

$$A_0 \left(\frac{d}{dx}\right)^n + A_1 \left(\frac{d}{dx}\right)^{n-1} + \ldots + A_n,$$

then the preceding formula holds for each of the component terms, and accordingly it holds for the sum of all the terms; ∴ &c.

The result admits also of being written in the form

$$\phi\left(a + \frac{d}{dx}\right) \cdot u = e^{-ax} \phi\left(\frac{d}{dx}\right)(e^{ax}u).$$

This symbolic equation is of importance in the solution of differential equations with constant coefficients. See "Boole's Differential Equations," chap. XVI.

51. If $y = \sin^{-1} x$, **to prove that**

$$(1 - x^2)\frac{d^{n+2}y}{dx^{n+2}} - (2n + 1) x \frac{d^{n+1}y}{dx^{n+1}} - n^2 \frac{d^n y}{dx^n} = 0. \quad (13)$$

Here $\quad \dfrac{dy}{dx} = \dfrac{1}{\sqrt{1 - x^2}}, \quad$ or $(1 - x^2)^{\frac{1}{2}} \dfrac{dy}{dx} = 1$;

hence, by differentiation,

$$(1 - x^2)^{\frac{1}{2}} \frac{d^2 y}{dx^2} - \frac{x \dfrac{dy}{dx}}{(1 - x^2)^{\frac{1}{2}}} = 0,$$

or $\qquad (1 - x^2) \dfrac{d^2 y}{dx^2} - x \dfrac{dy}{dx} = 0. \qquad (14)$

Again, by Leibnitz's Theorem, we have

$$\left(\frac{d}{dx}\right)^n \left\{(1 - x^2)\frac{d^2 y}{dx^2}\right\} = (1 - x^2)\frac{d^{n+2}y}{dx^{n+2}} - 2nx \frac{d^{n+1}y}{dx^{n+1}} - n(n-1)\frac{d^n y}{dx^n}.$$

Also $\qquad \left(\dfrac{d}{dx}\right)^n \left\{x \dfrac{dy}{dx}\right\} = x \dfrac{d^{n+1}y}{dx^{n+1}} + n \dfrac{d^n y}{dx^n}.$

On subtracting the latter expression from the former, we obtain the required result by (14).

If $x = 0$ in formula (13), it becomes

$$\left(\frac{d^{n+2}y}{dx^{n+2}}\right)_0 - n^2 \left(\frac{d^n y}{dx^n}\right)_0 = 0,$$

Applications of Leibnitz's Theorem.

where $\left(\dfrac{d^n y}{dx^n}\right)_0$ represents the value of $\dfrac{d^n y}{dx^n}$ when x becomes cypher.

Also, since $\left(\dfrac{dy}{dx}\right)_0 = 1$, we get, when n is an *odd* integer,

$$\left(\dfrac{d^{n+2} y}{dx^{n+2}}\right)_0 = 1^2 \cdot 3^2 \cdot 5^2 \ldots n^2.$$

Again we have $\left(\dfrac{d^2 y}{dx^2}\right)_0 = 0$; consequently, when n is an *even* integer, we have $\left(\dfrac{d^n y}{dx^n}\right)_0 = 0$.

52. **If** $y = (1+x^2)^{\frac{m}{2}} \sin(m \tan^{-1} x)$, **to prove that**

$$(1 + x^2)\dfrac{d^2 y}{dx^2} - 2(m - 1)x\dfrac{dy}{dx} + m(m - 1)y = 0. \quad (15)$$

Here
$$\dfrac{dy}{dx} = mx(1 + x^2)^{\frac{m}{2}-1} \sin(m \tan^{-1} x) + m(1 + x^2)^{\frac{m}{2}-1} \cos(m \tan^{-1} x),$$

or
$$(1 + x^2)\dfrac{dy}{dx} = mx(1 + x^2)^{\frac{m}{2}} \sin(m \tan^{-1} x) + m(1+x^2)^{\frac{m}{2}} \cos m(\tan^{-1} x)$$

$$= mxy + m(1 + x^2)^{\frac{m}{2}} \cos(m \tan^{-1} x);$$

$$\therefore (1 + x^2)^{\frac{m}{2}} \cos(m \tan^{-1} x) = \dfrac{1 + x^2}{m}\dfrac{dy}{dx} - xy.$$

The required result is obtained by differentiating the last equation, and eliminating $\cos(m \tan^{-1} x)$ and $\sin(m \tan^{-1} x)$ by aid of the two former.

Again, applying Leibnitz's Theorem as in the last Article, we get, in general—

$$(1 + x^2)\dfrac{d^{n+2} y}{dx^{n+2}} + 2(n - m + 1)x\dfrac{d^{n+1} y}{dx^{n+1}} + (n - m)(n - m + 1)\dfrac{d^n y}{dx^n} = 0.$$

Hence, when $x = 0$, we have

$$\left(\frac{d^{n+2}y}{dx^{n+2}}\right)_0 + (n - m)(n - m + 1)\left(\frac{d^n y}{dx^n}\right)_0 = 0.$$

Moreover, as when $x = 0$, we have $y = 0$, and $\dfrac{dy}{dx} = m$; it follows from the preceding that

$$\left(\frac{d^{2n}y}{dx^{2n}}\right)_0 = 0\;;\; \left(\frac{d^{2n+1}y}{dx^{2n+1}}\right)_0 = (-1)^n m(m-1)\ldots(m-2n). \quad (16)$$

For a complete discussion of this, and other analogous expressions, the student is referred to Bertrand, "Traité de Calcul Différentiel," p. 144, &c.

EXAMPLES.

1. $y = x^4 \log x$,   prove that $\dfrac{d^6y}{dx^6} = -\dfrac{\lfloor 4}{x^2}$.

2. $y = x \log x$,   ,, $\dfrac{d^n y}{dx^n} = (-1)^n \dfrac{1 \cdot 2 \ldots (n-2)}{x^{n-1}}$.

3. $y = x^x$,   ,, $\dfrac{d^2y}{dx^2} = x^x (1 + \log x)^2 + x^{x-1}$.

4. $y = \log (\sin x)$,   ,, $\dfrac{d^3y}{dx^3} = \dfrac{2 \cos x}{\sin^3 x}$.

5. $y = \tan^{-1} \dfrac{\sqrt{1+x^2} - 1}{x} + \tan^{-1} \dfrac{2x}{1-x^2}$,   ,, $\dfrac{d^2y}{dx^2} = -\dfrac{5x}{(1+x^2)^2}$.

6. $y = x^4 \log (x^3)$,   ,, $\dfrac{d^5y}{dx^5} = \dfrac{24}{x}$.

7. $y = \log \sqrt{\dfrac{1 + x\sqrt{2} + x^2}{1 - x\sqrt{2} + x^2}} + \tan^{-1} \dfrac{x\sqrt{2}}{1-x^2}$,   ,, $\dfrac{d^2y}{dx^2} = -\dfrac{8\sqrt{2} \cdot x^3}{(1+x^4)^2}$.

8. $y = e^{rx} \sin x$,   ,, $\dfrac{d^n y}{dx^n} = \dfrac{e^{rx} \sin (x + n\phi)}{\sin^n \phi}$,

where $\tan \phi = \dfrac{1}{r}$.

9. If $y = e^{ax} x^r$,   prove that

$$\dfrac{d^n y}{dx^n} = e^{ax} \left[a^n x^r + nra^{n-1} x^{r-1} + \dfrac{n(n-1)r(r-1)}{1 \cdot 2} a^{n-2} x^{r-2} + \ldots \right],$$

and $\left(\dfrac{d}{dx}\right)^n (e^{ax} x^r) = \left(\dfrac{a}{x}\right)^{n-r} \left(\dfrac{d}{dx}\right)^r (e^{ax} x^n)$.

10. If $y = a \cos (\log x) + b \sin (\log x)$,

prove that $x^2 \dfrac{d^2 y}{dx^2} + x \dfrac{dy}{dx} + y = 0$.

11. If $y = e^{a \sin^{-1} x}$,

prove that $(1 - x^2) \dfrac{d^2 y}{dx^2} - x \dfrac{dy}{dx} = a^2 y$.

Examples.

12. Prove that the equation
$$(1-x^2)\frac{d^2y}{dx^2} - x\frac{dy}{dx} + a^2y = 0$$
is satisfied by either of the following values of y:
$$y = \cos(a \sin^{-1} x), \text{ or } y = e^{a\sqrt{-1}\sin^{-1}x}.$$

13. Being given that $y = (x + \sqrt{x^2 - 1})^m$,

prove that
$$(x^2 - 1)\frac{d^2y}{dx^2} + x\frac{dy}{dx} - m^2 y = 0.$$

14. If $y = \sin(\sin x)$,

prove that
$$\frac{d^2y}{dx^2} + \frac{dy}{dx}\tan x + y\cos^2 x = 0.$$

15. In Fig. 3, Art. 37, if AB be regarded as a side of a regular polygon of an indefinitely great number of sides, show that the difference between the circumference of the circle and the perimeter of the polygon is represented by $\frac{\pi}{6} BD$, to the second order of infinitesimals.

16. If $y = A\cos nx + B\sin nx$, prove that $\left(\dfrac{d^2}{dx^2} + n^2\right)y = 0$.

17. If $y = \dfrac{1}{a^2 + x^2}$, prove that $\dfrac{d^n y}{dx^n} = (1-)^n \dfrac{\underline{|n}\,.\,\sin^{n+1}\phi \sin(n+1)\phi}{a^{n+2}}$,

where $\phi = \tan^{-1}\dfrac{a}{x}$.

This follows at once from Art. 46, since $\dfrac{d}{dx}\left(\tan^{-1}\dfrac{a}{x}\right) = \dfrac{-a}{a^2+x^2}$. It can also be proved otherwise, as follows:

$$\frac{1}{a^2+x^2} = \frac{1}{2a(-1)^{\frac{1}{2}}}\left[\frac{1}{x-a(-1)^{\frac{1}{2}}} - \frac{1}{x+a(-1)^{\frac{1}{2}}}\right];$$

$$\therefore \frac{d^n y}{dx^n} = \frac{1}{2a(-1)^{\frac{1}{2}}}\left(\frac{d}{dx}\right)^n \cdot \frac{1}{x-a(-1)^{\frac{1}{2}}} - \frac{1}{2a(-1)^{\frac{1}{2}}}\left(\frac{d}{dx}\right)^n \cdot \frac{1}{x+a(-1)^{\frac{1}{2}}}$$

$$= \frac{(-1)^n\,1\,.\,2\,\ldots\,n}{2a(-1)^{\frac{1}{2}}}\left[\frac{1}{(x-a(-1)^{\frac{1}{2}})^{n+1}} - \frac{1}{(x+a(-1)^{\frac{1}{2}})^{n+1}}\right]$$

$$= \frac{(-1)^n\,\underline{|n}}{2a(-1)^{\frac{1}{2}}}\left[\frac{(x+a(-1)^{\frac{1}{2}})^{n+1} - (x-a(-1)^{\frac{1}{2}})^{n+1}}{(x^2+a^2)^{n+1}}\right].$$

Examples.

Again, since $\dfrac{a}{x} = \tan \phi$, we have $a = \sqrt{a^2 + x^2} \sin \phi$, and $x = \sqrt{a^2 + x^2} \cos \phi$;

hence
$$(x + a(-1)^{\frac{1}{2}})^{n+1} = (a^2 + x^2)^{\frac{n+1}{2}} (\cos \phi + (-1)^{\frac{1}{2}} \sin \phi)^{n+1}$$

$$= (a^2 + x^2)^{\frac{n+1}{2}} \{\cos(n+1)\phi + (-1)^{\frac{1}{2}} \sin(n+1)\phi\},$$

and we get, finally,

$$\dfrac{d^n y}{dx^n} = (-1)^n \dfrac{\lfloor n \cdot \sin(n+1)\phi \cdot \sin^{n+1} \phi}{a^{n+2}}.$$

18. In like manner, if $y = \dfrac{x}{a^2 + x^2}$,

prove that
$$\dfrac{d^n y}{dx^n} = (-1)^n \dfrac{\lfloor n \cdot \sin^{n+1} \phi \cdot \cos(n+1)\phi}{a^{n+1}}.$$

19. If $u = xy$,

prove that
$$\dfrac{d^n u}{dx^n} = x \dfrac{d^n y}{dx^n} + n \dfrac{d^{n-1} y}{dx^{n-1}}.$$

✓ 20. If $u = (\sin^{-1} x)^2$,

prove that
$$(1 - x^2) \dfrac{d^2 u}{dx^2} - x \dfrac{du}{dx} = 2.$$

21. Prove, from the preceding, that

$$(1 - x^2) \dfrac{d^{n+2} u}{dx^{n+2}} - (2n+1) x \dfrac{d^{n+1} u}{dx^{n+1}} - n^2 \dfrac{d^n u}{dx^n} = 0;$$

and
$$\left(\dfrac{d^{n+2} u}{dx^{n+2}}\right)_0 = n^2 \left(\dfrac{d^n u}{dx^n}\right)_0.$$

✓ 22. If $y = e^{ax} \sin bx$, prove that $\dfrac{d^2 y}{dx^2} - 2a \dfrac{dy}{dx} + (a^2 + b^2) y = 0.$

23. Given $y = \dfrac{ax + b}{x^2 - c^2}$, find $\dfrac{d^n y}{dx^n}$.

Here
$$\dfrac{ax+b}{x^2-c^2} = \dfrac{ac+b}{2c} \dfrac{1}{x-c} + \dfrac{ac-b}{2c} \dfrac{1}{x+c}.$$

Hence
$$\dfrac{d^n y}{dx^n} = \dfrac{(-1)^n \lfloor n}{2c} \left(\dfrac{ac+b}{(x-c)^{n+1}} + \dfrac{ac-b}{(x+c)^{n+1}}\right).$$

CHAPTER III.

DEVELOPMENT OF FUNCTIONS.

53. Lemma.—If u be a function of $x + y$ which is finite and continuous for all values of $x + y$, between the limits a and b, then for all such values we shall have

$$\frac{du}{dx} = \frac{du}{dy}.$$

For, let $u = f(x + y)$, then if x become $x + h$,

$$\frac{du}{dx} = \text{limit of } \frac{f(x + y + h) - f(x + y)}{h},$$

when h is infinitely small.

Similarly, if y become $y + h$, we have

$$\frac{du}{dy} = \text{limit of } \frac{f(x + y + h) - f(x + y)}{h},$$

which is the same expression as before.

Hence
$$\frac{du}{dx} = \frac{du}{dy}.$$

Otherwise thus:—Let $z = x + y$, then $u = f(z)$,

$$\frac{dz}{dx} = 1, \text{ and } \frac{dz}{dy} = 1;$$

$$\frac{du}{dx} = \frac{du}{dz}\frac{dz}{dx} = f'(z);$$

$$\frac{du}{dy} = \frac{du}{dz}\frac{dz}{dy} = f'(z) = \frac{du}{dx}.$$

54. If $f(x + y)$ be a continuous function, which does not become infinite when $y = 0$, its expansion in powers of y can contain *no negative powers*; for, suppose it contains a term of the form My^{-m}, where M is independent of y, this term would become infinite when $y = 0$; but the given function in that case reduces to $f(x)$; hence we should have $f(x) = \infty$, which is contrary to our hypothesis. Consequently the expansion of $f(x + y)$ can contain only positive powers of y.

Again, if $f(x)$ and its successive derived functions be *finite* and continuous, the expansion of $f(x + y)$ can contain no *fractional* power of y. For, if it contain a term of the form $Py^{n+\frac{p}{q}}$, where $\frac{p}{q}$ is a proper fraction, then its $(n + 1)^{th}$ derived function with respect to y would contain y with a *negative* index, and, accordingly, would become *infinite* when $y = 0$; which is contrary to our hypothesis.

Hence, with the conditions expressed above, the expansion of $f(x + y)$ can contain only *positive integral* powers of y.

55. Taylor's Expansion of $f(x + y)$.*—Assuming that the function $f(x + y)$ is capable of being expanded in powers of y, then by the preceding this equation must be of the form

$$f(x + y) = P_0 + P_1 y + P_2 y^2 + \&c. + P_n y^n + \&c.,$$

in which $P_0, P_1, \ldots P_n$ are supposed to be finite and continuous functions of x.

When $y = 0$, this expansion reduces to $f(x) = P_0$.

Again, let $u = f(x + y)$; then by differentiation we have

$$\frac{du}{dx} = \frac{dP_0}{dx} + y \frac{dP_1}{dx} + y^2 \frac{dP_2}{dx} + \ldots + y^n \frac{dP_n}{dx} + \&c.;$$

$$\frac{du}{dy} = P_1 + 2 P_2 y + 3 P_3 y^2 + \&c.$$

---

* The investigation in this Article is introduced for the purpose of showing the beginner, in a simple manner, how Taylor's series can be arrived at. It is based on the assumption that the function $f(x + y)$ is capable of being expanded in a series of powers of y, and that it is also a continuous function. It demonstrates that whenever the function represented by $f(x + y)$ is capable of being expanded in a convergent series of positive ascending powers of y, the series must necessarily coincide with the form given in (1). An investigation of the conditions of convergency of the series, and of the applicability of the Theorem in general, will be introduced in a subsequent part of the Chapter. The particular case of this Theorem when $f(x)$ is a rational algebraic expression of the nth degree in x is already familiar to the student who has read the Theory of Equations.

Now, in order that these series should be *identical* for all values of y the coefficients of like powers must be equal. Accordingly, we must have

$$P_1 = \frac{dP_0}{dx} = \frac{df(x)}{dx} = f'(x),$$

$$P_2 = \frac{1}{1 \cdot 2}\frac{dP_1}{dx} = \frac{1}{1 \cdot 2}\frac{d^2f(x)}{dx^2} = \frac{1}{1 \cdot 2}f''(x),$$

$$P_3 = \frac{1}{3}\frac{dP_2}{dx} = \frac{1}{1 \cdot 2 \cdot 3}\frac{d^3f(x)}{dx^3} = \frac{1}{1 \cdot 2 \cdot 3}f'''(x);$$

and in general,

$$P_n = \frac{1}{1 \cdot 2 \ldots n}\frac{d^n f(x)}{dx^n} = \frac{1}{1 \cdot 2 \ldots n}f^{(n)}(x).$$

Accordingly, when $f(x)$ and its successive derived functions are finite and continuous we have

$$f(x+y) = f(x) + \frac{y}{1}f'(x) + \frac{y^2}{1 \cdot 2}f''(x) + \ldots + \frac{y^n}{\underline{|n}}f^{(n)}(x) + \ldots \quad (1)$$

This expansion is called Taylor's Theorem, having been first published, in 1715, by Dr. Brook Taylor in his *Methodus Incrementorum*.

It may also be written in the form

$$f(x+y) = f(x) + \frac{y}{1}\frac{df(x)}{dx} + \frac{y^2}{1 \cdot 2}\frac{d^2f(x)}{dx^2} \ldots + \frac{y^n}{\underline{|n}}\frac{d^nf(x)}{dx^n} + \ldots; \quad (2)$$

or, if $u = f(x)$, and $u_1 = f(x+y)$,

$$u_1 = u + \frac{y}{1}\frac{du}{dx} + \frac{y^2}{1 \cdot 2}\frac{d^2u}{dx^2} + \ldots + \frac{y^n}{\underline{|n}}\frac{d^nu}{dx^n} + \&c. \quad (3)$$

To complete the preceding proof it will be necessary to obtain an expression for the limit of the sum of the series after n terms, in order to determine whether the series is convergent or divergent. We postpone this discussion for the present, and shall proceed to illustrate the Theorem by showing that the expansions usually given in elementary treatises on Algebra and Trigonometry are particular cases of it.

56. The Binomial Theorem.—Let $u = (x + y)^n$; here $f(x) = x^n$, therefore, by Art. 41,

$$f'(x) = nx^{n-1}, \ldots f^{(r)}(x) = n(n-1)\ldots(n-r+1)x^{n-r}.$$

Hence the expansion becomes

$$(x+y)^n = x^n + \frac{n}{1}x^{n-1}y + \frac{n(n-1)}{1 \cdot 2}x^{n-2}y^2 + \ldots$$

$$+ \frac{n(n-1)\ldots(n-r-1)}{1 \cdot 2 \ldots r}x^{n-r}y^r + \&c. \qquad (4)$$

If n be a positive integer this consists of a finite number of terms; we shall subsequently examine the validity of the expansion when applied to the case where n is negative or fractional.

57. The Logarithmic Series.—To expand $\log(x+y)$.

Here $\quad f(x) = \log(x), \quad f'(x) = \frac{1}{x}, \quad f''(x) = -\frac{1}{x^2},$

$$f'''(x) = \frac{2}{x^3}, \ldots f^{(n)}(x) = (-1)^{n-1}\frac{1 \cdot 2 \ldots (n-1)}{x^n}.$$

Accordingly

$$\log(x+y) = \log x + \frac{y}{x} - \frac{1}{2}\frac{y^2}{x^2} + \frac{1}{3}\frac{y^3}{x^3} - \frac{1}{4}\frac{y^4}{x^4} + \&c.$$

If $x = 1$ this series becomes

$$\log(1+y) = \frac{y}{1} - \frac{y^2}{2} + \frac{y^3}{3} - \ldots (-1)^{n-1}\frac{y^n}{n} \ldots \&c. \qquad (5)$$

When taken to the base a, we get, by Art. 29,

$$\log_a(1+y) = M\left(\frac{y}{1} - \frac{y^2}{2} + \frac{y^3}{3} - \frac{y^4}{4} + \&c.\right). \qquad (6)$$

58. To expand $\sin(x+y)$.

Here $f(x) = \sin x, \quad f'(x) = \cos x,$

$f''(x) = -\sin x, \quad f'''(x) = -\cos x, \text{ &c.}$

Hence

$$\sin(x+y) = \sin x \left(1 - \frac{y^2}{1\cdot 2} + \frac{y^4}{1\cdot 2\cdot 3\cdot 4} - \text{&c.} \pm \frac{y^{2n}}{\underline{|2n}} \ldots \right)$$

$$+ \cos x \left(\frac{y}{1} - \frac{y^3}{1\cdot 2\cdot 3} + \frac{y^5}{1\cdot 2\cdot 3\cdot 4\cdot 5} \ldots \pm \frac{y^{2n-1}}{\underline{|2n-1}} \ldots \right). \quad (7)$$

As the preceding series is supposed to hold for all values, it must hold when $x = 0$, in which case it becomes

$$\sin y = \frac{y}{1} - \frac{y^3}{1\cdot 2\cdot 3} + \frac{y^5}{1\cdot 2\cdot 3\cdot 4\cdot 5} - \text{&c.} \quad (8)$$

Similarly, if $x = \frac{\pi}{2}$, we get

$$\cos y = 1 - \frac{y^2}{1\cdot 2} + \frac{y^4}{1\cdot 2\cdot 3\cdot 4} - \text{&c.} \quad (9)$$

We thus arrive at the well-known expansions* for the sine and cosine of an angle, in terms of its circular measure.

59. Maclaurin's Theorem.

—If we make $x = 0$, in Taylor's Expansion, it becomes

$$f(y) = f(0) + \frac{y}{1}f'(0) + \frac{y^2}{1\cdot 2}f''(0) + \ldots + \frac{y^n}{\underline{|n}}f^{(n)}(0) + \ldots, \quad (10)$$

where $f(0) \ldots f^{(n)}(0)$ represent the values which $f(x)$ and its successive derived functions assume when $x = 0$.

Substitute x for y in the preceding series and it becomes

$$f(x) = f(0) + \frac{x}{1}f'(0) + \frac{x^2}{1\cdot 2}f''(0) + \ldots + \frac{x^n}{\underline{|n}}f^{(n)}(0) + \text{&c.}$$

---

* These expansions are due to Newton, and were obtained by him by the method of reversion of series from the expansion of the arc in terms of its sine. This latter series he deduced from its derived function by a process analogous to integration (called by Newton the method of quadratures). See *Opuscula*, tom I., pp. 19, 21. Ed. Cast. Compare Art. 64, p. 68.

This result may be established otherwise thus; adopting the same limitation as in the case of Taylor's Theorem:—

Assume $f(x) = A + Bx + Cx^2 + Dx^3 + Ex^4 + \&c.$
then $\quad f'(x) = B + 2Cx + 3Dx^2 + 4Ex^3 + \&c.$
$\quad\quad f''(x) = 2C + 3.2Dx + 4.3Ex^2 + \&c.$
$\quad\quad f'''(x) = 3.2D + 4.3.2Ex + \&c.$

Hence, making $x = 0$ in each of these equations, we get

$$f(0) = A, \quad f'(0) = B, \quad \frac{f''(0)}{1 \cdot 2} = C, \quad \frac{f'''(0)}{1 \cdot 2 \cdot 3} = D, \&c.$$

whence we obtain the same series as before.

The preceding expansion is usually called Maclaurin's* Theorem; it was, however, previously given by Stirling, and is, as is shown already, but a particular case of Taylor's series. We proceed to illustrate it by a few examples.

60. **Exponential Series.**—Let $y = a^x$.

Here $\quad f(x) = a^x,\quad\quad$ hence $f(0) = 1,$
$\quad\quad f'(x) = a^x \log a, \quad\quad\quad$ „ $\quad f'(0) = \log a,$
$\quad\quad f''(x) = a^x (\log a)^2, \quad\quad$ „ $\quad f''(0) = (\log a)^2,$
$\quad\quad f^{(n)}(x) = a^x (\log a)^n, \quad$ „ $\quad f^{(n)}(0) = (\log a)^n;$

and the expansion is

$$a^x = 1 + \frac{(x \log a)}{1} + \frac{(x \log a)^2}{1 \cdot 2} + \ldots + \frac{(x \log a)^n}{1 \cdot 2 \ldots n} + \&c. \quad (11)$$

If e, the base of the Napierian system of Logarithms, be substituted for a, the preceding expansion becomes

$$e^x = 1 + \frac{x}{1} + \frac{x^2}{1 \cdot 2} + \ldots + \frac{x^n}{1 \cdot 2 \ldots n} + \ldots \quad (12)$$

---

* Maclaurin laid no claim to the theorem which is known by his name, for, after proving it, he adds—"This theorem was given by Dr. Taylor, *Method. Increm.*" See Maclaurin's *Fluxions*, vol. ii., Art. 751.

F

If $x = 1$ this gives for e the same value as that adopted in Art. 29, viz.:

$$e = 1 + \frac{1}{1} + \frac{1}{1.2} + \frac{1}{1.2.3} + \frac{1}{1.2.3.4} + \ldots$$

61. Expansion of $\sin x$ and $\cos x$ by Maclaurin's Theorem. Let $f(x) = \sin x$, then

$$f(0) = 0, \quad f'(0) = 1, \quad f''(0) = 0, \quad f'''(0) = -1, \text{ &c.,}$$

and we get

$$\sin x = \frac{x}{1} - \frac{x^3}{1.2.3} + \frac{x^5}{1.2.3.4.5} - \text{&c.} \ldots$$

In like manner

$$\cos x = 1 - \frac{x^2}{1.2} + \frac{x^4}{1.2.3.4} - \ldots;$$

the same expansions as already arrived at in Art. 58.

Since $\sin(-x) = -\sin x$, we might have inferred at once that the expansion for $\sin x$ in terms of x can only consist of *odd* powers of x. Similarly, as $\cos(-x) = \cos x$, the expansion of $\cos x$ can only contain *even* powers.

In general, if $F(x) = F(-x)$, the development of $F(x)$ can only consist of *even* powers of x. If $F(-x) = -F(x)$, the expansion can contain *odd* powers of x only.

Thus, the expansions of $\tan x$, $\sin^{-1} x$, $\tan^{-1} x$, &c., can contain no even powers of x; those of $\cos x$, $\sec x$, &c., no odd powers.

62. Huygens' Approximation to length of Circular Arc.*—If A be the chord of any circular arc, and B that of half the arc; then the length of the arc is equal to $\dfrac{8B - A}{3}$, q.p.

For, let R be the radius of the circle, and L the length of the arc: and we have

$$\frac{A}{R} = 2 \sin \frac{L}{2R}, \quad \frac{B}{R} = 2 \sin \frac{L}{4R},$$

---

* This important approximation is due to Huygens. The demonstration given above is that of Newton, and is introduced by him as an application of his expansion for the sine of an angle. *Vid.* "Epis. Prior ad Oldemburgium."

hence, by (8),

$$A = L - \frac{L^3}{2.3.4.R^2} + \frac{L^5}{2.3.4.5.16.R^4} - \&c.$$

$$8B = 4L - \frac{L^3}{2.3.4.R^2} + \frac{L^5}{2.3.4.5.64.R^4} - \&c.$$

consequently, neglecting powers of $\frac{L}{R}$ beyond the fourth, we get

$$\frac{8B - A}{3} = L\left(1 - \frac{L^4}{7680 R^4}\right). \qquad (13)$$

Hence, for an arc equal in length to the radius the error in adopting Huygens' approximation in less than $\frac{1}{7680}$th part of the whole arc; for an arc of half the length of the radius the proportionate error is one-sixteenth less; and so on.

In practice the approximation* is used in the form

$$L = 2B + \frac{1}{3}(2B - A).$$

This simple mode of finding approximately the length of an arc of a circle is much employed in practice. It may also be applied to find the approximate length of a portion of any continuous curve, by dividing it into an even number of suitable intervals, and regarding the intervals as approximately circular. See Rankine's Rules and Tables, Part I., Section 4.

---

* To show the accuracy of this approximation, let us apply it to find the length of an arc of 30° in a circle whose radius is 100,000 feet.

Here $\qquad B = 2R \sin 7° 30', \quad A = 2R \sin 15°;$

but, from the Tables,

$\qquad \sin 7° 30' = .1305268, \quad \sin 15° = .2588190.$

Hence $\qquad 2B + \dfrac{2B - A}{3} = 52359.71.$

The true value, assuming $\pi = 3.1415926$, is 52359.88; whence the error is but .17 of a foot, or about 2 inches.

63. Expansion of $\tan^{-1}x$.—Assume, according to Art. 61, the expansion of $\tan^{-1}x$ to be

$$Ax + Bx^3 + Cx^5 + Dx^7 + \&c.,$$

where A, B, C, &c., are undetermined coefficients:

then $\quad\dfrac{d.\tan^{-1}x}{dx} = A + 3Bx^2 + 5Cx^4 + 7Dx^6 + \&c.$;

but $\quad\dfrac{d.\tan^{-1}x}{dx} = \dfrac{1}{1+x^2} = 1 - x^2 + x^4 - x^6 + \&c.,$

when x lies between the limits ± 1.

Comparing coefficients, we have

$$A = 1, \quad B = -\frac{1}{3}, \quad C = \frac{1}{5}, \quad D = -\frac{1}{7}, \&c.$$

Hence

$$\tan^{-1}x = \frac{x}{1} - \frac{x^3}{3} + \frac{x^5}{5} - \ldots + (-1)^n\frac{x^{2n+1}}{2n+1} + \ldots; \quad (14)$$

when x is less than unity.

This expansion can be also deduced directly from Maclaurin's Theorem, by aid of the results given in Art. 46. This is left as an exercise for the student.

64. Expansion of $\sin^{-1}x$.—Assume, as before,

$$\sin^{-1}x = Ax + Bx^3 + Cx^5 + \&c.;$$

then $\quad\dfrac{1}{(1-x^2)^{\frac{1}{2}}} = A + 3Bx^2 + 5Cx^4 + \&c.$;

but $\quad\dfrac{1}{(1-x)^{\frac{1}{2}}} = (1-x^2)^{-\frac{1}{2}} = 1 + \dfrac{1}{2}x^2 + \dfrac{1\cdot 3}{2\cdot 4}x^4 + \ldots$

$$+ \frac{1\cdot 3 \ldots 2r-1}{2\cdot 4 \ldots 2r}x^{2r} + \ldots$$

Hence, comparing coefficients, we get

$$A = 1, \quad B = \frac{1}{2}\cdot\frac{1}{3}, \quad C = \frac{1\cdot 3}{2\cdot 4}\cdot\frac{1}{5}, \&c.$$

Finally,

$$\sin^{-1}x = \frac{x}{1} + \frac{1}{2}\cdot\frac{x^3}{3} + \frac{1\cdot 3}{2\cdot 4}\cdot\frac{x^5}{5} + \ldots + \frac{1\cdot 3\ldots 2r-1}{2\cdot 4\ldots 2r}\cdot\frac{x^{2r+1}}{2r+1} + \ldots \quad (15)$$

Since we have assumed that $\sin^{-1}x$ vanishes along with x we must in this expansion regard $\sin^{-1}x$ as being the circular measure of the *acute* angle whose sine is x.

There is no difficulty in determining the general formula for other values of $\sin^{-1}x$, if requisite.

A direct proof of the preceding result can be deduced from Maclaurin's expansion by aid of Art. 51. We leave this as an exercise for the student.

From the preceding expansion the value of π can be exhibited in the following series:

$$\frac{\pi}{6} = \frac{1}{2} + \frac{1}{2\cdot 3}\frac{1}{8} + \frac{1\cdot 3}{2\cdot 4\cdot 5}\frac{1}{32} + \&c.$$

For, since $\sin 30° = \frac{1}{2}$, we have $\frac{\pi}{6} = \sin^{-1}\frac{1}{2}$; $\therefore$ &c.

An approximate* value of π can be arrived at by the aid of this formula; at the same time it may be observed that many other expansions are better adapted for this purpose.

65. Euler's Expressions for Sine and Cosine.—In the exponential series (12), if $x\sqrt{-1}$ be substituted for x, we get

$$e^{x\sqrt{-1}} = 1 - \frac{x^2}{1\cdot 2} + \frac{x^4}{1\cdot 2\cdot 3\cdot 4} - \&c. \ldots$$

$$+ \sqrt{-1}\left[\frac{x}{1} - \frac{x^3}{1\cdot 2\cdot 3} + \&c. \ldots\right]$$

$$= \cos x + \sqrt{-1}\sin x; \text{ by Art. 59.}$$

Similarly, $e^{-x\sqrt{-1}} = \cos x - \sqrt{-1}\sin x.$

Hence $\left.\begin{array}{l} e^{x\sqrt{-1}} + e^{-x\sqrt{-1}} = 2\cos x, \\[4pt] e^{x\sqrt{-1}} - e^{-x\sqrt{-1}} = 2\sqrt{-1}\sin x. \end{array}\right\}$ (16)

A more complete development of these formulæ will be found in treatises on Algebra and Trigonometry.

---

* The expansion for $\sin^{-1}x$, and also this method of approximating to π, were given by Newton.

66. John Bernoulli's Series.—If, in Taylor's Expansion (1) we make $y = -x$, and transfer $f(x)$ to the other side of the equation, we get

$$f(x) = f(0) + xf'(x) - \frac{x^2}{1 \cdot 2} f''(x) + \frac{x^3}{1 \cdot 2 \cdot 3} f'''(x) - \&c. \quad (17)$$

This is equivalent to the series known as Bernoulli's,[*] and published by him in *Act. Lips.*, 1694.

As an example of this expansion, let $f(x) = e^x$; then

$$f(0) = 1, \quad f'(x) = e^x, \quad f''(x) = e^x, \&c.,$$

and we get

$$e^x = 1 + xe^x - \frac{x^2}{1 \cdot 2} e^x + \&c.,$$

Or, dividing by e^x, and transposing,

$$e^{-x} = 1 - x + \frac{x^2}{1 \cdot 2} - \&c.,$$

which agrees with Art. 60.

67. Symbolic Form of Taylor's Theorem.—The expansion

$$f(x + y) = f(x) + y \frac{d}{dx} \cdot f(x) + \frac{y^2}{1 \cdot 2} \left(\frac{d}{dx}\right)^2 \cdot f(x) + \&c.$$

may be written in the form

$$f(x + y) = \left\{ 1 + y \frac{d}{dx} + \frac{y^2}{1 \cdot 2} \left(\frac{d}{dx}\right)^2 + \ldots + \frac{y^n}{\underline{|n}} \left(\frac{d}{dx}\right)^n + \ldots \right\} f(x), \quad (18)$$

in which the student will perceive that the terms within the brackets proceed according to the law of the exponential series (12); the equation may accordingly be written in the shape

$$f(x + y) = e^{y \frac{d}{dx}} f(x), \quad (19)$$

---

[*] In his *Reduc. Quad. ad long. curv.*, John Bernoulli introduces this theorem again, adding—"Quam eandem scriem postea Taylorus, interjecto viginti annorum intervallo, in librum quem edidit, A.D. 1715, *de methodo incrementorum*, transferre dignatus est sub alio tantum characterum habitu." The great injustice of this statement need not be insisted on; for while Taylor's Theorem is one of the most important in the entire range of analysis, that of Bernoulli is comparatively of little use; and is, as shown above, but a simple case of Taylor's Expansion.

where $e^{y\frac{d}{dx}}$ is supposed to be expanded as in the exponential theorem, and $\dfrac{y^n}{\underline{|n}} \dfrac{d^n f(x)}{dx^n}$ written for $\dfrac{y^n}{\underline{|n}} \left(\dfrac{d}{dx}\right)^n f(x)$, &c.

This form of Taylor's Theorem is of extensive application in the Calculus of Finite Differences.

68. Other Forms derived from Taylor's Series.—In the expansion (3), Art. 55, substitute h for y,

then $\quad u_1 = u + \dfrac{h}{1}\dfrac{du}{dx} + \dfrac{h^2}{1 \cdot 2}\dfrac{d^2 u}{dx^2} + \ldots \dfrac{h^n}{1 \cdot 2 \ldots n}\dfrac{d^n u}{dx^n} +$ &c.

If now h be diminished indefinitely, it may be represented by dx, and the series becomes

$$u_1 = u + \frac{du}{dx}\frac{dx}{1} + \frac{d^2 u}{dx^2}\frac{dx^2}{1 \cdot 2} + \ldots + \frac{d^n u}{dx^n}\frac{dx^n}{1 \cdot 2 \ldots n}\ldots,$$

or $\quad u_1 - u = \dfrac{f'(x)}{1} dx + \dfrac{f''(x)}{1 \cdot 2} dx^2 + \dfrac{f'''(x)}{1 \cdot 2 \cdot 3} dx^3 +$ &c., $\quad$ (20)

in which $u_1 - u$ is the *complete* increment of u, corresponding to the increment dx in x.

Again, since each term in this expansion is infinitely small in comparison with the preceding one, if all the terms after the first be neglected (by Art. 38) as being infinitely small in comparison with it, we get

$$du = f'(x)\, dx,$$

the same result as given in Art. 7.

Another form of the preceding expansion is

$$u_1 - u = \frac{du}{1} + \frac{d^2 u}{1 \cdot 2} + \frac{d^3 u}{1 \cdot 2 \cdot 3} + \ldots + \frac{d^n u}{1 \cdot 2 \ldots n} + \&c. \quad (21)$$

69. Theorem.—*If a function of x become infinite for any finite value of x then all its successive derived functions become infinite at the same time.*

If the function be algebraic, the only way that it can become infinite for a finite value of x is by its containing a term of the form $\dfrac{P}{Q}$, in which Q vanishes for one or more values of x for which P remains finite. Accordingly, let $u = \dfrac{P}{Q}$: then $\dfrac{du}{dx} = \dfrac{\dfrac{dP}{dx} - \dfrac{P}{Q}\dfrac{dQ}{dx}}{Q}$; this also becomes infinite when $Q = 0$.

Similarly, $\dfrac{d^2u}{dx^2}, \dfrac{d^3u}{dx^3}$, &c., each become infinite when $Q = 0$.

Again, certain transcendental functions, such as $e^{\frac{1}{(x-a)^2}}$, cosec $(x - a)$, &c., become infinite when $x = a$; but it can be easily shown, by differentiation, that their derived functions also become infinite at the same time. Similar remarks apply in all other cases.

The student who desires a more general investigation is referred to De Morgan's Calculus, page 179.

70. Remarks on Taylor's Expansion.—In the preceding applications of Taylor's Theorem, the series arrived at (Art. 56 excepted) each consisted of an *infinite number* of terms; and it has been assumed in our investigation that the sum of these infinite series has, in each case, a *finite limiting value*, represented by the original function, $f(x + y)$, or $f(x)$. In other words, we have assumed that the *remainder* of the series after n terms, in each case, becomes infinitely small when n is taken sufficiently large—or, that the series is *convergent*. The meaning of this term will be explained in the next Article.

71. Convergent and Divergent Series.—A series, $u_1, u_2, u_3, \ldots u_n, \ldots$ consisting of an indefinite number of terms, which succeed each other according to some fixed law, is said to be *convergent*, when the sum of its first n terms approaches nearer and nearer to a finite limiting value, according as n is taken greater and greater; and this limiting value is called the sum of the series, from which it can be made to differ by an amount less than any assigned quantity, on taking a sufficient number of terms. It is evident that in the case of a convergent series the terms become indefinitely small when n is taken indefinitely great.

If the sum of the first n terms approximates to no finite limit the series is said to be *divergent*.

In general, a series consisting of real and positive terms is convergent whenever the sum of its first n terms does not increase indefinitely with n. For, if this sum do not become indefinitely great as n increases, it cannot be greater than a certain *finite value*, to which it constantly approaches as n is increased indefinitely.

72. Application to Geometrical Progression.— The preceding statements will be best understood by applying them to the case of the ordinary progression

$$1 + x + x^2 + x^3 + \ldots + x^n + \ldots$$

The sum of the first n terms of this series is $\dfrac{1 - x^n}{1 - x}$ in all cases.

(1). Let $x < 1$; then the terms become smaller and smaller as n increases; and if n be taken sufficiently great the value of x^n can be made as small as we please.

Hence, the sum of the first n terms tends to the limiting value $\dfrac{1}{1 - x}$; also the remainder after n terms is represented by $\dfrac{x^n}{1 - x}$, which becomes smaller and smaller as n increases, and may be regarded as vanishing ultimately.

(2). Let $x > 1$. The series is in this case an increasing one, and x^n becomes infinitely great along with n. Hence the sum of n terms, $\dfrac{1 - x^n}{1 - x}$ or $\dfrac{x^n - 1}{x - 1}$, as well as the remainder after n terms, becomes *infinite* along with n. Accordingly the statement that the limit of the sum of the series

$$1 + x + x^2 + \ldots + x^n + \ldots \; ad\ infinitum$$

is $\dfrac{1}{1 - x}$ holds only when x is less than unity, i.e. when the series is a convergent one.

In like manner the sum of n terms of the series

$$1 - x + x^2 - x^3 + \&c.$$

is

$$\dfrac{1 - (-1)^n x^n}{1 + x}.$$

As before, when $x < 1$, the limit of the sum is $\dfrac{1}{1+x}$; but when $x > 1$, x^n becomes infinitely great along with n, and the limit of the sum of an *even* number of terms is $-\infty$; while that of an *odd* number is $+\infty$. Hence the series in this case has no limit.

73. **Theorem.**—*If, in a series of positive terms represented by*

$$u_1 + u_2 + \ldots + u_n + \&c.,$$

the ratio $\dfrac{u_{n+1}}{u_n}$ *be less than a certain limit smaller than unity, for all values of n beyond a certain number, the series is convergent, and has a finite limit.*

Suppose k to be a fraction less than unity, and greater than the greatest of the ratios $\dfrac{u_{n+1}}{u_n}$ … (beyond the number n), then we have

$$\dfrac{u_{n+1}}{u_n} < k, \qquad \therefore u_{n+1} < k u_n.$$

$$\dfrac{u_{n+2}}{u_{n+1}} < k, \qquad \therefore u_{n+2} < k^2 u_n.$$

$$\dfrac{u_{n+r}}{u_{n+r-1}} < k, \qquad \therefore u_{n+r} < k^r u_n.$$

Hence, the limit of the remainder of the series after u_n is less than the sum of the series

$$k u_n + k^2 u_n + \ldots + k^r u_n \ldots \quad \textit{ad infinitum};$$

therefore, by Art. 72, less than

$$\dfrac{k u_n}{1-k}, \text{ since } k < 1.$$

Hence, since u_n decreases as n increases, and becomes infinitely small ultimately, the remainder after n terms becomes also infinitely small when n is taken sufficiently great; and consequently, the series is convergent, and has a finite limit.

Again, if the ratio $\dfrac{u_{n+1}}{u_n}$ be > 1, for all values of n beyond a certain number, the series is divergent, and has no finite limit. This can be established by a similar process; for, assuming $k > 1$, and less than the least of the fractions $\dfrac{u_{n+1}}{u_n}, \ldots$ then by Art. 72 the series

$$u_n + ku_n + k^2 u_n + \&c.\ ad\ infinitum$$

has an infinite value; but each term of the series

$$u_n + u_{n+1} + u_{n+2} + \&c.$$

is greater than the corresponding term in the above geometrical progression; hence, its sum must be also infinite, &c. These results hold also if the terms of the series be alternately positive and negative; for in this case k becomes negative, and the series will be convergent or divergent according as $-k$ is $<$ or >1; as can be readily seen.

In order to apply the preceding principles to Taylor's Theorem it will be necessary to determine a general expression for the remainder after n terms in that expansion; in order to do so, we commence with the following :—

74. **Lemma.**—*If a continuous function $\phi(x)$ vanish when $x = a$, and also when $x = b$, then its derived function $\phi'(x)$, if also continuous, must vanish for some value of x between a and b.*

Suppose b greater than a; then if $\phi'(x)$ *do not vanish between a and b*, it must be either always positive or always negative for *all* values of x between these limits; and consequently, by Art. 6, $\phi(x)$ must constantly increase, or constantly diminish, as x increases from a to b, which is impossible, since $\phi(x)$ vanishes for both limits. Accordingly, $\phi'(x)$ cannot be either always positive or always negative; and hence it must change its sign between the limits, and, being a continuous function, it must vanish for some intermediate value.

This result admits of being illustrated from geometry. For, let $y = \phi(x)$ represent a continuous curve; then, since $\phi(a) = 0$, and $\phi(b) = 0$, we have $y = 0$, when $x = a$, and also when $x = b$; therefore the curve cuts the axis of x at distances a and b from the origin; and accordingly at some intermediate point it must have its tangent parallel to the axis of x. Hence, by Art. 10, we must have $\phi'(x) = 0$ for some value of x between a and b.

75. Lagrange's Theorem on the Limits of Taylor's Series.—Suppose R_n to represent the remainder after n terms in Taylor's expansion, then writing X for $x + y$ in (1), we shall have

$$f(X) = f(x) + \frac{(X-x)}{1} f'(x) + \frac{(X-x)^2}{1 \cdot 2} f''(x) + \ldots$$
$$+ \frac{(X-x)^{n-1}}{\underline{|n-1}} f^{(n-1)}(x) + R_n, \qquad (22)$$

in which $f(x)$, $f'(x)$ . . . . . $f^{(n)}(x)$ are supposed finite and continuous for all values of the variable between X and x.

From the form of the terms included in R_n it evidently may be written in the shape

$$R_n = \frac{(X-x)^n}{\underline{|n}} P,$$

where P is some function of X and x.

Consequently we have

$$f(X) - \left\{ f(x) + \frac{(X-x)}{1} f'(x) + \ldots + \frac{(X-x)^{n-1}}{\underline{|n-1}} f^{(n-1)}(x) \right.$$
$$\left. + \frac{(X-x)^n}{\underline{|n}} P \right\} = 0. \qquad (23)$$

Now, let z be substituted for x in every term in the preceding, *with the exception of* P, and let $F(z)$ represent the resulting expression: we shall have

$$F(z) = f(X) - \left\{ f(z) + \frac{(X-z)}{1} f'(z) + \ldots + \frac{(X-z)^n}{\underline{|n}} P \right\}, \quad (24)$$

in which P has the same value as before.

Again, the right-hand side in this equation vanishes when $z = X$; $\therefore F(X) = 0$.

Also, from (23), the right-hand side vanishes when $z = x$; $\therefore F(x) = 0$.

Limits of Taylor's Series.

Accordingly, since the function $F(z)$ vanishes when $z = X$, and also when $z = x$, it follows from Art. 74 that its derived function $F'(z)$ also vanishes for some value of z between the limits X and x.

Proceeding to obtain $F'(z)$ by differentiation from equation (24), it can be easily seen that the terms destroy each other in pairs, with the exception of the two last. Thus we shall have

$$F'(z) = - \frac{(X-z)^{n-1}}{\underline{|n-1}} f^{(n)}(z) + \frac{(X-z)^{n-1}}{\underline{|n-1}} P.$$

Consequently, for some value of z between x and X we must have

$$f^{(n)}(z) = P.$$

Again, if θ be a positive quantity less than unity it is easily seen that the expression

$$x + \theta (X - x),$$

by assigning a suitable value to θ, can be made equal to any number intermediate between x and X.

Hence, finally,

$$P = f^{(n)} \{x + \theta (X - x)\},$$

where θ is some quantity > 0 and < 1.

Consequently, the remainder after n terms of Taylor's series can be represented by

$$R_n = \frac{(X-x)^n}{\underline{|n}} f^{(n)} \{x + \theta (X - x)\}. \qquad (25)$$

Making this substitution, the equation (22) becomes

$$f(X) = f(x) + \frac{(X-x)}{1} f'(x) + \frac{(X-x)^2}{1 \cdot 2} f''(x) + \ldots$$
$$+ \frac{(X-x)^{n-1}}{\underline{|n-1}} f^{(n-1)}(x) + \frac{(X-x)^n}{\underline{|n}} f^{(n)} \{x + \theta (X-x)\}. \qquad (26)$$

The preceding demonstration is taken, with some slight modifications, from Bertrand's "Traité de Calcul Différentiel" (273).

Again, if h be substituted for $X - x$, the series becomes
$$f(x + h) = f(x) + hf'(x) + \&c.$$
$$+ \frac{h^{n-1}}{\underline{|n-1}} f^{(n-1)}(x) + \frac{h^n}{\underline{|n}} f^{(n)}(x + \theta h). \tag{27}$$

In this expression n may be any positive integer.

If $n = 1$ the result becomes
$$f(x + h) = f(x) + hf'(x + \theta h). \tag{28}$$

When $n = 2$,
$$f(x + h) = f(x) + hf'(x) + \frac{h^2}{1 \cdot 2} f''(x + \theta h). \tag{29}$$

The student should observe that θ has in general different values in each of these functions, but that they are all subject to the same condition, viz., $\theta > 0$ and < 1.

It will be a useful exercise on the preceding method for the student to investigate the formulæ (28) and (29) independently, by aid of the Lemma of Art. 74.

The preceding investigation may be regarded as furnishing a *complete and rigorous proof of Taylor's Theorem, and formula* (27) *as representing its most general expression.*

76. Geometrical Illustration.—The equation
$$f(X) = f(x) + (X - x) f' \{ x + \theta (X - x) \}$$
admits of a simple geometrical verification; for, let $y = f(x)$ represent a curve referred to rectangular axes, and suppose (X, Y), (x, y) to be two points P_1, P_2 on it: then
$$\frac{f(X) - f(x)}{X - x} = \frac{Y - y}{X - x}.$$

But $\dfrac{Y - y}{X - x}$ is the tangent of the angle which the chord $P_1 P_2$ makes with the axis of x; also, since the curve cuts the chord in the points P_1, P_2, it is obvious that, when the point on the curve and the direction of the tangent alter continuously, the *tangent to the curve at some point between* P_1 *and* P_2 must be parallel to the chord $P_1 P_2$; but by Art. 10, $f'(x_1)$ is the trigonometrical tangent of the angle which the tangent at the point (x_1, y_1) makes with the axis of x. Hence, for some value, x_1, between X and x, we must have

$$f'(x_1) = \frac{Y-y}{X-x} = \frac{f(X)-f(x)}{X-x},$$

or, writing x_1 in the form $x + \theta(X - x)$,

$$f(X) = f(x) + (X - x) f'\{x + \theta(X - x)\}.$$

77. Second Form of Remainder.—The remainder after n terms in Taylor's Series may also be written in the form

$$R_n = \frac{(1 - \theta)^{n-1}}{\underline{|n-1|}} h^n f^{(n)}(x + \theta h).$$

For it is evident that R_n may be written in the form $(X - x) P_1$;

$$\therefore f(X) = f(x) + (X - x) f'(x) + \ldots + \frac{(X-x)^{n-1}}{\underline{|n-1|}} f^{(n-1)}(x)$$

$$+ (X - x) P_1.$$

Substitute z for x, as before, in every term *except* P_1; and the same reasoning is applicable, word for word, as that employed in Art. 75. The value of $F'(z)$ becomes, however, in this case

$$F'(z) = -\frac{(X-z)^{n-1}}{\underline{|n-1|}} f^{(n)}(z) + P_1,$$

and, as $F'(z)$ must vanish for some value of z between x and X, we must have, representing that value by $x + \theta(X - x)$,

$$P_1 = \frac{(X-x)^{n-1}(1-\theta)^{n-1}}{\underline{|n-1|}} f^{(n)} \{x + \theta(X - x)\}, \qquad (30)$$

where θ, as before, is > 0 and < 1.

If h be introduced instead of $X - x$, the preceding result becomes

$$R_n = \frac{(1 - \theta)^{n-1}}{\underline{|n-1|}} h^n f^{(n)}(x + \theta h), \qquad (31)$$

which is of the required form.

Hence, Taylor's Theorem admits of being written in the form

$$f(x+h) = f(x) + \frac{h}{1}f'(x) + \frac{h^2}{1\cdot 2}f''(x) + \ldots + \frac{h^{n-1}}{\underline{|n-1}}f^{(n-1)}(x)$$

$$+ \frac{h^n}{\underline{|n-1}}(1-\theta)^{n-1}f^{(n)}(x+\theta h). \qquad (32)$$

The same remarks are applicable to this form* as were made with respect to (27).

From these formulæ we see that the essential conditions for the application of Taylor's Theorem to the expansion of any function in a series consisting of an infinite number of terms are, that none of its derived functions shall become infinite, and that the quantity

$$\frac{h^n}{\underline{|n}}f^{(n)}(x+\theta h)$$

shall become infinitely small, when n is taken sufficiently large; as otherwise the series does not admit of a finite limit.

78. Limit of $\dfrac{h^n}{1\cdot 2\ldots n}$ **when n is indefinitely great.**

Let $u_n = \dfrac{h^n}{1\cdot 2\ldots n}$, then $\dfrac{u_{n+1}}{u_n} = \dfrac{h}{n+1}$; $\therefore \dfrac{u_{n+1}}{u_n}$ becomes smaller and smaller as n increases; hence, when n is taken sufficiently great, the series $u_{n+1}, u_{n+2}, \ldots$ &c., diminishes rapidly, and the terms become ultimately infinitely small. Consequently, *whenever the nth derived function $f^{(n)}(x)$ continues to be finite for all values of n, however great, the remainder after n terms in Taylor's Expansion becomes infinitely small, and the series has a finite limit.*

---

* This second form is in some cases more advantageous than that in (27). An example of this will be found in Art. 83.

79. General Form of Maclaurin's Series.—The expansion (27) becomes, on making $x = 0$, and substituting x afterwards instead of h,

$$f(x) = f(0) + \frac{x}{1}f'(0) + \frac{x^2}{1 \cdot 2}f''(0) + \ldots + \frac{x^{n-1}}{\underline{|n-1}}f^{(n-1)}(0)$$

$$+ \frac{x^n}{\underline{|n}}f^{(n)}(\theta x). \qquad (33)$$

Hence the remainder after n terms is represented by

$$\frac{x^n}{\underline{|n}}f^{(n)}(\theta x);$$

where θ is > 0 and < 1.

This remainder becomes infinitely small for any function $f(x)$ whenever $\frac{x^n}{\underline{|n}}f^{(n)}(\theta x)$ becomes evanescent for infinitely great values of n.

We shall now proceed to examine the remainders in the different elementary expansions which were given in the commencement of this chapter.

80. Remainder in the Expansion of a^x.—Our formula gives for R_n in this case

$$\frac{x^n}{\underline{|n}}(\log a)^n a^{\theta x}.$$

Now, $a^{\theta x}$ is finite, being less than a^x; and it has been proved in Art. 78 that $\frac{(x \log a)^n}{\underline{|n}}$ becomes infinitely small for large values of n. Hence the remainder in this case becomes evanescent when n is taken sufficiently large. Accordingly the series is a convergent one, and the expansion by Taylor's Theorem is always applicable.

81. Remainder in the Expansion of sin x.—In this case

$$R_n = \frac{x^n}{\underline{|n}} \sin\left(\frac{n\pi}{2} + \theta x\right).$$

This value of R_n ultimately vanishes by Art. 78, and the series is accordingly convergent.

The same remarks apply to the expansion of cos x. Accordingly, both of these series hold for all values of x.

82. Remainder in the Expansion of log $(1 + x)$.— The series

$$\frac{x}{1} - \frac{x^2}{2} + \frac{x^3}{3} - \frac{x^4}{4} + \&c.,$$

when x is > 1, is no longer convergent; for the ratio of any term to the preceding one tends to the limit $- x$; consequently the terms form an increasing series, and become ultimately infinitely great. Hence the expansion is inapplicable in this case.

Again, since $f^n(x) = (-1)^{n-1} \dfrac{1 \cdot 2 \ldots n-1)}{(1+x)^n}$, the remainder R_n is denoted by $\dfrac{(-1)^{n-1}}{n} \left(\dfrac{x}{1+\theta x}\right)^n$; hence, if x be positive and less than unity, $\dfrac{x}{1+\theta x}$ is a proper fraction, and the value of R_n evidently tends to become infinitely small for large values of n; accordingly the series is convergent, and the expansion holds in this case.

83. Binomial Theorem for Fractional and Negative Indices.—In the expansion

$$(1+x)^m = 1 + \frac{m}{1}x + \frac{m(m-1)}{1 \cdot 2}x^2 + \ldots$$
$$+ \frac{m(m-1)\ldots(m-n+1)x^n}{1 \cdot 2 \ldots n} + \&c.$$

if u_n denote the nth term, we have

$$\frac{u_{n+1}}{u_n} = \frac{m-n+1}{n} x,$$

the value of which, when n increases indefinitely, tends to become $- x$; the series, accordingly, is convergent if $x < 1$, but is not convergent if $x > 1$.

Accordingly, the Binomial Expansion does not hold when x is greater than unity.

Again, as

$$f^{(n)}(x) = m(m-1) \ldots (m-n+1)(1+x)^{m-n},$$

the remainder, by formula (25), is

$$\frac{m(m-1) \ldots (m-n+1)}{1 . 2 \ldots n} x^n (1+\theta x)^{m-n},$$

or

$$\frac{m(m-1) \ldots (m-n+1)}{1 . 2 \ldots n} \frac{x^n}{(1+\theta x)^{n-m}}.$$

Now, suppose x *positive and less than unity*; then, when n is very great, the expression

$$\frac{m(m-1) \ldots (m-n+1)}{1 . 2 \ldots n} x^n$$

becomes indefinitely small; also $\dfrac{1}{(1+\theta x)^{n-m}}$ is less than unity; hence, the expansion by the Binomial Theorem holds in this case.

Again, suppose x *negative and less than unity*. We employ the form for the remainder given in Art. 77, which becomes in this case

$$(-1)^n \frac{m(m-1) \ldots (m-n+1) x^n}{1 . 2 \ldots (n-1)} (1-\theta)^{n-1} (1-\theta x)^{m-n};$$

or

$$(-1)^n \frac{m(m-1) \ldots (m-n+1)(1-\theta)^{m-1} x^n}{1 . 2 \ldots (n-1)} \left| \frac{1-\theta}{1-\theta x} \right|^{n-m}.$$

Also, since $x < 1$, $\theta x < \theta$; $\therefore 1 - \theta x > 1 - \theta$; hence $\dfrac{1-\theta}{1-\theta x}$ is a proper fraction; $\therefore$ any integral power of it is less than unity; hence, by the preceding, the remainder, when n is sufficiently great, tends ultimately to vanish.

In general $(x+y)^m$ may be written in either of the forms

$$x^m\left(1+\frac{y}{x}\right)^m \text{ or } y^m\left(1+\frac{x}{y}\right)^m:$$

now, if the index m be fractional or negative, and $x > y$, or $\frac{y}{x}$ a proper fraction, the Binomial Expansion holds for the series

$$(x+y)^m = x^m\left(1+\frac{y}{x}\right)^m = x^m + \frac{m}{1}x^{m-1}y + \frac{m(m-1)}{1\cdot 2}x^{m-2}y^2 + \&c.,$$

but does not hold for the series

$$(x+y)^m = y^m\left(1+\frac{x}{y}\right)^m = y^m + \frac{m}{1}y^{m-1}x + \frac{m(m-1)}{1\cdot 2}y^{m-2}x^2 + \&c.,$$

since the former series is convergent and the latter divergent.

We conclude that in all cases one or other of the expansions of the Binomial series holds; but never both, except when m is a positive integer, in which case the number of terms is finite.

84. Remainder in the Expansion of $\tan^{-1}x$.—The series

$$\tan^{-1}x = \frac{x}{1} - \frac{x^3}{3} + \frac{x^5}{5} - \&c.,$$

is evidently convergent or divergent, according as $x <$ or > 1. To find an expression for the remainder when $x < 1$, we have, by (8), p. 50—

$$f^{(n)}(x) = \left(\frac{d}{dx}\right)^n \cdot \tan^{-1}x = (-1)^{n-1}\frac{\lfloor n-1 \cdot \sin\left(n\frac{\pi}{2} - n\tan^{-1}x\right)}{(1+x^2)^{\frac{n}{2}}}.$$

Hence we have, in this case,

$$R_n = (-1)^{n-1}\frac{x^n \sin\left\{n\frac{\pi}{2} - n\tan^{-1}(\theta x)\right\}}{n(1+\theta^2x^2)^{\frac{n}{2}}};$$

which, when x lies between $+1$ and -1, evidently becomes infinitely small as n increases, and accordingly the series holds for such values of x.

Expansion by aid of Differential Equations.

85. Expansion of $\sin^{-1}x$.—Since the function $\sin^{-1}x$ is impossible unless x be < 1, it is easily seen that the series given in Art. 64 is always convergent; for its terms are each less than the corresponding terms in the geometrical progression

$$x + x^3 + x^5 + \&c.$$

Consequently, the limit of the series is always less than the limit of the preceding progression.

A similar mode of demonstration is applicable to the expansion of $\tan^{-1}x$ when $x < 1$, as well as to other analogous series.

In every case, the value of R_n, the remainder after n terms, furnishes us with the degree of approximation in the evaluation of an expansion on taking its first n terms for its value.

86. Expansion by aid of Differential Equations.—In many cases we are enabled to find the relation between the coefficients in the expansion of a function of x by aid of differential* equations; and thus to find the form of the series.

For example, let $y = e^x$, then

$$\frac{dy}{dx} = e^x = y.$$

Now suppose that we have

$$y = a_0 + a_1 x + a_2 x^2 + \ldots a_n x^n + \ldots,$$

then
$$\frac{dy}{dx} = a_1 + 2a_2 x + \ldots n a_n x^{n-1} + \&c.$$

Accordingly we have

$$a_1 + 2a_2 x + 3a_3 x^2 + \ldots = a_0 + a_1 x + a_2 x^2 + \&c.,$$

---

* This method is indicated by Newton, and there can be little doubt that it was by aid of it he arrived at the expansion of $\sin(m \sin^{-1} x)$, as well as other series.—Vide *Ep. posterior ad Oldemburgium*. It is worthy of observation that Newton's letters to Oldemburg were written for the purpose of transmission to Leibnitz.

hence, equating coefficients, we have

$$a_1 = a_0, \quad a_2 = \frac{a_1}{2} = \frac{a_0}{2}, \quad a_3 = \frac{a_2}{3} = \frac{a_0}{2 \cdot 3}, \quad \&c.$$

Moreover, if we make $x = 0$, we get $a_0 = 1$,

$$\therefore e^x = 1 + \frac{x}{1} + \frac{x^2}{1 \cdot 2} + \frac{x^3}{1 \cdot 2 \cdot 3} + \&c.,$$

the same series as before.

Again, let

$$y = \sin(m \sin^{-1} x).$$

Here, by Art. 47, we have

$$(1 - x^2)\frac{d^2y}{dx^2} - x\frac{dy}{dx} + m^2 y = 0.$$

Now, if we suppose y developed in the form

$$y = a_0 + a_1 x + a_2 x^2 + \ldots + a_n x^n + \&c.,$$

then

$$\frac{dy}{dx} = a_1 + 2a_2 x + 3a_3 x^2 + \ldots + n a_n x^{n-1} + \&c.,$$

$$\frac{d^2y}{dx^2} = 2a_2 + 3 \cdot 2a_3 x + \ldots + n(n-1)a_n x^{n-2} + \&c.$$

Substituting and equating the coefficients of x^n we get

$$a_{n+2} = \frac{n^2 - m^2}{(n+1)(n+2)} a_n. \tag{34}$$

Again, when $x = 0$ we have $y = 0$; $\therefore a_0 = 0$.

Hence we see that the series consists only of odd powers of x; a result which might have been anticipated from Art. 61.

To find a_1. When $x = 0$, $\cos(m \sin^{-1} x) = 1$, hence $\left(\dfrac{dy}{dx}\right) = m$; accordingly $a_1 = m$;

$$\therefore a_3 = -\frac{m^2 - 1}{2 \cdot 3} a_1 = -\frac{m(m^2 - 1)}{1 \cdot 2 \cdot 3},$$

$$a_5 = -\frac{m^2 - 9}{4 \cdot 5} a_3 = \frac{m(m^2 - 1)(m^2 - 9)}{1 \cdot 2 \cdot 3 \cdot 4 \cdot 5}:$$

hence we get

$$\sin^* (m \sin^{-1} x) = \frac{m}{1} x - \frac{m(m^2 - 1)}{1 \cdot 2 \cdot 3} x^3$$

$$+ \frac{m(m^2 - 1)(m^2 - 9)}{1 \cdot 2 \cdot 3 \cdot 4 \cdot 5} x^5 - \&c. \qquad (35)$$

In the preceding, we have assumed that $\sin^{-1} x$ is an acute angle, as otherwise both it, and also $\sin(m \sin^{-1} x)$, would admit of an indefinite number of values.—See Art. 26.

87. Expansion of $\sin mz$ and $\cos mz$.—If, in (35), z be substituted for $\sin^{-1} x$, the formula becomes

$$\sin mz = m \sin z \left\{ \frac{1}{1} - \frac{m^2 - 1}{1 \cdot 2 \cdot 3} \sin^2 z \right.$$

$$\left. + \frac{(m^2 - 1)(m^2 - 9)}{1 \cdot 2 \cdot 3 \cdot 4 \cdot 5} \sin^4 z - \&c. \right\} \qquad (36)$$

In a similar manner it can be proved that

$$\cos mz = 1 - \frac{m^2 \sin^2 z}{1 \cdot 2} + \frac{m^2(m^2 - 4)}{1 \cdot 2 \cdot 3 \cdot 4} \sin^4 z - \&c. \qquad (37)$$

If m be an *odd* integer the expansion for $\sin mz$ consists of a finite number of terms, while that for $\cos mz$ contains an infinite number. If m be an *even* integer the number of terms in the series for $\cos mz$ is finite, while that in $\sin mz$ is infinite.

The preceding series hold equally when m is a fraction.

A more complete exposition of these important expansions will be found in Bertrand's "Calcul Différentiel."

In general, in the expansion (36), the ratio of any term to that which precedes it is $\dfrac{n^2 - m^2}{(n + 1)(n + 2)} \sin^2 z$, which, when n is very great, approaches to $\sin^2 z$. Hence, since $\sin z$ is less than unity, the series is *convergent* in all cases. Similar observations apply to expansion (37).

---

* This expansion is erroneously attributed to Euler by M. Bertrand; it was originally given by Newton. See preceding note.

Development of Functions.

The expansion

$$e^{a\sin^{-1}x} = 1 + \frac{ax}{1} + \frac{a^2 x^2}{1\cdot 2} + \frac{a(a^2+1^2)}{1\cdot 2\cdot 3}x^3 + \frac{a^2(a^2+2^2)}{1\cdot 2\cdot 3\cdot 4}x^4 + \ldots$$

can be easily arrived at by a similar process.

88. Arbogast's Method of Derivations.

If $\quad u = a + b\dfrac{x}{1} + c\dfrac{x^2}{1\cdot 2} + d\dfrac{x^3}{1\cdot 2\cdot 3} + \&c.,$

to find the coefficients in the expansion of $\phi(u)$ in ascending powers of x—

Let $\quad f(x) = \phi(u),$

and suppose $f(x) = A + \dfrac{B}{1}x + \dfrac{C}{1\cdot 2}x^2 + \&c.$

$$= f(0) + \frac{x}{1}f'(0) + \frac{x^2}{1\cdot 2}f''(0) + \&c.,$$

then we have evidently

$$A = f(0) = \phi(a).$$

Also, writing u', u'', u''', &c. instead of

$$\frac{du}{dx},\ \frac{d^2u}{dx^2},\ \frac{d^3u}{dx^3},\ \&c.,$$

by successive differentiation of the equation $f(x) = \phi(u)$, we obtain

$f'(x) = \phi'(u)\cdot u',$
$f''(x) = \phi'(u)\cdot u'' + \phi''(u)\cdot (u')^2,$
$f'''(x) = \phi'(u)\cdot u''' + 3\phi''(u)\cdot u'\cdot u'' + \phi'''(u)(u')^3,$
$f^{iv}(x) = \phi'(u)\cdot u^{iv} + \phi''(u)[4u'u''' + 3(u'')^2] + 6\phi'''(u)\cdot (u')^2\cdot u''$
$\qquad\qquad\qquad\qquad\qquad\qquad\qquad\qquad\qquad + \phi^{iv}(u)\cdot (u')^4.$

Now, when $x = 0$, u, u', u'', u''', ... obviously become $a, b, c, d, \ldots$ respectively.

Accordingly,

$B = f'(0) = \phi'(a) \cdot b,$

$C = f''(0) = \phi'(a) \cdot c + \phi''(a) \cdot b^2,$

$D = f'''(0) = \phi'(a) \cdot d + 3\phi''(a) \cdot bc + \phi'''(a) \cdot b^3,$

$E = f^{iv}(0) = \phi'(a) \cdot e + \phi''(a)(4bd + 3c^2) + 6\phi'''(a) \cdot b^2 c$
$\qquad\qquad\qquad\qquad\qquad\qquad\qquad + \phi^{iv}(a) \cdot b^4.$

From the mode of formation of these terms, they are seen to be each deduced from the preceding one by an analogous law by that to which the derived functions are deduced one from the other; and, as $f'(x), f''(x) \ldots$ are deduced from $f(x)$ by successive differentiation, so in like manner, $B, C, D, \ldots$ are deduced from $\phi(u)$ by successive *derivation*; where, after differentiation, a, b, c, &c., are substituted for

$$u, \frac{du}{dx}, \frac{d^2u}{dx^2}, \ldots \text{&c.}$$

If this process of *derivation* be denoted by the letter δ, then

$$B = \delta \cdot A, \quad C = \delta \cdot B, \quad D = \delta \cdot C, \text{&c.} \qquad (38)$$

From the preceding, we see that in forming the term $\delta \cdot \phi(a)$, we take the derived function $\phi'(a)$, and multiply it by the next letter b, and similarly in other cases.

Thus $\quad \delta \cdot b = c, \qquad \delta \cdot c = d, \ldots$

$\qquad\quad \delta \cdot b^m = m b^{m-1} c, \quad \delta \cdot c^m = m c^{m-1} d \ldots$

Also $\quad \delta \cdot \phi'(a) b = \phi'(a) c + \phi''(a) b^2.$

This gives the same value for C as that found before; D is derived from C in accordance with the same law; and so on.

The preceding method is due to Arbogast: for its complete discussion the student is referred to his "Calcul des Dérivations." The Rules there arrived at for forming the successive coefficients in the simplest manner are given in "Galbraith's Algebra," page 342.

As an illustration of this method, we shall apply it to find a few terms in the expansion of

$$\sin\left(a + b\frac{x}{1} + c\frac{x^2}{1.2} + d\frac{x^3}{1.2.3} + \&c.\right).$$

Here $A = \sin a$, $B = \delta . \sin a = b \cos a$,

$C = \delta . b \cos a = c \cos a - b^2 \sin a$,

$D = \delta . C = d \cos a - 3bc \sin a - b^3 \cos a$,

$E = \delta . D = e \cos a - (4bd + 3c^2) \sin a - 6b^2c \cos a$
$\qquad\qquad\qquad\qquad\qquad\qquad\qquad + b^4 \sin a$.

If the series $a + bx + c\dfrac{x^2}{1.2} + \&c.$ consist of a finite number of terms the *derivative* of the last letter is zero—thus, if d be the last letter, $\delta . d = 0$, and d is regarded as a constant with respect to the symbol of derivation δ.

If the expansion of $\phi(u)$ be required when u is of the form

$$\alpha + \beta x + \gamma x^2 + \delta x^3 + \&c.,$$

the result can be attained from the preceding method by substituting a, b, c, d, &c. instead of α, β, $1.2\,\gamma$, $1.2.3\,.\,\delta$, &c., and proceeding as before.

The student will observe that in the expression for the terms D, E, &c., the *coefficients* of the derived functions $\phi'(a)$, $\phi''(a)$, &c., are completely *independent of the form* of the function ϕ, and are expressed in terms of the letters, b, c, d, &c. solely; so that, if *calculated once for all*, they can be applied to the determination of the coefficients in every particular case, by finding the different derived functions $\phi'(a)$, $\phi''(a)$, &c., for that case, and multiplying by the respective coefficients, determined as stated above.

EXAMPLES.

1. If $u = f(ax + by)$, then $\dfrac{1}{a}\dfrac{du}{dx} = \dfrac{1}{b}\dfrac{du}{dy}$. This furnishes the condition that a given function of x and y should be a function of $ax + by$.

2. Find, by Maclaurin's theorem, the first three terms in the expansion of $\tan x$.

$$\text{Ans. } x + \frac{x^3}{3} + \frac{2x^5}{15}.$$

3. Find the first four terms in the expansion of $\sec x$.

$$\text{Ans. } 1 + \frac{x^2}{2} + \frac{5x^4}{24} + \frac{61 x^6}{720}.$$

4. Find, by Maclaurin's theorem, as far as x^4, the expansion of $\log(1 + \sin x)$ in ascending powers of x.

Let $f(x) = \log(1 + \sin x)$,

then $f'(x) = \dfrac{\cos x}{1 + \sin x} = \dfrac{1 - \sin x}{\cos x} = \sec x - \tan x$,

$f''(x) = \sec x \tan x - \sec^2 x = -f'(x) \sec x$;

$\therefore f'''(x) = -f''(x) \sec x - f'(x) \sec x \tan x$,

$f^{\text{iv}}(x) = -f'''(x) \sec x - 2f''(x) \sec x \tan x - f'(x)(2 \sec^3 x - \sec x)$;

$\therefore f(0) = 0, \ f'(0) = 1, \ f''(0) = -1, \ f'''(0) = 1, \ f^{\text{iv}}(0) = -2$;

$\therefore \log(1 + \sin x) = x - \dfrac{x^2}{2} + \dfrac{x^3}{6} - \dfrac{x^4}{12} + \&c.$

5. Find six terms of the development of $\dfrac{e^x}{\cos x}$ in ascending powers of x.

$$\text{Ans. } 1 + x + x^2 + \frac{2x^3}{3} + \frac{x^4}{2} + \frac{3x^5}{10} \ldots$$

6. Apply the method of Art. 86, to find the expansions of $\sin x$ and $\cos x$.

7. Prove that

$$\tan^{-1}(x + h) = \tan^{-1} x + h \sin z \frac{\sin z}{1} - (h \sin z)^2 \frac{\sin 2z}{2} + (h \sin z)^3 \frac{\sin 3z}{3} - \&c.,$$

where $z = \cot^{-1} x$.

Here $f(x) = \tan^{-1} x = \dfrac{\pi}{2} - z$; and by Art. 46, $\dfrac{d^n z}{dx^n} = (-1)^n \underline{|n-1}\ \sin^n z \sin nz$; $\therefore$ &c.

8. Hence prove the expansion

$$\frac{\pi}{2} = z + \frac{\sin z}{1}\cos z + \frac{\sin 2z}{2}\cos^2 z + \frac{\sin 3z}{3}\cos^3 z + \&c.$$

Let $h = -\cot z = -x$, &c.

9. Prove that

$$\frac{\pi}{2} = \frac{z}{2} + \frac{\sin z}{1} + \frac{\sin 2z}{2} + \frac{\sin 3z}{3} + \&c.$$

Let $h \sin z = -1$ in Example 7; then $h + x = \dfrac{\cos z - 1}{\sin z} = -\tan\dfrac{z}{2}$; ∴ &c.

10. Prove the expansion

$$\frac{\pi}{2} = \frac{\sin z}{\cos z} + \frac{1}{2}\frac{\sin 2z}{\cos^2 z} + \frac{1}{3}\frac{\sin 3z}{\cos^3 z} + \&c.$$

Assume $h = -\dfrac{1}{\sin z \cos z}$, then

$$x + h = -\tan z = \tan(\pi - z); \therefore \pi - z = \tan^{-1}(x + h), \&c.$$

Substituting in Example 7, we get the result required.

The preceding expansions were first given by Euler.

11. Prove the equations

$$\sin 9x = 9\sin x - 120\sin^3 x + 432\sin^5 x - 576\sin^7 x + 256\sin^9 x,$$
$$\cos 6x = 32\cos^6 x - 48\cos^4 x + 18\cos^2 x - 1.$$

These follow from the formulæ of Article 78.

12. If $m = 2$, Newton's formula, Art. 87, gives

$$\sin 2x = 2\left\{\sin x - \frac{\sin^3 x}{2} - \frac{\sin^5 x}{2.4} - \&c.\right\};$$

verify this result by aid of the elementary equation $\sin 2x = 2\sin x \cos x$.

13. If $\phi(x + h) + \phi(x - h) = \phi(x)\phi(h)$, for all values of x and h, prove that

$$\frac{\phi''(x)}{\phi(x)} = \frac{\phi^{\text{iv}}(x)}{\phi''(x)} = \&c. = \text{constant};$$

and also $\phi'(0) = 0, \quad \phi'''(0) = 0, \&c.$

14. If, in the last, $\dfrac{\phi''(x)}{\phi(x)} = a^2$; prove that $\phi(x) = e^{ax} + e^{-ax}$.

If $\dfrac{\phi''(x)}{\phi(x)} = -a^2$; prove that $\phi(x) = 2\cos(ax)$.

Examples.

15. Apply Arbogast's method to find the first four terms in the expansion of

$$(a + bx + cx^2 + dx^3 + \&c.)^n.$$

$$Ans. \ a^n + na^{n-1}bx + \left(\frac{n(n-1)}{1.2}b^2 + nac\right)a^{n-2}x^2$$

$$+ n\left\{\frac{(n-1)(n-2)}{2.3}a^{n-3}b^3 + (n-1)a^{n-2}bc + a^{n-1}d\right\}x^3 + \&c.$$

16. Prove that the expansion of $\dfrac{e^x + 1}{e^x - 1} \cdot x$ can contain no odd powers of x. For if the sign of x be changed, the function remains unaltered.

17. Hence, show that the expansion of $\dfrac{x}{e^x - 1}$ contains no odd powers of x beyond the first.

Here $\dfrac{x}{e^x - 1} + \dfrac{x}{2} = \dfrac{x}{2} \cdot \dfrac{e^x + 1}{e^x - 1}$; $\therefore$ &c.

18. If $u = \dfrac{x}{e^x - 1}$, prove that

$$\frac{n}{1}\left(\frac{d^{n-1}u}{dx^{n-1}}\right)_0 + \frac{n(n-1)}{1.2}\left(\frac{d^{n-2}u}{dx^{n-2}}\right)_0 + \ldots + n\left(\frac{du}{dx}\right)_0 + (u)_0 = 0;$$

and hence calculate the coefficients of the first five terms in the expansion of u. Here $e^x u = x + u$, and by Art. 48, we have

$$e^x\left(u + n\frac{du}{dx} + \frac{n(n-1)}{1.2}\frac{d^2u}{dx^2} + \ldots + \frac{d^n u}{dx^n}\right) = \frac{d^n u}{dx^n}, \ \therefore \ \&c.$$

19. If $\dfrac{x}{e^x - 1} = 1 - \dfrac{x}{2} + \dfrac{B_1}{1.2}x^2 - \dfrac{B_2}{1.2.3.4}x^4 + \dfrac{B_3}{1.2\ldots 6}x^6 - \ldots$

prove that

$$B_1 = \frac{1}{6}, \quad B_2 = \frac{1}{30}, \quad B_3 = \frac{1}{42}, \quad B_4 = \frac{1}{30}, \ \&c.$$

These are called Bernoulli's numbers, and are of importance in connexion with the expansion of a large number of functions.

20. Prove that

$$\frac{x}{e^x + 1} = \frac{x}{2} - \frac{B_1 x^2}{1.2}(2^2 - 1) + \frac{B_2 x^4}{1.2.3.4}(2^4 - 1) - \frac{B_3 x^6}{1.2\ldots 6}(2^6 - 1) + \ldots$$

Examples.

21. Hence, prove that

$$\frac{e^x - 1}{e^x + 1} = B_1 x (2^2 - 1) + \frac{B_2 x^3}{3 \cdot 4}(2^4 - 1) + \frac{B_3 x^5}{3 \cdot 4 \cdot 5 \cdot 6}(2^6 - 1) + \&c.$$

$$= \frac{x}{2} - \frac{x^3}{24} + \frac{x^5}{240} - \&c.$$

22. Prove that

$$x \cot x = 1 - \frac{2^2 B_1 x^2}{1 \cdot 2} - \frac{2^4 B_2 x^4}{1 \cdot 2 \cdot 3 \cdot 4} - \frac{2^6 B_3 x^6}{1 \cdot 2 \ldots 6} - \&c.$$

23. Also, $\tan \dfrac{x}{2} = B_1 x (2^2 - 1) + \dfrac{B_2 x^3}{3 \cdot 4}(2^4 - 1) + \&c.$

24. Prove that

$$\frac{x}{2} \cot \frac{x}{2} = 1 - B_1 \frac{x^2}{\lfloor 2} - B_2 \frac{x^4}{\lfloor 4} - \frac{B_3 x^6}{\lfloor 6} - \ldots$$

This follows immediately by substituting $\dfrac{x}{2}$ for x in Ex. 22.

25. Given $u(u - x) = 1$; find the four first terms in the expansion of u in terms of x, by Maclaurin's Theorem.

26. If

$$x \frac{d^2 y}{dx^2} + \frac{dy}{dx} + y = 0,$$

expand y in powers of x by the method of indeterminate coefficients.

27. Show that the series

$$\frac{x}{1^m} + \frac{x^2}{2^m} + \frac{x^3}{3^m} + \frac{x^4}{4^m} + \ldots$$

is convergent when $x < 1$, and divergent when $x > 1$, for all values of m.

28. Prove the expansion

$$\frac{f(x)}{(x - a)^m \phi(x)} = \frac{1}{(x - a)^m} \frac{f(a)}{\phi(a)} + \frac{1}{(x - a)^{m-1}} \frac{d}{da} \left\{ \frac{f(a)}{\phi(a)} \right\}$$

$$+ \frac{1}{1 \cdot 2 \cdot (x - a)^{m-2}} \left(\frac{d}{da}\right)^2 \left\{ \frac{f(a)}{\phi(a)} \right\} + \&c. \ldots$$

29. Find, by Maclaurin's Theorem, the first four terms in the expansion of $(1 + x)^{\frac{1}{x}}$ in ascending powers of x.

Let

$$f(x) = (1 + x)^{\frac{1}{x}},$$

then $f'(x) = f(x) \left(\dfrac{1}{x(1+x)} - \dfrac{\log(1+x)}{x^2} \right)$

$= -f(x) \left\{ \dfrac{1}{2} - \dfrac{2}{3}x + \dfrac{3}{4}x^2 - \&c. \right\}.$

$\therefore f''(x) = -f'(x) \left(\dfrac{1}{2} - \dfrac{2}{3}x + \dfrac{3}{4}x^2 - \&c. \right) + f(x) \left(\dfrac{2}{3} - \dfrac{3}{2}x + \&c. \right);$

$f'''(x) = -f''(x) \left(\dfrac{1}{2} - \dfrac{2}{3}x + \dfrac{3}{4}x^2 - \&c. \right) + 2f'(x) \left(\dfrac{2}{3} - \dfrac{3}{2}x + \&c. \right).$

But, by Art. 29, $\qquad f(0) = e;$

$$\therefore f'(0) = -\dfrac{e}{2}, \quad f''(0) = \dfrac{11e}{12}, \quad f'''(0) = -\dfrac{21}{8}e.$$

Hence $\qquad (1+x)^{\frac{1}{x}} = e - \dfrac{ex}{2} + \dfrac{11ex^2}{24} - \dfrac{7e}{16}x^3 + \&c.$

This result can be verified by direct development, as follows:

let $\qquad u = (1+x)^{\frac{1}{x}},$

then $\qquad \log u = \dfrac{1}{x} \log(1+x) = 1 - \dfrac{x}{2} + \dfrac{x^2}{3} - \dfrac{x^3}{4} + \ldots;$

$\therefore u = e^{1 - \frac{x}{2} + \frac{x^2}{3} - \frac{x^3}{4} \cdots} = e \cdot e^{-\frac{x}{2} + \frac{x^2}{3} - \frac{x^3}{4} \cdots}$

$= e \left[1 - \left(\dfrac{x}{2} - \dfrac{x^2}{3} + \dfrac{x^3}{4} \ldots \right) + \dfrac{x^2}{2} \left(\dfrac{1}{2} - \dfrac{x}{3} + \dfrac{x^2}{4} \ldots \right)^2 - \dfrac{x^3}{2 \cdot 3} \left(\dfrac{1}{2} - \dfrac{x}{3} + \ldots \right)^3 \ldots \right]$

$= e \left[1 - \dfrac{x}{2} + \dfrac{11x^2}{24} - \dfrac{7x^3}{16} \ldots \right].$

30. In Art. 76, if $f(x)$ and $f'(x)$ be not both continuous between the points P_1, P_2, show that there is not necessarily a tangent between those points, parallel to the chord.

31. Find the development of $\dfrac{x \sin 3x}{\sin x \sin 2x}$ in ascending powers of x, the coefficients being expressed in Bernoullian numbers. "Camb. Math. Trip., 1878."

Since $\dfrac{x \sin 3x}{\sin x \sin 2x} = x \cot x + x \cot 2x$, the expansion in question, by (22), is

$$\dfrac{3}{2} - \dfrac{2^2 B_2 x^2}{\underline{|2}} (2+1) - \dfrac{2^4 B_4 x^4}{\underline{|4}} (2^3 + 1) - \dfrac{2^6 B_6 x^6}{\underline{|6}} (2^5 + 1) - \&c.$$

CHAPTER IV.

INDETERMINATE FORMS.

89. Indeterminate Forms.—Algebraic expressions sometimes become indeterminate for particular values of the variable on which they depend; thus, if the same value a when substituted for x makes both the numerator and the denominator of the fraction $\dfrac{f(x)}{\phi(x)}$ vanish, then $\dfrac{f(a)}{\phi(a)}$ becomes of the form $\dfrac{0}{0}$, and its value is said to be *indeterminate*.

Similarly, the fraction becomes indeterminate if $f(x)$ and $\phi(x)$ both become *infinite* for a particular value of x. We proceed to show how its *true* value is to be found in such cases. By its *true* value we mean the *limiting value which the fraction assumes when x differs by an infinitely small amount from the particular value which renders the expression indeterminate*.

It will be observed that the determination of the differential coefficient of any expression $f(x)$ may be regarded as a case of finding an indeterminate form, for it reduces to the determination of $\dfrac{f(x+h) - f(x)}{h}$ when $h = 0$.

In many cases the true values of indeterminate forms can be best found by ordinary algebraical and trigonometrical processes.

We shall illustrate this statement by a few examples.

Examples.

1. The fraction $\dfrac{ax^2 - 2acx + ac^2}{bx^2 - 2bcx + bc^2}$ becomes of the form $\dfrac{0}{0}$ when $x = c$; but since it can be written in the shape $\dfrac{a(x-c)^2}{b(x-c)^2}$, its true value in all cases is $\dfrac{a}{b}$.

Examples. 97

2. The fraction $\dfrac{x}{\sqrt{a+x}-\sqrt{a-x}}$ becomes $\dfrac{0}{0}$ when $x = 0$.

To find its true value, multiply its numerator and denominator by the *complementary* surd, $\sqrt{a+x}+\sqrt{a-x}$, and the fraction becomes

$$\frac{x(\sqrt{a+x}+\sqrt{a-x})}{2x} \text{ or } \frac{\sqrt{a+x}+\sqrt{a-x}}{2};$$

the true value of which is $\sqrt{a}$ when $x = 0$.

3. $\dfrac{\sqrt{a^2+ax+x^2}-\sqrt{a^2-ax+x^2}}{\sqrt{a+x}-\sqrt{a-x}}$, when $x = 0$.

Multiply by the two complementary surd forms, and the fraction becomes

$$\frac{2ax\{\sqrt{a+x}+\sqrt{a-x}\}}{2x\{\sqrt{a^2+ax+x^2}+\sqrt{a^2-ax+x^2}\}},$$

or

$$\frac{a(\sqrt{a+x}+\sqrt{a-x})}{\sqrt{a^2+ax+x^2}+\sqrt{a^2-ax+a^2}},$$

the true value of which evidently is $\sqrt{a}$ when $x = 0$. From the preceding examples we infer that when an expression of a surd form becomes indeterminate, its true value can usually be determined by multiplying by the complementary surd form or forms.

4. $\dfrac{2x-\sqrt{5x^2-a^2}}{x-\sqrt{2x^2-a^2}}$ when $x = a$. Ans. $\dfrac{1}{2}$.

5. $\dfrac{a-\sqrt{a^2-x^2}}{x^2}$ when $x = 0$. Ans. $\dfrac{1}{2a}$.

6. $\dfrac{a\sin\theta - \sin a\theta}{\theta(\cos\theta - \cos a\theta)}$ becomes $\dfrac{0}{0}$ when $\theta = 0$.

To find its true value, substitute their expansions for the sines and cosines, and the fraction becomes

$$\frac{a\left(\theta - \dfrac{\theta^3}{1.2.3}+\ldots\right) - \left(a\theta - \dfrac{a^3\theta^3}{1.2.3}+\ldots\right)}{\theta\left\{-\dfrac{\theta^2}{1.2}+\ldots+\dfrac{a^2\theta^2}{1.2}-\ldots\right\}}$$

or

$$\frac{\dfrac{\theta^3}{6}(a^3-a)+\ldots}{\dfrac{\theta^3}{2}(a^2-1)-\ldots}.$$

H

Divide by $\theta^3(a^2-1)$, and since all the terms after the first in the new numerator and denominator vanish when $\theta = 0$, the true value of the fraction is $\dfrac{a}{3}$ in this case.

7. The fraction
$$\frac{A_0 x^m + A_1 x^{m-1} + A_2 x^{m-2} + \ldots A_m}{a_0 x^n + a_1 x^{n-1} + \ldots + a_n} \text{ becomes } \frac{\infty}{\infty} \text{ when } x = \infty:$$
its true value can, however, be easily determined, for it is evidently equal to that of
$$x^{m-n} \frac{A_0 + \dfrac{A_1}{x} + \dfrac{A_2}{x^2} + \ldots}{a_0 + \dfrac{a_1}{x} + \dfrac{a_2}{x^2} + \ldots}.$$

Moreover, when $x = \infty$, the fractions $\dfrac{A_1}{x}, \dfrac{A_2}{x^2} \ldots \dfrac{a_1}{x} \ldots$, all vanish; hence, the true value of the given fraction is that of
$$x^{m-n} \frac{A_0}{a_0} \text{ when } x = \infty.$$

The value of this expression depends on the sign of $m - n$.

(1.) If $m > n$, $x^{m-n} = \infty$ when $x = \infty$; or the fraction is infinite in this case.

(2.) If $m = n$, the true value is $\dfrac{A_0}{a_0}$.

(3.) If $m < n$, then $x^{m-n} = 0$ when $x = \infty$; and the true value of the fraction is zero.

Accordingly, the proposed expression, when $x = \infty$, is infinite, finite, or zero, according as m is greater than, equal to, or less than n. Compare Art. 39.

8. $u = \sqrt{x+a} - \sqrt{x+b}$, when $x = \infty$.

Here $u = \dfrac{a-b}{\sqrt{x+a} + \sqrt{x+b}} = 0$ when $x = \infty$.

9. $\sqrt{x^2 + ax} - x$, when $x = \infty$. *Ans.* $\dfrac{a}{2}$.

10. $u = a^x \sin\left(\dfrac{c}{a^x}\right)$, when $x = \infty$.

(1.) If $a < 1$, $a^x = 0$ when $x = \infty$, and therefore the true value of u is zero in this case.

(2.) If $a > 1$, then a^x becomes infinite along with x; but as $\dfrac{c}{a^x}$ is infinitely small at the same time, we have $\sin\dfrac{c}{a^x} = \dfrac{c}{a^x}$. Hence, the true value of u is c in this case.

11. $u = \sqrt{a^2 - x^2} \cot \dfrac{\pi}{2} \sqrt{\dfrac{a-x}{a+x}}$ is of the form $0 \times \infty$ when $x = a$.

Here
$$u = \dfrac{\sqrt{a^2 - x^2}}{\tan \dfrac{\pi}{2} \sqrt{\dfrac{a-x}{a+x}}},$$

but, when $a - x$ is infinitely small,

$$\tan \dfrac{\pi}{2} \sqrt{\dfrac{a-x}{a+x}} = \dfrac{\pi}{2} \sqrt{\dfrac{a-x}{a+x}};$$

$$\therefore u = \dfrac{\sqrt{a^2 - x^2}}{\dfrac{\pi}{2} \sqrt{\dfrac{a-x}{a+x}}} = \dfrac{a+x}{\dfrac{\pi}{2}} = \dfrac{4a}{\pi} \text{ when } x = a.$$

12. $u = \dfrac{x \sin(\sin x) - \sin^2 x}{x^6}$, when $x = 0$.

Substitute the ordinary expansion for $\sin x$, neglecting powers beyond the sixth, and u becomes

$$\dfrac{x \left\{ \sin x - \dfrac{\sin^3 x}{\underline{3}} + \dfrac{\sin^5 x}{\underline{5}} \right\} - \left(x - \dfrac{x^3}{\underline{3}} + \dfrac{x^5}{\underline{5}} \right)^2}{x^6}$$

$$= \dfrac{x - \dfrac{x^3}{\underline{3}} + \dfrac{x^5}{\underline{5}} - \dfrac{1}{6} \left(x - \dfrac{x^3}{\underline{3}} \right)^3 + \dfrac{x^5}{\underline{5}} - x \left(1 - \dfrac{x^2}{\underline{3}} + \dfrac{x^4}{\underline{5}} \right)^2}{x^5}.$$

Hence we get, on dividing by x^5, the true value of the fraction to be $\dfrac{1}{18}$ when $x = 0$.

13. $\dfrac{(a \sin^2 \phi + \beta \cos^2 \phi)^n - \beta^n}{a^n - \beta^n}$, when $a = \beta$. *Ans.* $\sin^2 \phi$.

Similar processes may be applied to other cases; there are, however, many indeterminate forms in which such processes would either fail altogether, or else be very laborious.

We now proceed to show how the Differential Calculus furnishes us with a general method for evaluating indeterminate forms.

90.—**Method of the Differential Calculus.**—Suppose $\dfrac{f(x)}{\phi(x)}$ to be a fraction which becomes of the form $\dfrac{0}{0}$ when $x = a$;

i. e. $f(a) = 0$, and $\phi(a) = 0$;

substitute $a + h$ for x and the fraction becomes

$$\frac{f(a+h)}{\phi(a+h)}, \text{ or } \frac{\dfrac{f(a+h)-f(a)}{h}}{\dfrac{\phi(a+h)-\phi(a)}{h}};$$

but when h is infinitely small the numerator and denominator in this expression become $f'(a)$ and $\phi'(a)$, respectively; hence, in this case,

$$\frac{f(a+h)}{\phi(a+h)} = \frac{f'(a)}{\phi'(a)}.$$

Accordingly, $\dfrac{f'(a)}{\phi'(a)}$ represents the *limiting* or *true* value of the fraction $\dfrac{f(a)}{\phi(a)}$.

(1.) If $f'(a) = 0$, and $\phi'(a)$ be not zero, the true value of $\dfrac{f(a)}{\phi(a)}$ is zero.

(2.) If $f'(a)$ be not zero, and $\phi'(a) = 0$, the true value of $\dfrac{f(a)}{\phi(a)}$ is ∞.

(3.) If $f'(a) = 0$, and $\phi'(a) = 0$, our new fraction $\dfrac{f'(a)}{\phi'(a)}$ is still of the indeterminate form $\dfrac{0}{0}$. Applying the preceding process of reasoning to it, it follows that its true value is that of $\dfrac{f''(a)}{\phi''(a)}$.

If this fraction be also of the form $\dfrac{0}{0}$, we proceed to the next derived functions.

In general, if the first derived functions which do not vanish be $f^{(n)}(a)$ and $\phi^{(n)}(a)$, then the true value of $\dfrac{f(a)}{\phi(a)}$ is that of $\dfrac{f^{(n)}(a)}{\phi^{(n)}(a)}$.

Examples.

1. $u = \dfrac{x \sin x - \dfrac{\pi}{2}}{\cos x}$, when $x = \dfrac{\pi}{2}$.

Here $f(x) = x \sin x - \dfrac{\pi}{2}$,

$\phi(x) = \cos x$;

$\therefore f'(x) = x \cos x + \sin x,$ $\qquad f'\left(\dfrac{\pi}{2}\right) = 1,$

$\phi'(x) = -\sin x,$ $\qquad \phi'\left(\dfrac{\pi}{2}\right) = -1.$

Hence $u = -1$, when $x = \dfrac{\pi}{2}$.

2. $u = \dfrac{e^{mx} - e^{ma}}{(x-a)^r}$, when $x = a$.

Here $f(x) = e^{mx} - e^{ma},$

$\phi(x) = (x-a)^r$;

$\therefore f'(x) = me^{mx},$ $\qquad f'(a) = me^{ma}.$

$\phi'(x) = r(x-a)^{r-1},$ $\qquad \phi'(a)$ is 0 or ∞, as $r >$ or < 1.

Hence the true value of u is ∞ or 0, according as $r >$ or < 1.

This result can also be arrived at by writing the fraction in the form

$$\dfrac{\{e^{m(x-a)} - 1\}e^{ma}}{(x-a)^r} = \dfrac{e^{mh} - 1}{h^r} e^{ma}, \text{ where } h = x - a;$$

hence, expanding e^h, and making $h = 0$, we evidently get the same result as before.

3. $\dfrac{x - \sin x}{x^3}$ when $x = 0$.

Here $f'(x) = 1 - \cos x,$ $\qquad f'(0) = 0.$
$\phi'(x) = 3x^2,$ $\qquad \phi'(0) = 0.$
$f''(x) = \sin x,$ $\qquad f''(0) = 0.$
$\phi''(x) = 6x,$ $\qquad \phi''(0) = 0.$
$f'''(x) = \cos x,$ $\qquad f'''(0) = 1.$
$\phi'''(x) = 6,$ $\qquad \phi'''(0) = 6.$

Hence, the true value is $\frac{1}{6}$, as can also be immediately arrived at by substituting $x - \frac{x^3}{6} + \&c.$ instead of $\sin x$.

4. $\dfrac{a^x - 1}{x}$ when $x = 0$;   Ans. $\log a$.

5. $\dfrac{e^x f(x) - e^a f(a)}{e^x \phi(x) - e^a \phi(a)}$ when $x = a$.   ,, $\dfrac{f(a) + f'(a)}{\phi(a) + \phi'(a)}$.

It may be observed that each of these examples can be exhibited in the form $\infty - \infty$, that is, as the difference of two functions each of which becomes infinite for the particular value of x in question.

91. Form $0 \times \infty$.—The expression $f(x) \times \phi(x)$ becomes indeterminate for any value of x which makes one of its factors zero and the other infinite. The function in this case is easily reducible to the form $\dfrac{0}{0}$; for suppose $f(a) = 0$, and $\phi(a) = \infty$, then the expression can be written $\dfrac{f(a)}{\frac{1}{\phi(a)}}$, which is of the required form.

EXAMPLES.

1. Find the value of $(1 - x) \tan \dfrac{\pi x}{2}$ when $x = 1$.

This expression becomes $\dfrac{1 - x}{\cot \frac{\pi x}{2}}$, the true value of which is $\dfrac{2}{\pi}$ when $x = 1$.

2. $\sec x \left(x \sin x - \dfrac{\pi}{2} \right)$,   when $x = \dfrac{\pi}{2}$.

This becomes $\dfrac{x \sin x - \frac{\pi}{2}}{\cos x}$, a form already discussed.

3. $\operatorname{Tan}(x - a) \cdot \log(x - a)$,   when $x = a$.   Ans. 0.

4. $\operatorname{Cosec}^2 \beta x \cdot \log(\cos ax)$,   ,, $x = 0$.   ,, $-\dfrac{a^2}{2\beta^2}$.

92. Form $\frac{\infty}{\infty}$. As stated before, the fraction $\frac{f(x)}{\phi(x)}$ also becomes indeterminate for the value $x = a$, if

$$f(a) = \infty, \text{ and } \phi(a) = \infty.$$

It can, however, be reduced to the form $\frac{0}{0}$ by writing it in the shape

$$\frac{\dfrac{1}{\phi(x)}}{\dfrac{1}{f(x)}}.$$

The true value of the latter fraction, by Art. 90, is that of

$$\frac{\dfrac{\phi'(x)}{\{\phi(x)\}^2}}{\dfrac{f'(x)}{f(x)}}, \text{ or } \frac{\phi'(x)}{f'(x)}\left\{\frac{f(x)}{\phi(x)}\right\}^2.$$

Now, suppose A represents the limiting value of $\frac{f(x)}{\phi(x)}$ when $x = a$, then we have

$$A = \frac{\phi'(a)}{f'(a)} A^2, \text{ or } A = \frac{f'(a)}{\phi'(a)}:$$

that is, the true value of the indeterminate form $\frac{\infty}{\infty}$ is found in the same manner as that of the form $\frac{0}{0}$.

In the preceding demonstration, in dividing both sides of our equation by A, we have assumed that A is neither zero nor infinity; so that the proof would fail in either of these cases.

It can, however, be completed as follows:—

Suppose the real limit of $\frac{f(a)}{\phi(a)}$ to be zero, then that of $\frac{f(a) + k\phi(a)}{\phi(a)}$ is k, where k may be any constant; but as the latter fraction has a finite limit, its value by the preceding method is

$$\frac{f'(a) + k\phi'(a)}{\phi'(a)}, \text{ or } \frac{f'(a)}{\phi'(a)} + k;$$

therefore $\frac{f'(a)}{\phi'(a)} = 0$; *i.e.* when A is zero, $\frac{f'(a)}{\phi'(a)}$ is also zero, and *vice versa*.

Similarly, if the true value of $\frac{f(x)}{\phi(x)}$ be infinity when $x = a$, then $\frac{\phi(a)}{f(a)}$ is really zero; we have, therefore, $\frac{\phi'(a)}{f'(a)} = 0$, by what has been just established; $\therefore \frac{f'(a)}{\phi'(a)} = \infty$.

Accordingly, in all cases the value of $\frac{f'(a)}{\phi'(a)}$* determines that of $\frac{f(a)}{\phi(a)}$ for either of the indeterminate forms $\frac{0}{0}$ or $\frac{\infty}{\infty}$.

---

* On referring to Art. 69, the student will observe that $\frac{f'(x)}{\phi'(x)}$ is of the form $\frac{\infty}{\infty}$ whenever $\frac{f(x)}{\phi(x)} = \frac{\infty}{\infty}$, so that the process given above would not seem to assist us towards determining the true value of the fraction in this case; however, we generally find a common factor, or else some simple transformation, by which we are enabled to exhibit our expression, after differentiation, in the form $\frac{0}{0}$.

For example $\dfrac{\tan x}{\log\left(x - \dfrac{\pi}{2}\right)}$ is of the form $\frac{\infty}{-\infty}$ when $x = \frac{\pi}{2}$: here $f'(x) = \sec^2 x$, $\phi'(x) = \dfrac{1}{x - \dfrac{\pi}{2}}$, and the fraction $\frac{f'(x)}{\phi'(x)}$ is still of the form $\frac{\infty}{\infty}$, but it can be transformed into $\dfrac{x - \dfrac{\pi}{2}}{\cos^2 x}$, which is of the form $\frac{0}{0}$: the true value of the latter fraction can be easily shown to be $-\infty$ when $x = \frac{\pi}{2}$.

In some instances an expression becomes indeterminate from an infinite value of x. The student can easily see, on substituting $\frac{1}{y}$ for x, that our rules apply equally to this case.

93. Indeterminate Expressions of the Form $\{f(x)\}^{\phi(x)}$.

Let $u = \{f(x)\}^{\phi(x)}$, then $\log u = \phi(x) \log f(x)$. This latter product is indeterminate whenever one of its factors becomes zero and the other infinite for the same value of x.

(1.) Let $\phi(x) = 0$, and $\log \{f(x)\} = \pm \infty$; the latter requires either $f(x) = \infty$, or $f(x) = 0$.

Hence, $\{f(x)\}^{\phi(x)}$ becomes indeterminate when it is of the form 0^0, or ∞^0.

(2.) Let $\phi(x) = \pm \infty$, and $\log \{f(x)\} = 0$, or $f(x) = 1$; this gives the indeterminate forms

$$1^\infty \text{ and } 1^{-\infty}.$$

Hence, the indeterminate forms of this class are

$$0^0, \infty^0, \text{ and } 1^{\pm\infty}.$$

EXAMPLES.

1. $(\sin x)^{\tan x}$ is of the form 0^0, when $x = 0$.

Here
$$\log u = \tan x \log (\sin x) = \frac{\log (\sin x)}{\cot x}.$$

The true value of this fraction is that of

$$\frac{\cot x}{-\operatorname{cosec}^2 x} = -\cos x \sin x, \text{ or } 0 \text{ when } x = 0.$$

Hence the value of $(\sin x)^{\tan x} = e^0 = 1$ at the same time.

2. $(\sin x)^{\tan x}$, when $x = \dfrac{\pi}{2}$.

This is of the form 1^∞, but its true value is easily found to be unity.

3. $\left(\dfrac{\tan x}{x}\right)^{\frac{1}{x^2}}$, when $x = 0$.

Here
$$\log u = \frac{\log\left(\dfrac{\tan x}{x}\right)}{x^2};$$

but
$$\frac{\tan x}{x} = 1 + \frac{x^2}{3} + \&c.$$

$$\therefore \log \frac{\tan x}{x} = \log\left(1 + \frac{x^2}{3} + \&c.\right) = \frac{x^2}{3} + \&c.$$

Hence, the true value of $\log u$ is $\frac{1}{3}$ when $x = 0$; and, accordingly, the value of u is $e^{\frac{1}{3}}$ at the same time.

4. $$u = \left(1 + \frac{a}{x}\right)^x, \text{ when } x = 0.$$

Let $x = \dfrac{1}{z}$, then $\log u = \dfrac{\log(1 + az)}{z}$;

$\therefore$ by Art. 92, the true value of $\log u$ when $z = \infty$ is that of $\dfrac{a}{1 + az}$, or is zero. Hence, the value of u is 1 at the same time.

5. $$u = \left(1 + \frac{a}{x}\right)^x, \text{ when } x = \infty.$$

Let $x = \dfrac{1}{z}$, then $\log u = \dfrac{\log(1 + az)}{z}$,

the true value of which is a when z is zero.
Hence, the true value of u is e^a; as also follows immediately from Art. 29.

6. $\left(\dfrac{1}{x}\right)^{\tan x}$, when $x = 0$.   *Ans.* 1.

7. $\left(2 - \dfrac{x}{a}\right)^{\tan \frac{\pi x}{2a}}$, when $x = a$.   ,, $e^{\frac{2}{\pi}}$.

94. Compound Indeterminate Forms.—If an indeterminate form be the product of two or more expressions, each of which becomes indeterminate for the same value of x, its true value can be determined by considering the limiting value of each of the expressions separately; also when the value of any indeterminate form is known, that of any power of it can be determined. These are evident principles: at the same time the student will find them of importance in the evaluation of indeterminate functions of complex form. We will illustrate their use by a few elementary applications.

EXAMPLES.

1. Find the value of
$$x^m (\sin x)^{\tan x} \left(\frac{\pi - 2x}{2 \sin 2x}\right)^n, \text{ when } x = \frac{\pi}{2}.$$

The value of x^m is $\left(\dfrac{\pi}{2}\right)^m$, and that of $(\sin x)^{\tan x}$ is unity: see p. 105.

Again, $\dfrac{\pi - 2x}{2 \sin 2x}$ becomes $\dfrac{2z}{2 \sin 2z}$ on substituting $\dfrac{\pi}{2} - z$ for x: hence its true value is $\dfrac{1}{2}$ when $z = 0$.

Accordingly, the true value of the proposed expression when $x = \dfrac{\pi}{2}$ is $\dfrac{\pi^m}{2^{m+n}}$.

2. $\qquad\qquad\qquad\dfrac{x^n}{e^x}$ when $x = \infty$.

This fraction can be written in the form $\left(\dfrac{x}{e^{\frac{x}{n}}}\right)^n$. The true value of $\dfrac{x}{e^{\frac{x}{n}}}$, by the method of Art. 92, is that of $\dfrac{1}{\frac{1}{n}e^{\frac{x}{n}}}$; but the value of the latter fraction is zero when $x = \infty$; hence the true value of the proposed fraction is also zero at the same time.

3. $\qquad u = x^n (\log x)^m$, when $x = 0$, and m and n are *positive*.

Here $\qquad\qquad\qquad u = (x^{\frac{n}{m}} \log x)^m$,

$\dfrac{\log x}{x^{-\frac{n}{m}}}$ is of the form $\dfrac{\infty}{\infty}$ when $x = 0$;

its true value is that of $\qquad \dfrac{\frac{1}{x}}{-\frac{n}{m} x^{-\frac{n}{m}-1}}$, or $-\dfrac{mx^{\frac{n}{m}}}{n}$.

Hence, the true value of the given expression is zero.
This form is immediately reducible to the preceding, by assuming $x^n = e^{-y}$.

4. $\qquad\qquad\qquad u = \dfrac{a^{x^m}}{b^{x^n}}$ when $x = \infty$.

Here $\qquad\qquad\qquad u = \left(\dfrac{a}{b^{x^{n-m}}}\right)^{x^m}$:

but if $b > 1$, and $n > m$, $b^{x^{n-m}} = \infty$ when $x = \infty$. Consequently the value of u is of the form 0^∞, or is zero in this case.

Again, if $m > n$, $b^{x^{n-m}} = 0$ when $x = \infty$, and the true value of u is ∞.

5. $$u = \frac{a\, x^{-\frac{1}{n}}}{b\, x^{-\frac{1}{m}}} \text{ when } x = 0.$$

Let $x = \dfrac{1}{z}$, and this fraction is immediately reducible to the form discussed in the previous Example.

6. $$\frac{(1-\cos x)^n \{\log(1+x)\}^m}{x^{2n+m}}, \text{ when } x = 0. \qquad\qquad Ans.\ \frac{1}{2^n}.$$

7. $$u = \frac{(1+x)^{\frac{1}{x}} - e}{x}, \text{ when } x = 0.$$

From Art. 29, this is of the form $\dfrac{0}{0}$; to find its true value, proceed by the method of Art. 90, and it becomes

$$(1+x)^{\frac{1}{x}}\left\{\frac{x - (1+x)\log(1+x)}{x^2(1+x)}\right\}.$$

Again, substituting for $(1+x)^{\frac{1}{x}}$ its limiting value e, we get

$$e\left\{\frac{x - (1+x)\log(1+x)}{x^2(1+x)}\right\};$$

the true value of which is readily found to be $-\dfrac{e}{2}$ when $x = 0$. Compare Ex. 29, p. 94.

8. $$\left|\frac{m^x - 1}{\sin x}\right| \left\{\frac{a \sin x - \sin ax}{x(\cos x - \cos ax)}\right\}^n, \text{ when } x = 0.$$

The true value of $\dfrac{m^x - 1}{\sin x}$, when $x = 0$, is $\log m$;

and that of $\dfrac{a \sin x - \sin ax}{x(\cos x - \cos ax)}$, when $x = 0$,

has been found in Example 6, Art 89, to be $\dfrac{a}{3}$; hence the true value of the given expression when $x = 0$, is $\left(\dfrac{a}{3}\right)^n \log m$.

EXAMPLES.

1. $\dfrac{f(x)-f(a)}{\phi(x)-\phi(a)}$, when $x=a$. Ans. $\dfrac{f'(a)}{\phi'(a)}$.

2. $\left(\dfrac{\sin nx}{x}\right)^m$, $x=0$. n^m.

3. $\dfrac{\cos x\theta - \cos n\theta}{(x^2-n^2)^r}$, $x=n$. ∞.

4. $\dfrac{\sqrt{a+x}-\sqrt{2x}}{\sqrt{a+3x}-2\sqrt{x}}$, $x=a$. $\sqrt{2}$.

5. $\dfrac{x^{n+1}-a^{n+1}}{n+1}$, $n=-1$. $\log\left(\dfrac{x}{a}\right)$.

6. $\dfrac{e^x - e^{-x} - 2x}{(e^x-1)^3}$, $x=0$. $\dfrac{1}{3}$.

7. $\dfrac{1-\sin x+\cos x}{\sin x+\cos x-1}$, $x=\dfrac{\pi}{2}$. 1.

8. $\dfrac{\tan x - \sin x}{\sin^3 x}$, $x=0$. $\dfrac{1}{2}$.

9. $\dfrac{(a^2-x^2)^{\frac{1}{2}}+(a-x)^{\frac{3}{2}}}{(a^3-x^3)^{\frac{1}{3}}+(a-x)^{\frac{1}{3}}}$, $x=a$. $\dfrac{\sqrt{2a}}{1+a\sqrt{3}}$.

10. $\dfrac{x^{\frac{1}{2}}\tan x}{(e^x-1)^{\frac{3}{2}}}$, $x=0$. 1.

11. $\dfrac{a^{\sin x}-a}{\log \sin x}$, $x=\dfrac{\pi}{2}$. $a\log a$.

12. $\dfrac{n}{x}-\cot\left(\dfrac{x}{n}\right)$, $x=0$. 0.

13. $\dfrac{x^2+2\cos x-2}{x^4}$, $x=0$. $\dfrac{1}{12}$.

14. $\dfrac{\left(x+\sin 2x - 6\sin\dfrac{x}{2}\right)^2}{\left(4+\cos x - 5\cos\dfrac{x}{2}\right)^3}$, $x=0$. $8\left(\dfrac{29}{3}\right)^2$.

15. $\dfrac{\sqrt{2 + \cos 2x - \sin x}}{x \sin 2x + x \cos x}$, when $x = \dfrac{\pi}{2}$. *Ans.* $\dfrac{\sqrt{10}}{3\pi}$.

16. $\dfrac{x^a \sin na - n^a \sin xa}{\tan na - \tan xa}$, $x = n$.

$n^{a-1}(n \cos na - \sin na) \cos^2 na$.

17. $\dfrac{x^2}{1 - \cos mx} \cdot \dfrac{\tan nx - n \tan x}{n \sin x - \sin nx}$, $x = 0$. $\dfrac{4}{m^2}$.

18. $\dfrac{(2 \sin x - \sin 2x)^2}{(\sec x - \cos 2x)^3}$, $x = 0$. $\dfrac{8}{125}$.

19. $x^{\frac{1}{1-x}}$, $x = 1$. $\dfrac{1}{e}$.

20. $\dfrac{(x-y)\{\phi'(x) + \phi'(y)\} - 2\phi(x) + 2\phi(y)}{(x-y)^3}$, $x = y$. $\dfrac{\phi'''(y)}{6}$.

21. $\dfrac{x \log(1+x)}{1 - \cos x}$, $x = 0$. 2.

22. $x \cdot e^{\frac{1}{x}}$, $x = 0$. ∞.

23. $\dfrac{e^x - e^{-x}}{\log(1+x)}$, $x = 0$. 2.

24. $\dfrac{\pi x - 1}{2x^2} + \dfrac{\pi}{(e^{2\pi x} - 1)x}$, $x = 0$. $\dfrac{\pi^2}{2}$.

25. $\dfrac{\log(\tan 2x)}{\log(\tan x)}$, $x = 0$. 1.

26. $\dfrac{e^x + \log(1-x) - 1}{\tan x - x}$, $x = 0$. $-\dfrac{1}{2}$.

27. $\dfrac{2m}{(1-m)\sqrt{1-m^2}} \tan^{-1} \sqrt{\dfrac{1-m^2}{m}} \cos \phi - \dfrac{1+m}{1-m} \cos \phi$, $m = 1$.

$-\dfrac{\cos 3\phi}{3}$.

28. $\dfrac{\log(1 + x + x^2) + \log(1 - x + x^2)}{\sec x - \cos x}$, $x = 0$. 1.

29. $\left(\dfrac{a_1{}^x + a_2{}^x + \ldots a_n{}^x}{n}\right)^{\frac{n}{x}}$, $x = 0$. $a_1 a_2 \ldots a_n$.

30. $\left(\dfrac{\log x}{x}\right)^{\frac{1}{x}}$, when $x = \infty$. Ans. 1.

31. $\dfrac{(1+x)^{\frac{1}{x}} - e + \dfrac{ex}{2}}{x^2}$, $x = 0$. $\dfrac{11e}{24}$.

32. $\dfrac{\sin x - \log (e^x \cos x)}{x^3}$, $x = 0$. $\dfrac{1}{2}$.

33. $x^2 \left(1 + \dfrac{1}{x}\right)^x - ex^3 \log\left(1 + \dfrac{1}{x}\right)$, $x = \infty$. $\dfrac{e}{8}$.

34. $\dfrac{1 - x + \log x}{1 - \sqrt{2x - x^2}}$, $x = 1$. -1.

35. $\dfrac{x^2 - x}{1 - x + \log x}$, $x = 1$. ∞.

36. $\dfrac{x^x - x}{1 - x + \log x}$, $x = 1$. -2.

37. $\dfrac{\cos x - \log(1+x) + \sin x - 1}{e^x - (1+x)}$, $x = 0$. 0.

38. $\dfrac{e^x + \sin x - 1}{\log(1+x)}$, $x = 0$. 2.

39. $\dfrac{e^x - e^{-x} - 2x}{\tan x - x}$, $x = 0$. 1.

40. $\dfrac{d}{dx}\left(\dfrac{ax^2 + bx + c}{a_1 x + b_1}\right)$, $x = \infty$. $\dfrac{a}{a_1}$.

41. $\dfrac{a - x - a \log\left(\dfrac{a}{x}\right)}{a - \sqrt{2ax - x^2}}$, $x = a$. -1.

42. $\dfrac{\tan(a+x) - \tan(a-x)}{\tan^{-1}(a+x) - \tan^{-1}(a-x)}$, $x = 0$. $\dfrac{1 + a^2}{\cos^2 a}$.

43. $\dfrac{x^3 - 3x + 2}{x^4 - 6x^2 + 8x - 3}$, $x = 1$. ∞.

Examples.

44. $(\sin x)^{\sin x}$, when $x = 0$. Ans. 1.

45. $(\sec x)^{n \sec x}$, $x = 0$. 1.

46. $(\sin x)^{\tan x}$, $x = \dfrac{\pi}{2}$. 1.

47. Find the value of

$$\frac{(x-y)a^n + (y-a)x^n + (a-x)y^n}{(x-y)(y-a)(a-x)}, \qquad -\frac{n \cdot \overline{n-1}}{1 \cdot 2} a^{n-2}.$$

when $x = y = a$.

Substitute $a + h$ for x, and $a + k$ for y, and after some easy transformations we get the answer, on making $h = 0$, and $k = 0$.

48. $\dfrac{x + \tan x - \tan 2x}{2x + \tan x - \tan 3x}$, $x = 0$. Ans. $\dfrac{7}{26}$.

49. $\dfrac{x + \sin x - \sin 2x}{2x + \tan x - \tan 3x}$, $x = 0$. $\dfrac{-7}{52}$.

50. $\dfrac{\sqrt{x} - \sqrt{a} + \sqrt{x-a}}{\sqrt{x^2 - a^2}}$, $x = a$. $\dfrac{1}{\sqrt{2a}}$.

51. $\dfrac{x - \dfrac{2}{3}\sin x - \dfrac{1}{3}\tan x}{x^5}$, $x = 0$. $\dfrac{-1}{20}$.

CHAPTER V.

PARTIAL DIFFERENTIAL COEFFICIENTS AND DIFFERENTIATION OF FUNCTIONS OF TWO OR MORE VARIABLES.

95. Partial Differentiation.—In the preceding chapters we have regarded the functions under consideration as depending on one variable solely; thus, such expressions as

$$e^{ax},\ \sin bx,\ x^m,\ \&c.,$$

have been treated as functions of x only; the quantities a, b, m, ... being regarded as constants. We may, however, conceive these quantities as also capable of change, and as receiving small increments; then, if we regard x as constant, we can, by the methods already established, find the differential coefficients of these expressions with regard to the quantities, a, b, m, &c., *considered as variable*.

In this point of view, e^{ax} is regarded as a function of a as well as of x, and its differential coefficient with regard to a is represented by $\dfrac{d(e^{ax})}{da}$, or $x\, e^{ax}$ by Art. 30; in the derivation of which *x is regarded as a constant.*

In like manner, $\sin(ax+by)$ may be considered as a function of the four quantities, x, y, a, b, and we can find its differential coefficient with respect to any one of them, *the others being regarded as constants*. Let these derived functions be denoted by

$$\frac{du}{dx},\quad \frac{du}{dy},\quad \frac{du}{da},\quad \frac{du}{db},$$

respectively, where u stands for the expression under consideration, and we have

$$\frac{du}{dx} = a\cos(ax+by),\qquad \frac{du}{dy} = b\cos(ax+by),$$

$$\frac{du}{da} = x\cos(ax+by),\qquad \frac{du}{db} = y\cos(ax+by).$$

These expressions are called the *partial differential coefficients* of u with respect to x, y, a, b, respectively.

More generally, if
$$f(x, y, z)$$
denotes a function of three variables, x, y, z, its differential coefficient, when x *alone* is supposed to change, is called the *partial* differential coefficient of the function *with respect to x*; and similarly for the other variables y and z. If the function be represented by u, its partial differential coefficients are denoted by
$$\frac{du}{dx}, \frac{du}{dy}, \frac{du}{dz},$$
and from the preceding it follows that the partial derived functions of any expression are formed by the same rules as the derived functions in the case of a single variable.

Examples.

1. $u = (ax^2 + by^2 + cz^2)^n$.

Here
$$\frac{du}{dx} = 2nax(ax^2 + by^2 + cz^2)^{n-1},$$

$$\frac{du}{dy} = 2nby(ax^2 + by^2 + cz^2)^{n-1},$$

$$\frac{du}{dz} = 2ncz(ax^2 + by^2 + cz^2)^{n-1}.$$

2. $u = \sin^{-1}\frac{x}{y}$.

$$\frac{du}{dx} = \frac{1}{\sqrt{y^2 - x^2}}, \quad \frac{du}{dy} = \frac{-x}{y\sqrt{y^2 - x^2}}.$$

3. $u = x^y$, $\quad \dfrac{du}{dx} = yx^{y-1}, \quad \dfrac{du}{dy} = x^y \log x.$

4. $u = x^2 \phi(xy)$.

$$\frac{du}{dx} = 2x\phi(xy) + x^2 y\phi'(xy).$$

$$\frac{du}{dy} = x^3 \phi'(xy).$$

96. Differentiation of a Function of Two Variables.

—Let $u = \phi(x, y)$, and suppose x and y to receive the increments h, k, respectively, and let Δu denote the corresponding increment of u, then

$$\Delta u = \phi(x + h, y + k) - \phi(x, y)$$
$$= \phi(x + h, y + k) - \phi(x, y + k) + \phi(x, y + k) - \phi(x, y)$$
$$= \frac{\phi(x + h, y + k) - \phi(x, y + k)}{h} h + \frac{\phi(x, y + k) - \phi(x, y)}{k} k.$$

If now h and k be supposed to become infinitely small, by Art. 6 we have

$$\frac{\phi(x + h, y + k) - \phi(x, y + k)}{h} = \frac{d \cdot \phi(x, y + k)}{dx},$$

and $\quad \dfrac{\phi(x, y + k) - \phi(x, y)}{k} = \dfrac{d \cdot \phi(x, y)}{dy}.$

In the limit, when k is infinitely small, $\phi(x, y + k)$ becomes $\phi(x, y)$, and

$$\frac{d \cdot \phi(x, y + k)}{dx} \text{ becomes } \frac{d \cdot \phi(x, y)}{dx};$$

hence we get, *neglecting infinitely small quantities of the second order*,

$$du = \frac{du}{dx} h + \frac{du}{dy} k,$$

where h and k are infinitely small.

If dx, dy, be substituted for h and k, the preceding becomes

$$du = \frac{du}{dx} dx + \frac{du}{dy} dy. \tag{1}$$

In this equation du is called the *total differential* of u, where both x and y are supposed to vary.

The student should carefully observe the different meanings given to the infinitely small quantity du in this equation. Thus, in the expression $\dfrac{du}{dx} dx$, du stands for the infinitely small change in u arising from the increment dx in x, y being regarded as constant. Similarly, in $\dfrac{du}{dy} dy$, du stands for the infinitely small change arising from the increment dy in y, x being regarded as constant. If these partial increments be represented by $d_x u$, $d_y u$, the preceding result may be written in the form

$$du = d_x u + d_y u.$$

That is, the total increment in a function of two variables is found by adding its partial increments, arising from the differentials of each of the variables taken separately.

EXAMPLES.

1. Let $x = r \cos \theta$, in which r and θ are considered variables, to find the total differential of x.

Here $\quad \dfrac{dx}{dr} = \cos \theta, \quad \dfrac{dx}{d\theta} = -r \sin \theta.$

Hence $\quad dx = \cos \theta\, dr - r \sin \theta\, d\theta.$

2. $\quad u = \dfrac{x^2}{a^2} + \dfrac{y^2}{b^2}.$

Here $\quad \dfrac{du}{dx} = \dfrac{2x}{a^2}, \quad \dfrac{du}{dy} = \dfrac{2y}{b^2};$

$\therefore du = \dfrac{2x}{a^2} dx + \dfrac{2y}{b^2} dy.$

3. $u = \phi\left(\dfrac{x}{y}\right).$ Let $\dfrac{x}{y} = z$, then $u = \phi(z)$,

$$\dfrac{du}{dx} = \dfrac{du}{dz}\dfrac{dz}{dx} = \dfrac{\phi'\left(\dfrac{x}{y}\right)}{y},$$

$$\dfrac{du}{dy} = \dfrac{du}{dz}\dfrac{dz}{dy} = \dfrac{-x\phi'\left(\dfrac{x}{y}\right)}{y^2};$$

$$\therefore du = \phi'\left(\dfrac{x}{y}\right) \dfrac{y\, dx - x\, dy}{y^2}.$$

Again, multiplying the former of the two preceding equations by x, and the latter by y, and adding, we get

$$x \dfrac{du}{dx} + y \dfrac{du}{dy} = 0.$$

97. Differentiation of a Function of Three or more Variables.—Suppose

$$u = \phi(x, y, z),$$

and let h, k, l represent infinitely small increments in x, y, z, respectively; then

$$\Delta u = \phi(x+h, y+k, z+l) - \phi(x, y, z)$$

$$= \frac{\phi(x+h, y+k, z+l) - \phi(x, y+k, z+l)}{h} h$$

$$+ \frac{\phi(x, y+k, z+l) - \phi(x, y, z+l)}{k} k + \frac{\phi(x, y, z+l) - \phi(x, y, z)}{l} l,$$

which becomes in the limit, by the same argument as before, when dx, dy, dz, are substituted for h, k, l,

$$du = \frac{du}{dx} dx + \frac{du}{dy} dy + \frac{du}{dz} dz. \qquad (2)$$

Or, the infinitely small increment in u is the sum of its infinitely small increments arising from the variation of each variable considered separately.

A similar process of reasoning can be easily extended to a function of any number of variables; hence, in general, if u be a function of n variables, $x_1, x_2, x_3, \ldots x_n$,

$$du = \frac{du}{dx_1} dx_1 + \frac{du}{dx_2} dx_2 + \ldots + \frac{du}{dx_n} dx_n. \qquad (3)$$

98. If

$$u = f(v, w),$$

where v, w, *are both functions of* x; then, from Art. 96, it is easily seen that

$$\frac{du}{dx} = \frac{df(v, w)}{dv} \frac{dv}{dx} + \frac{df(v, w)}{dw} \cdot \frac{dw}{dx}.$$

This result is usually written in the form

$$\frac{du}{dx} = \frac{du}{dv} \frac{dv}{dx} + \frac{du}{dw} \frac{dw}{dx}. \qquad (4)$$

In general, if

$$u = \phi(y_1, y_2, \ldots y_n),$$

where $y_1, y_2, \ldots y_n$, are each functions of x, we have

$$\frac{du}{dx} = \frac{du}{dy_1}\frac{dy_1}{dx} + \frac{du}{dy_2}\frac{dy_2}{dx} + \ldots + \frac{du}{dy_n}\frac{dy_n}{dx}. \qquad (5)$$

Also, if y_1, y_2, &c., y_n, be at the same time functions of another variable z, we have

$$\frac{du}{dz} = \frac{du}{dy_1}\frac{dy_1}{dz} + \frac{du}{dy_2}\frac{dy_2}{dz} + \&c. + \frac{du}{dy_n}\frac{dy_n}{dz},$$

and so on.

EXAMPLES.

1. Let $u = \phi(X, Y)$,

where $X = ax + by, \quad Y = a'x + b'y$;

then
$$\frac{du}{dx} = \frac{du}{dX}\frac{dX}{dx} + \frac{du}{dY}\frac{dY}{dx},$$

$$\frac{du}{dy} = \frac{du}{dX}\frac{dX}{dy} + \frac{du}{dY}\frac{dY}{dy};$$

but $\quad \dfrac{dX}{dx} = a, \quad \dfrac{dX}{dy} = b, \quad \dfrac{dY}{dx} = a', \quad \dfrac{dY}{dy} = b'.$

Hence
$$\frac{du}{dx} = a\frac{du}{dX} + a'\frac{du}{dY},$$

$$\frac{du}{dy} = b\frac{du}{dX} + b'\frac{du}{dY}.$$

2. More generally, let

$$u = \phi(X, Y, Z),$$

where
$$X = ax + by + cz,$$
$$Y = a'x + b'y + c'z,$$
$$Z = a''x + b''y + c''z.$$

When these substitutions are made, u becomes a function of x, y, z, and we have

$$\frac{du}{dx} = a\frac{du}{dX} + a'\frac{du}{dY} + a''\frac{du}{dZ},$$

$$\frac{du}{dy} = b\frac{du}{dX} + b'\frac{du}{dY} + b''\frac{du}{dZ},$$

$$\frac{du}{dz} = c\frac{du}{dX} + c'\frac{du}{dY} + c''\frac{du}{dZ}.$$

98*. Differentiation of a Function of Differences.

If u be a function of the differences of the variables, a, β, γ: to prove that

$$\frac{du}{da} + \frac{du}{d\beta} + \frac{du}{d\gamma} = 0.$$

Let $a - \beta = x$, $\beta - \gamma = y$, $\gamma - a = z$; then, u is a function of x, y, z; and, accordingly, we may write

$$u = \phi(x, y, z).$$

Hence
$$\frac{du}{da} = \frac{du}{dx}\frac{dx}{da} + \frac{du}{dy}\frac{dy}{da} + \frac{du}{dz}\frac{dz}{da} = \frac{du}{dx} - \frac{du}{dz}.$$

Similarly,
$$\frac{du}{d\beta} = \frac{du}{dy} - \frac{du}{dx}, \quad \frac{du}{d\gamma} = \frac{du}{dz} - \frac{du}{dy};$$

$$\therefore \frac{du}{da} + \frac{du}{d\beta} + \frac{du}{d\gamma} = 0.$$

This result admits of obvious extension to a function of the differences of any number of variables.

EXAMPLES.

1. If
$$\Delta = \begin{vmatrix} 1, & 1, & 1, & 1, \\ a, & \beta, & \gamma, & \delta, \\ a^2, & \beta^2, & \gamma^2, & \delta^2, \\ a^3, & \beta^3, & \gamma^3, & \delta^3, \end{vmatrix}, \text{ prove that}$$

$$\frac{d\Delta}{da} + \frac{d\Delta}{d\beta} + \frac{d\Delta}{d\gamma} + \frac{d\Delta}{d\delta} = 0.$$

2. If
$$\Delta = \begin{vmatrix} 1, & 1, & 1, & 1, \\ a, & \beta, & \gamma, & \delta, \\ a^2, & \beta^2, & \gamma^2, & \delta^2, \\ a^4, & \beta^4, & \gamma^4, & \delta^4, \end{vmatrix}, \text{ prove that}$$

$$\frac{d\Delta}{da} + \frac{d\Delta}{d\beta} + \frac{d\Delta}{d\gamma} + \frac{d\Delta}{d\delta} = 4 \begin{vmatrix} 1, & 1, & 1, & 1, \\ a, & \beta, & \gamma, & \delta, \\ a^2, & \beta^2, & \gamma^2, & \delta^2, \\ a^3, & \beta^3, & \gamma^3, & \delta^3, \end{vmatrix}.$$

99. Definition of an Implicit Function.—Suppose that y, instead of being given explicitly as a function of x, is determined by an equation of the form

$$f(x, y) = 0,$$

then y is said to be an *implicit* function of x; for its value, or values, are given implicitly when that of x is known.

100. Differentiation of an Implicit Function.—Let k denote the increment of y corresponding to the increment h in x, and denote $f(x, y)$ by u.

Then, since the equation $f(x, y) = 0$ is supposed to hold for all values of x and the corresponding values of y, we must have

$$f(x + h, y + k) = 0.$$

Hence $du = 0$; and accordingly, by Art. 96, we have, when h and k are infinitely small,

$$\frac{du}{dx} h + \frac{du}{dy} k = 0;$$

hence in the limit
$$\frac{k}{h} = \frac{dy}{dx} = -\frac{\dfrac{du}{dx}}{\dfrac{du}{dy}}. \qquad (6)$$

This result enables us to determine the differential coefficient of y with respect to x whenever the form of the equation $f(x, y) = 0$ is given.

In the case of implicit functions we may regard x as being a function of y, or y a function of x, whichever we please—in the former case y is treated as the *independent* variable, and, in the latter, x: when y is taken as the independent variable, we have

$$\frac{dx}{dy} = -\frac{\dfrac{du}{dy}}{\dfrac{du}{dx}} = \frac{1}{\dfrac{dy}{dx}}.$$

This is the extension of the result given in Art. 20, and might have been established in a similar manner.

Differentiation of an Implicit Function.

EXAMPLES.

1. If $x^3 + y^3 - 3axy = c$, to find $\dfrac{dy}{dx}$.

Here $\dfrac{du}{dx} = 3(x^2 - ay)$, $\dfrac{du}{dy} = 3(y^2 - ax)$;

$$\therefore \dfrac{dy}{dx} = \dfrac{x^2 - ay}{ax - y^2}.$$

See Art. 38.

2. If $\dfrac{x^m}{a^m} + \dfrac{y^m}{b^m} = 1$, to find $\dfrac{dy}{dx}$.

Here $\dfrac{du}{dx} = \dfrac{mx^{m-1}}{a^m}$, $\dfrac{du}{dy} = \dfrac{my^{m-1}}{b^m}$; $\therefore \dfrac{dy}{dx} = -\left(\dfrac{x}{y}\right)^{m-1}\left(\dfrac{b}{a}\right)^m$.

3. $x \log y - y \log x = 0$. $\dfrac{dy}{dx} = \dfrac{y}{x}\left(\dfrac{x \log y - y}{y \log x - x}\right)$.

101. If $u = \phi(x, y)$, where x and y are connected by the equation $f(x, y) = 0$, to find the total differential of u with respect to x; y being regarded as a function of x.

Here, by Art. 98, we get

$$\dfrac{du}{dx} = \dfrac{d\phi}{dx} + \dfrac{d\phi}{dy}\dfrac{dy}{dx}.$$

Also

$$\dfrac{df}{dx} + \dfrac{df}{dy}\dfrac{dy}{dx} = 0.$$

Hence, eliminating $\dfrac{dy}{dx}$, we get

$$\dfrac{du}{dx} = \dfrac{\dfrac{d\phi}{dx}\dfrac{df}{dy} - \dfrac{df}{dx}\dfrac{d\phi}{dy}}{\dfrac{df}{dy}}. \tag{7}$$

This result can also be written in the following determinant form:

$$\frac{du}{dx} = \frac{\begin{vmatrix} \dfrac{d\phi}{dx}, & \dfrac{d\phi}{dy} \\ \dfrac{df}{dx}, & \dfrac{df}{dy} \end{vmatrix}}{\dfrac{df}{dy}}.$$

More generally, let $u = \phi(x, y, z)$, where $x, y, z,$ are connected by two equations,

$$f_1(x, y, z) = 0, \quad f_2(x, y, z) = 0;$$

then, as in the preceding case, we have

$$\frac{du}{dx} = \frac{d\phi}{dx} + \frac{d\phi}{dy}\frac{dy}{dx} + \frac{d\phi}{dz}\frac{dz}{dx},$$

and also

$$\frac{df_1}{dx} + \frac{df_1}{dy}\frac{dy}{dx} + \frac{df_1}{dz}\frac{dz}{dx} = 0,$$

$$\frac{df_2}{dx} + \frac{df_2}{dy}\frac{dy}{dx} + \frac{df_2}{dz}\frac{dz}{dx} = 0.$$

Hence, we get

$$\frac{du}{dx} = \frac{\begin{vmatrix} \dfrac{d\phi}{dx}, & \dfrac{d\phi}{dy}, & \dfrac{d\phi}{dz} \\ \dfrac{df_1}{dx}, & \dfrac{df_1}{dy}, & \dfrac{df_1}{dz} \\ \dfrac{df_2}{dx}, & \dfrac{df_2}{dy}, & \dfrac{df_2}{dz} \end{vmatrix}}{\begin{vmatrix} \dfrac{df_1}{dy}, & \dfrac{df_1}{dz} \\ \dfrac{df_2}{dy}, & \dfrac{df_2}{dz} \end{vmatrix}}. \quad (8)$$

This result easily admits of generalization.

102. Euler's Theorem of Homogeneous Functions.—If

$$u = Ax^p y^q + Bx^{p'} y^{q'} + Cx^{p''} y^{q''} + \&c.,$$

where

$$p + q = p' + q' = p'' + q'' = \&c. = n,$$

to prove that

$$x \frac{du}{dx} + y \frac{du}{dy} = nu. \qquad (9)$$

Here $\quad x \dfrac{du}{dx} = Apx^p y^q + Bp' x^{p'} y^{q'} + \&c.;$

$$y \frac{du}{dy} = Aqx^p y^q + Bq' x^{p'} y^{q'} + \&c.;$$

$$\therefore x \frac{du}{dx} + y \frac{du}{dy} = A(p + q) x^p y^q + B(p' + q') x^{p'} y^{q'} + \&c.$$
$$= n A x^p y^q + n B x^{p'} y^{q'} + \&c. = nu.$$

Hence, if u be any homogeneous expression of the nth degree in x and y, not involving fractions, we have

$$x \frac{du}{dx} + y \frac{du}{dy} = nu.$$

Again, suppose u to be a homogeneous function of a fractional form, represented by $\dfrac{\phi_1}{\phi_2}$; where ϕ_1, ϕ_2, are homogeneous expressions of the nth and m^{th} degrees, respectively, in x and y; then, from the equation

$$u = \frac{\phi_1}{\phi_2}$$

we have $\qquad \dfrac{du}{dx} = \dfrac{\phi_2 \dfrac{d\phi_1}{dx} - \phi_1 \dfrac{d\phi_2}{dx}}{(\phi_2)^2},$

and $\qquad \dfrac{du}{dy} = \dfrac{\phi_2 \dfrac{d\phi_1}{dy} - \phi_1 \dfrac{d\phi_2}{dy}}{(\phi_2)^2};$

accordingly we get

$$x\frac{du}{dx} + y\frac{du}{dy} = \frac{\phi_2\left(x\dfrac{d\phi_1}{dx} + y\dfrac{d\phi_1}{dy}\right) - \phi_1\left(x\dfrac{d\phi_2}{dx} + y\dfrac{d\phi_2}{dy}\right)}{(\phi_2)^2};$$

but, by the preceding,

$$x\frac{d\phi_1}{dx} + y\frac{d\phi_1}{dy} = n\phi_1, \qquad x\frac{d\phi_2}{dx} + y\frac{d\phi_2}{dy} = m\phi_2;$$

hence
$$x\frac{du}{dx} + y\frac{du}{dy} = \frac{n\phi_1\phi_2 - m\phi_1\phi_2}{(\phi_2)^2}$$

$$= (n-m)\frac{\phi_1}{\phi_2} = (n-m)u;$$

which proves the theorem for homogeneous expressions of a fractional form.

This result admits of being established in a more general manner, as follows:

It is easily seen that a homogeneous expression of the nth degree in x and y, since the sum of the indices of x and of y in each term is n, is capable of being represented in the general form of

$$x^n \phi\left(\frac{y}{x}\right).$$

Accordingly, let
$$u = x^n \phi\left(\frac{y}{x}\right) = x^n v,$$

where
$$v = \phi\left(\frac{y}{x}\right).$$

Then
$$\frac{du}{dx} = nx^{n-1}v + x^n \frac{dv}{dx},$$

and
$$\frac{du}{dy} = x^n \frac{dv}{dy}:$$

multiply the former equation by x, and the latter by y, and add; then

$$x\frac{du}{dx} + y\frac{du}{dy} = nx^n v + x^n\left(x\frac{dv}{dx} + y\frac{dv}{dy}\right);$$

but (by Ex. 3, Art. 96),

$$x\frac{dv}{dx} + y\frac{dv}{dy} = 0;$$

hence
$$x\frac{du}{dx} + y\frac{du}{dy} = nx^n v = nu,$$

which proves the theorem in general.

In the case of three variables, x, y, z,
suppose $\quad u = Ax^p y^q z^r,$
then we have

$$x\frac{du}{dx} = Ap\, x^p y^q z^r, \quad y\frac{du}{dy} = Aq\, x^p y^q z^r, \quad z\frac{du}{dz} = Ar\, x^p y^q z^r;$$

$$\therefore x\frac{du}{dx} + y\frac{du}{dy} + z\frac{du}{dz} = A(p+q+r)\, x^p y^q z^r = (p+q+r)\, u;$$

and the same method of proof can be extended to any homogeneous function of three or more variables.

Hence, if u be a homogeneous function of the nth degree in x, y, z, we have

$$x\frac{du}{dx} + y\frac{du}{dy} + z\frac{du}{dz} = nu. \qquad (10)$$

It may be observed that the preceding result holds also if n be a *fractional* or *negative* number, as can be easily seen.

This result can also be proved in general, by the same method as in the case of two variables, from the consideration that a homogeneous function of the nth degree in x, y, z admits of being written in the general form

$$u = x^n \phi\left(\frac{y}{x}, \frac{z}{x}\right),$$

or in the form

$$u = x^n \phi(v, w), \text{ where } v = \frac{y}{x}, \text{ and } w = \frac{z}{x}.$$

Proceeding, as in the former case, the student can show,

without difficulty, that we shall have
$$x\frac{du}{dx} + y\frac{du}{dy} + z\frac{du}{dz} = nu.$$

Another proof will be found in a subsequent chapter, along with the extension of the theorem to differentiations of a higher order.

EXAMPLES.

Verify Euler's Theorem in the following cases by direct differentiation:—

1. $u = \dfrac{x^3 + y^3}{(x+y)^{\frac{1}{2}}}$; prove $x\dfrac{du}{dx} + y\dfrac{du}{dy} = \dfrac{5u}{2}$.

2. $u = \dfrac{x^3 + ax^2y + by^3}{a'x^2 + b'y^2}$, ,, $x\dfrac{du}{dx} + y\dfrac{du}{dy} = u$.

3. $u = \sin^{-1}\dfrac{x^2 - y^2}{x^2 + y^2}$, ,, $x\dfrac{du}{dx} + y\dfrac{du}{dy} = 0$.

4. $u = x\phi\left(\dfrac{x^3 - y^3}{x^3 + y^3}\right)$, ,, $x\dfrac{du}{dx} + y\dfrac{du}{dy} = u$.

103. Theorem.—If $U = u_0 + u_1 + u_2 \ldots + u_n$, where u_0 is a constant, and $u_1, u_2, \ldots u_n$, are homogeneous functions of x, y, z, &c., of the 1st, 2nd, $\ldots n^{th}$ degrees, respectively, then

$$x\frac{dU}{dx} + y\frac{dU}{dy} + z\frac{dU}{dz} + \ldots = u_1 + 2u_2 + 3u_3 + \ldots + nu_n. \quad (11)$$

For, by Euler's Theorem, we have

$$x\frac{du_r}{dx} + y\frac{du_r}{dy} + z\frac{du_r}{dz} + \&c. = ru_r,$$

since u_r is homogeneous of the rth degree in the variables.

Cor. If $U = 0$, then

$$x\frac{dU}{dx} + y\frac{dU}{dy} + z\frac{dU}{dz} \ldots = -\{u_{n-1} + 2u_{n-2} + \ldots + nu_0\}. \quad (12)$$

This follows on subtracting

$$nu_0 + nu_1 + \ldots + nu_n = 0$$

from the preceding result.

104. Remarks on Euler's Theorem.—In the application of Euler's Theorem the student should be careful to see that the functions to which it is applied are *really* homogeneous expressions. For instance, at first sight the expression $\sin^{-1}\left(\dfrac{x+y}{x^{\frac{1}{2}}+y^{\frac{1}{2}}}\right)$ might appear to be a homogeneous function in x and y; but if the function be expanded, it is easily seen that the terms thus obtained are of different degrees, and, consequently, Euler's Theorem cannot be directly applied to it. However, if the equation be written in the form $\dfrac{x+y}{x^{\frac{1}{2}}+y^{\frac{1}{2}}} = \sin u$, we have, by Euler's formula,

$$x\frac{d\sin u}{dx} + y\frac{d\sin u}{dy} = \frac{\sin u}{2},$$

or
$$\cos u\left(x\frac{du}{dx} + y\frac{du}{dy}\right) = \frac{\sin u}{2};$$

hence
$$x\frac{du}{dx} + y\frac{du}{dy} = \frac{\tan u}{2} = \frac{1}{2}\frac{x+y}{\sqrt{(x^{\frac{1}{2}}+y^{\frac{1}{2}})^2 - (x+y)^2}}.$$

When, however, the degrees in the numerator and the denominator are the same, the function is of the degree zero, and in all such cases we have

$$x\frac{du}{dx} + y\frac{du}{dy} = 0.$$

For example, $\sin^{-1}\left(\dfrac{x^{\frac{1}{2}}+y^{\frac{1}{2}}}{x^{\frac{1}{4}}-y^{\frac{1}{4}}}\right)$, $\tan^{-1}\dfrac{x+y}{x-y}$, $e^{\frac{hx}{y}}$, &c., may be treated as homogeneous expressions, whose degree of homogeneity is zero. The same remark applies to all expressions which are reducible to the form $\phi\left(\dfrac{y}{x}\right)$; as already shown in Ex. 3, Art. 96.

105. If $\quad x = r\cos\theta,\ y = r\sin\theta,$

to prove that $\quad xdy - ydx = r^2 d\theta.$ \hfill (13)

In Ex. 1, Art. 96, we found
$$dx = \cos\theta dr - r\sin\theta d\theta;$$
similarly
$$dy = \sin\theta dr + r\cos\theta d\theta.$$
Hence
$$xdy = r\cos\theta\sin\theta dr + r^2\cos^2\theta d\theta,$$
$$ydx = r\cos\theta\sin\theta dr - r^2\sin^2\theta d\theta;$$
$$\therefore xdy - ydx = r^2 d\theta.$$

106. If x and y have the same values as in the last, to prove that
$$(dx)^2 + (dy)^2 = (dr)^2 + r^2(d\theta)^2. \tag{14}$$

Square and add the expressions for dx, dy, found above, and the required result follows immediately.

The two preceding formulæ are of importance in the theory of plane curves, and admit of being easily established from geometrical considerations.

107. If $\quad u = ax^2 + by^2 + cz^2 + 2fyz + 2gzx + 2hxy,$
to find the condition among the constants that the same values of x, y, z should satisfy the three equations
$$\frac{du}{dx} = 0, \quad \frac{du}{dy} = 0, \quad \frac{du}{dz} = 0.$$

Here
$$\frac{du}{dx} = 2ax + 2hy + 2gz = 0,$$
$$\frac{du}{dy} = 2hx + 2by + 2fz = 0,$$
$$\frac{du}{dz} = 2gx + 2fy + 2cz = 0.$$

Hence, eliminating x, y, z between these three equations, the required condition is
$$abc - af^2 - bg^2 - ch^2 + 2fgh = 0;$$
or, in the determinant form,
$$\begin{vmatrix} a & h & g \\ h & b & f \\ g & f & c \end{vmatrix} = 0.$$

The preceding determinant is called the *discriminant* of the quadratic expression, and is an *invariant* of the function; it also expresses the condition that the conic represented by the equation $u = 0$ should break up into two right lines. (*Salmon's Conic Sections*, Art. 76.)

The foregoing result can be verified easily from the latter point of view; for, suppose the quadratic expression, u, to be the product of two linear factors, X and Y;

or
$$u = XY,$$

where
$$X = lx + my + nz, \quad Y = l'x + m'y + n'z;$$

then
$$\frac{du}{dx} = X\frac{dY}{dx} + Y\frac{dX}{dx} = l'X + lY,$$

$$\frac{du}{dy} = X\frac{dY}{dy} + Y\frac{dX}{dy} = m'X + mY,$$

$$\frac{du}{dz} = X\frac{dY}{dz} + Y\frac{dX}{dz} = n'X + nY.$$

Here the expressions at the right-hand side become zero for the values of x, y, z, which satisfy the equations $X = 0$, $Y = 0$,

or
$$lx + my + nz = 0, \quad l'x + m'y + n'z = 0.$$

Hence in this case the equations

$$\frac{du}{dx} = 0, \quad \frac{du}{dy} = 0, \quad \frac{du}{dz} = 0$$

are also satisfied simultaneously by the same values.

We shall next proceed to illustrate the principles of partial differentiation by applying them to a few elementary questions in plane and spherical triangles. In such cases we may regard *any three** of the parts, a, b, c, A, B, C, as being

---

* The case of the three angles of a plane triangle is excepted, as they are equivalent to only two independent data.

K

independent variables, and each of the others as a function of the three so chosen.

108. Equation connecting the Variations of the three Sides and one Angle.—If two sides, a, b, and the contained angle, C, in a plane triangle, receive indefinitely small increments, to find the corresponding increment in the third side c, we have

$$c^2 = a^2 + b^2 - 2ab\cos C;$$

$$\therefore cdc = (a - b\cos C)\,da + (b - a\cos C)\,db + ab\sin C\,dC;$$

but $\quad a = b\cos C + c\cos B, \quad b = a\cos C + c\cos A.$

Hence, dividing by c, and substituting $c\sin B$ for $b\sin C$, we get

$$dc = \cos B\,da + \cos A\,db + a\sin B\,dC. \tag{15}$$

Otherwise thus, geometrically.

By equation (2), Art. 97, we have

$$dc = \frac{dc}{da}da + \frac{dc}{db}db + \frac{dc}{dC}dC.$$

Now, in the determination of $\dfrac{dc}{da}$ we must regard b and C as constants; accordingly, let us suppose the side CB, or a, to receive a small increment, BB' or Δa, as in the figure. Join AB', and draw $B'D$ perpendicular to AB, produced if necessary; then, by Art. 37, $AB' = AD$ when BB' is infinitely small, neglecting infinitely small quantities of the second order.

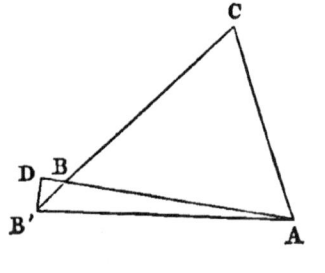

Fig. 4.

Hence

$$\Delta c = AB' - AB = AD - AB = BD;$$

$$\therefore \frac{dc}{da} = \text{limit of } \frac{\Delta c}{\Delta a} = \frac{BD}{BB'} = \cos B.$$

Similarly, $\dfrac{dc}{db} = \cos A$; which results agree with those arrived at before by differentiation.

Again, to find $\dfrac{dc}{dC}$. Suppose the angle C to receive a small increment ΔC, represented by BCB' in the accompanying figure; take $CB' = CB$, join AB', and draw BD perpendicular to AB'.
Then

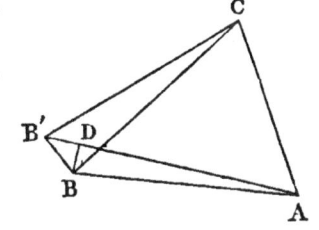

$\Delta c = AB' - AB = B'D$ (in the limit)

$= BB' \cos AB'B = BB' \sin ABC$ (q.p.).

Fig. 5.

Also, in the limit, $BB' = B'C \sin BCB' = a \, \Delta C$.

Hence $\dfrac{dc}{dC} =$ limiting value of $\dfrac{\Delta c}{\Delta C} = a \sin B$;

the same result as that arrived at by differentiation.

In the investigation in Fig. 5 it has been assumed that $AB - AD$ is infinitely small in comparison with BD; or that the fraction $\dfrac{AB - AD}{BD}$ vanishes in the limit. For the proof of this the student is referred to Art. 37.

When the base of a plane triangle is calculated from the observed lengths of its sides and the magnitude of its vertical angle, the result in (15) shows how the *error* in the computed value of the base can be *approximately* found in terms of the small errors in observation of the sides and of the contained angle.

109. **To find** $\dfrac{dC}{dA}$ **when** a **and** b **are considered Constant.**—In the preceding figure, BAB' represents the change in the angle A arising from the change ΔC in C; moreover, as the angle A is diminished in this case, we must denote BAB' by $-\Delta A$, and we have

$$BB' = -\dfrac{AB \Delta A}{\sin AB'B} = -\dfrac{AB \Delta A}{\cos B} = -\dfrac{c \Delta A}{\cos B}.$$

K 2

Also, $$BB' = a\,\Delta C;$$

$$\therefore \frac{dC}{dA} = \frac{\Delta C}{\Delta A} \text{ (in the limit)} = -\frac{c}{a\cos B}. \tag{16}$$

This result admits of another easy proof by differentiation.
For $$a \sin B = b \sin A;$$
hence, when a and b are constants, we have
$$a \cos B\, dB = b \cos A\, dA;$$
also, since $A + B + C = \pi$, we have
$$dA + dB + dC = 0.$$
Substitute for dB in the former its value deduced from the latter equation, and we get
$$(a \cos B + b \cos A)\, dA = -a \cos B\, dC;$$
or $$c\, dA = -a \cos B\, dC, \text{ as before.}$$

110. Equation connecting the Variations of two Sides and the opposite Angles.—In general, if we take the logarithmic differential of the equation
$$a \sin B = b \sin A,$$
regarding a, b, A, B, as variables, we get
$$\frac{da}{a} + \frac{dB}{\tan B} = \frac{db}{b} + \frac{dA}{\tan A}. \tag{17}$$

111. Landen's Transformation.—The result in equation (16) admits of being transformed into
$$\frac{dA}{a \cos B} = -\frac{dC}{c};$$
but
$$c = \sqrt{a^2 + b^2 - 2ab \cos C}, \text{ and } a \cos B = \sqrt{a^2 - b^2 \sin^2 A};$$
hence we get
$$\frac{dA}{\sqrt{a^2 - b^2 \sin^2 A}} = -\frac{dC}{\sqrt{a^2 + b^2 - 2ab \cos C}}.$$

If C be denoted by $180° - 2\phi_1$, the angle at A by ϕ, and $\dfrac{b}{a}$ by k, the preceding equation becomes

$$\frac{d\phi}{\sqrt{1 - k^2 \sin^2 \phi}} = \frac{2 d\phi_1}{\sqrt{1 + 2k \cos 2\phi_1 + k^2}} = \frac{2 d\phi_1}{\sqrt{(1+k)^2 - 4k \sin^2 \phi_1}}$$

$$= \frac{2}{(1+k)} \frac{d\phi_1}{\sqrt{1 - k_1^2 \sin^2 \phi_1}}; \qquad (18)$$

where $\qquad k_1 = \dfrac{2\sqrt{k}}{1+k}.$

Also, the equation $a \sin B = b \sin A$ becomes

$$\sin(2\phi_1 - \phi) = k \sin \phi.$$

The result just established furnishes a proof of Landen's[*] transformation in Elliptic Functions.

We shall next investigate some analogous formulæ in Spherical Trigonometry.

112. **Relation connecting the Variations of Three Sides and One Angle.**—Differentiating the well-known relation

$$\cos c = \cos a \cos b + \sin a \sin b \cos C,$$

regarding a and b *as constants*, we get

$$\frac{dc}{dC} = \frac{\sin a \sin b \sin C}{\sin c} = \sin a \sin B.$$

Again, the value of $\dfrac{dc}{da}$, when b and C are constants, can be easily determined geometrically as follows:—

---

[*] This transformation is often attributed to Lagrange; it had, however, been previously arrived at by Landen. (See *Philosophical Transactions*, 1771 and 1775.)

In the spherical triangle ABC, making a construction similar to that of Fig. 4, Art. 108, we have

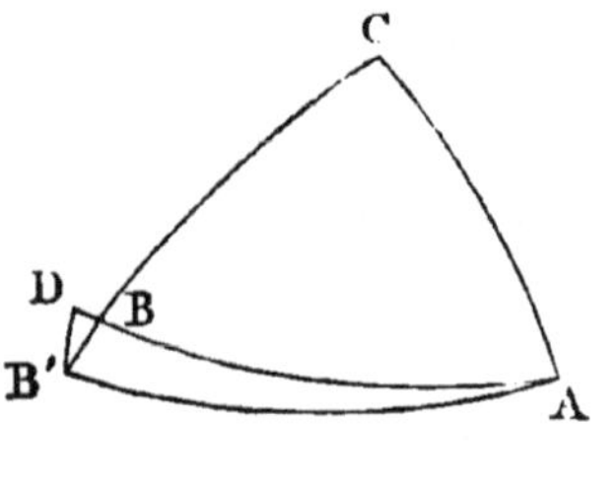

$$BB' = \Delta a;\ \therefore \frac{dc}{da} = \text{limit of } \frac{\Delta c}{\Delta a} = \frac{BD}{BB'}$$

(in the limit) $= \cos B$.

Similarly, when a and C are constants, $\dfrac{dc}{db} = \cos A$.

Fig. 6.

Hence, finally,

$$dc = \cos B\, da + \cos A\, db + \sin a \sin B\, dC. \qquad (19)$$

This result can also be obtained by a process of differentiation. This method is left as an exercise for the student.

As, in the corresponding case of plane triangles, we have assumed that $AB' = AD$ in the limit; *i.e.*, that $\dfrac{AB' - AD}{B'D}$ is infinitely small in comparison with AD in the limit; this assumption may be stated otherwise, thus:—

If the angle A of a right-angled spherical triangle be very small, then the ratio $\dfrac{c - b}{A}$ becomes very small at the same time, where c and b have their usual significations.

This result is easily established, for by Napier's rules we have

$$\cos A = \frac{\tan b}{\tan c} = \frac{\sin b \cos c}{\cos b \sin c};$$

$$\therefore \frac{1 - \cos A}{1 + \cos A} = \frac{\sin c \cos b - \cos c \sin b}{\sin c \cos b + \cos c \sin b} = \frac{\sin (c - b)}{\sin (c + b)};$$

or

$$\sin (c - b) = \tan^2 \frac{A}{2} \sin (c + b);\ \therefore \frac{\sin (c - b)}{\tan \dfrac{A}{2}} = \sin (c + b) \tan \frac{A}{2}.$$

But the right-hand side of this equation becomes very small along with A, and consequently $c - b$ becomes at the same time very small in comparison with that angle.

The formula (19) can also be written in the form

$$dC = \frac{dc}{\sin a \sin B} - \frac{da}{\sin a \tan B} - \frac{db}{\sin b \tan A}. \qquad (20)$$

The corresponding formulæ for the differentials of A and B are obtained by an interchange of letters.

Again, from any equation in Spherical Trigonometry another can be derived by aid of the *polar triangle*.

Thus, by this transformation, formula (19) becomes

$$dC = -\cos b \, dA - \cos a \, dB + \sin A \sin b \, dc. \qquad (21)$$

These, and the analogous formulæ, are of importance in Astronomy in determining the *errors* in a computed angular distance arising from small *errors* in observation. They also enable us to determine the most favourable positions for making certain observations; viz., those in which small errors in observation produce the *least error* in the required result.

113. Remarks on Partial Differentials.—The beginner must be careful to attach their proper significations to the expressions $\frac{dC}{da}$, $\frac{da}{dC}$, &c., in each case. Thus when a and b are constants, we have $\frac{dc}{dC} = \sin a \sin B$; but when A and a are constants, we have $\frac{dc}{dC} = \frac{\tan c}{\tan C}$; these are quite different quantities represented by the same expression $\frac{dc}{dC}$.

The reason is, that in the former case we investigate the ultimate ratio of the simultaneous increments of a side and its opposite angle, when the *other two sides are considered as constant*; while in the latter we investigate the similar ratio when *one side and its opposite angle are constant*.

Similar remarks apply in all cases of *partial* differentiation.

When our formulæ are applied to the case of small *errors* in the sides and angles of a triangle, it is usual to designate these errors by Δa, Δb, Δc, ΔA, ΔB, ΔC; and when these expressions are substituted for da, db, &c., in our formulæ, they give approximate results.

For instance (19) becomes in this case

$$\Delta c = \Delta a \cos B + \Delta b \cos A + \Delta C \sin a \sin B; \qquad (22)$$

and similarly in other cases.

It is easily seen that the *error* arising in the application of these formulæ to such cases is a small quantity of the *second* order; that is, it involves the squares and products of the small quantities Δa, Δb, Δc, &c. This will also appear more fully from the results arrived at in a subsequent chapter.

114. Theorem.—If the base c, and the vertical angle C, of a spherical triangle be constant, formula (19) becomes

$$\frac{da}{\cos A} + \frac{db}{\cos B} = 0.$$

Now, writing ϕ instead of a, ψ instead of b, and k for $\dfrac{\sin C}{\sin c}$, this equation becomes

$$\left(\text{since } k = \frac{\sin A}{\sin a} = \frac{\sin B}{\sin b}\right)$$

$$\frac{d\phi}{\sqrt{1 - k^2 \sin^2 \phi}} + \frac{d\psi}{\sqrt{1 - k^2 \sin^2 \psi}} = 0. \qquad (23)$$

where ϕ and ψ are connected by the following* relation:—

$$\cos c = \cos \phi \cos \psi + \sin \phi \sin \psi \cos C,$$

or $\quad \cos c = \cos \phi \cos \psi + \sin \phi \sin \psi \sqrt{1 - k^2 \sin^2 c}.$

115. In a Spherical Triangle, to prove that

$$\frac{da}{\cos A} + \frac{db}{\cos B} + \frac{dc}{\cos C} = 0, \qquad (24)$$

when $\dfrac{\sin C}{\sin c}$ **is constant.**

---

* This mode of establishing the connexion between Elliptic Functions by aid of Spherical Trigonometry is due to Lagrange.

Examples in Spherical Trigonometry.

Let $\sin C = k \sin c$, and we get

$$dC = \frac{k \cos c}{\cos C} dc = \frac{\sin A \cos c}{\sin a \cos C} dc:$$

substitute this value for dC in (19), and it becomes

$$dc = \cos A\, db + \cos B\, da + \frac{\cos c \sin A \sin B}{\cos C} dc;$$

or $\quad \cos A\, db + \cos B\, da = \left(1 - \dfrac{\cos c \sin A \sin B}{\cos C}\right) dc$

$$= -\frac{\cos A \cos B}{\cos C} dc;$$

since $\quad \sin A \sin B \cos c = \cos C + \cos A \cos B.$

Hence $\quad \dfrac{da}{\cos A} + \dfrac{db}{\cos B} + \dfrac{dc}{\cos C} = 0.$

Again, since $\cos A = \sqrt{1 - \sin^2 A} = \sqrt{1 - k^2 \sin^2 a}$, &c., the preceding result may be written in the form

$$\frac{da}{\sqrt{1 - k^2 \sin^2 a}} + \frac{db}{\sqrt{1 - k^2 \sin^2 b}} + \frac{dc}{\sqrt{1 - k^2 \sin^2 a}} = 0, \quad (25)$$

where a, b, c, are connected by the equation

$$\cos c = \cos a \cos b + \sin a \sin b \sqrt{1 - k^2 \sin^2 c}.$$

116. Theorem of Legendre.—We get from (24)

$$\cos B \cos C\, da + \cos A \cos C\, db + \cos B \cos A\, dc = 0,$$

or $(\cos A - \sin B \sin C \cos a)\, da + (\cos B - \sin A \sin C \cos b)\, db$

$$+ (\cos C - \sin A \sin B \cos c)\, dc = 0;$$

$\therefore \cos A\, da + \cos B\, db + \cos C\, dc$

$= \sin B \sin C d(\sin a) + \sin A \sin C d (\sin b) + \sin A \sin B d (\sin c)$

$= k^2 \{\sin b \sin c\, d (\sin a) + \sin a \sin c\, d (\sin b) + \sin a \sin b\, d (\sin c)\}$

$= k^2 d (\sin a \sin b \sin c);$

or $\sqrt{1-k^2\sin^2 a}\, da + \sqrt{1-k^2\sin^2 b}\, db + \sqrt{1-k^2\sin^2 c}\, dc$

$$= k^2 d\, (\sin a\, \sin b\, \sin c). \tag{26}$$

This furnishes a proof of Legendre's formula for the comparison of Elliptic Functions of the second species.

The most important application of these results has place when one of the angles, C suppose, is obtuse; in this case $\cos C$ is negative, and formula (25) becomes

$$\frac{da}{\sqrt{1-k^2\sin^2 a}} + \frac{db}{\sqrt{1-k^2\sin^2 b}} = \frac{dc}{\sqrt{1-k^2\sin^2 c}}:$$

where the relation connecting a, b, c is

$$\cos c = \cos a \cos b - \sin a \sin b \sqrt{1-k^2\sin^2 c}.$$

In like manner, equation (26) becomes, in this case,

$$\sqrt{1-k^2\sin^2 a}\, da + \sqrt{1-k^2\sin^2 b}\, db$$

$$= \sqrt{1-k^2\sin^2 c}\, dc + k^2 d\, (\sin a\, \sin b\, \sin c).$$

117. **If $u = \phi(x + at, y + \beta t)$, where x, y, a, β, are independent of t, and of each other, to prove that**

$$\frac{du}{dt} = a\frac{du}{dx} + \beta\frac{du}{dy}. \tag{27}$$

Let $x' = x + at,\quad y' = y + \beta t;$

then $u = \phi(x', y'),$

and $\dfrac{dx'}{dx} = 1,\quad \dfrac{dy'}{dy} = 1,\quad \dfrac{dx'}{dt} = a,\quad \dfrac{dy'}{dt} = \beta.$

Also, since y' is independent of x, we have

$$\frac{du}{dx} = \frac{du}{dx'}\frac{dx'}{dx} = \frac{du}{dx'}, \text{ and } \frac{du}{dy} = \frac{du}{dy'}.$$

Hence $\dfrac{du}{dt} = \dfrac{du}{dx'}\dfrac{dx'}{dt} + \dfrac{du}{dy'}\dfrac{dy'}{dt} = a\dfrac{du}{dx} + \beta\dfrac{du}{dy}.$

Partial Differentiation.

In like manner, if x', y', z', be substituted for $x + at$, $y + \beta t$, $z + \gamma t$, in the equation

$$u = \phi(x + at,\ y + \beta t,\ z + \gamma t),$$

it becomes $\quad u = \phi(x',\ y',\ z')\ ;$

also $\quad \dfrac{du}{dx} = \dfrac{du}{dx'}\dfrac{dx'}{dx} + \dfrac{du}{dy'}\dfrac{dy'}{dx} + \dfrac{du}{dz'}\dfrac{dz'}{dx}\ ;$

but $\quad \dfrac{dx'}{dx} = 1,\ \dfrac{dy'}{dx} = 0,\ \dfrac{dz'}{dx} = 0,$

$\therefore \dfrac{du}{dx} = \dfrac{du}{dx'},\ \text{also}\ \dfrac{du}{dy} = \dfrac{du}{dy'},\ \dfrac{du}{dz} = \dfrac{du}{dz'}.$

Again $\quad \dfrac{du}{dt} = \dfrac{du}{dx'}\dfrac{dx'}{dt} + \dfrac{du}{dy'}\dfrac{dy'}{dt} + \dfrac{du}{dz'}\dfrac{dz'}{dt}\ ;$

but $\quad \dfrac{dx'}{dt} = a,\ \dfrac{dy'}{dt} = \beta,\ \dfrac{dz'}{dt} = \gamma.$

Hence $\quad \dfrac{du}{dt} = a\dfrac{du}{dx} + \beta\dfrac{du}{dy} + \gamma\dfrac{du}{dz}. \qquad (28)$

This result can be easily extended to any number of variables.

EXAMPLES.

1. If $u = \sin^{-1}\left(\dfrac{x}{a}\right) + \sin^{-1}\left(\dfrac{y}{b}\right)$, prove that $du = \dfrac{dx}{\sqrt{a^2-x^2}} + \dfrac{dy}{\sqrt{b^2-y^2}}$.

2. If $u = xy\phi\left(\dfrac{y}{x}\right)$, ,, $x\dfrac{du}{dx} + y\dfrac{du}{dy} = 2u$.

3. Find the conditions that u, a function of x, y, z, should be a function of $x+y+z$.

$$\text{Ans. } \dfrac{du}{dx} = \dfrac{du}{dy} = \dfrac{du}{dz}.$$

4. If $f(ax+by) = 0$, find $\dfrac{dy}{dx}$. ,, $-\dfrac{a}{b}$.

5. If $f(u) = \phi(v)$, where u and v are each functions of x and y, prove that

$$\dfrac{du}{dx}\dfrac{dv}{dy} = \dfrac{dv}{dx}\dfrac{du}{dy}.$$

6. Find the values of $x\dfrac{du}{dx} + y\dfrac{du}{dy}$, when

$$(a)\ u = \dfrac{ax^{\frac{1}{2}} + by^{\frac{1}{2}}}{mx^2 + ny^2},$$

$$(\beta)\ u = \tan^{-1}\left(\dfrac{x-y}{x+y}\right)^{\frac{2}{3}}.$$

7. If $u = \sin ax + \sin by + \tan^{-1}\left(\dfrac{y}{z}\right)$, prove that

$$du = a\cos ax\,dx + b\cos by\,dy + \dfrac{zdy - ydz}{y^2 + z^2}.$$

8. If $u = \log_y x$, find $\dfrac{du}{dx}$ and $\dfrac{du}{dy}$. Ans. $\dfrac{du}{dx} = \dfrac{1}{x\log y}$, $\dfrac{du}{dy} = \dfrac{-\log x}{y(\log y)^2}$.

9. If $\theta = \tan^{-1}\dfrac{x}{y}$, prove that

$$(x^2+y^2)d\theta = ydx - xdy.$$

10. If $u = y^{xz}$, prove that

$$du = y^{xz-1}(xzdy + yz\log y\,dx + xy\log y\,dz).$$

11. If $a + \sqrt{a^2 - y^2} = y e^{\frac{x + \sqrt{a^2 - y^2}}{a}}$, prove that

$$\frac{dy}{dx} = \frac{-y}{\sqrt{a^2 - y^2}}.$$

12. In a spherical triangle, when a, b are constant, prove that

$$\frac{dA}{dB} = \frac{\tan A}{\tan B}, \text{ and } \frac{dC}{dB} = -\frac{\sin C}{\sin B \cos A}.$$

13. In a plane triangle, if the angles and sides receive small variations, prove that

$c \Delta B + b \cos A \Delta C = 0$; a, b being constant,
$\cos C \Delta b + \cos B \Delta c = 0$; a, A being constant,
$\tan A \Delta b = b \Delta C$; a, B being constant.

14. The base c of a spherical triangle is measured, and the two adjacent base angles A, B are found by observation. Suppose that small errors dA, dB are committed in the observations of A and B; show that the corresponding error in the computed value of C is

$$- \cos a\, dB - \cos b\, dA.$$

15. If the base c and the area of a spherical triangle be given, prove that

$$\sin^2 \frac{a}{2} dB + \sin^2 \frac{b}{2} dA = 0.$$

16. Given the base and the vertical angle of a spherical triangle, prove that the variation of the perpendicular p is connected with the variations of the sides by the relation

$$\sin C dp = \sin s' da + \sin s db,$$

s and s' being the segments into which the perpendicular divides the vertical angle.

17. In a plane triangle, if the sides a, b be constant, prove that the variations of its base angles are connected by the equation

$$\frac{dA}{\sqrt{a^2 - b^2 \sin^2 A}} = \frac{dB}{\sqrt{b^2 - a^2 \sin^2 B}}.$$

18. Prove the following relation between the small increments in two sides and the opposite angles of a spherical triangle,

$$\frac{da}{\tan a} + \frac{dB}{\tan B} = \frac{dA}{\tan A} + \frac{db}{\tan b}.$$

19. In a right-angled spherical triangle, prove that, if A be invariable $\sin 2c\, db = \sin 2b\, dc$; and if c be invariable, $\tan a\, da + \tan b\, db = 0$.

20. If a be one of the equal sides of an isosceles spherical triangle, whose vertical angle is very small, and represented by $d\omega$, prove that the quantity by which either base angle falls short of a right angle is $\dfrac{1}{2}\cos a\,d\omega$.

21. In a spherical triangle, if one angle C be given, as well as the sum of the other angles, prove that

$$\frac{da}{\sin a} + \frac{db}{\sin b} = 0.$$

22. If all the parts of a spherical triangle vary, then will

$$\cos A\,da + \cos B\,db + \cos C\,dc = kd\,(k\sin a \sin b \sin c);$$

where

$$k = \frac{\sin A}{\sin a} = \frac{\sin B}{\sin b} = \frac{\sin C}{\sin c}.$$

Also

$$\frac{da}{\cos A} + \frac{db}{\cos B} + \frac{dc}{\cos C} = \tan A \tan B \tan C\, d\left(\frac{1}{k}\right).$$

These theorems can be transformed by aid of the polar triangle?—*M'Cullagh, Fellowship Examination*, 1837.

These are more general than the theorems contained in Arts. 115 and 116, and can be deduced by the same method without difficulty.

23. If $z = \phi(x^2 - y^2)$, prove that

$$y\frac{dz}{dx} + x\frac{dz}{dy} = 0.$$

24. If $z = \dfrac{1}{x}f\left(\dfrac{y}{x}\right)$, prove that

$$x\frac{dz}{dx} + y\frac{dz}{dy} + z = 0.$$

25. Find $\dfrac{dy}{dx}$ and $\dfrac{dz}{dx}$ when x, y, z are connected by two equations of the form

$$f(x, y, z) = 0, \qquad \phi(x, y, z) = 0.$$

$$\text{Ans. } \frac{dy}{dx} = \frac{\dfrac{df}{dx}\dfrac{d\phi}{dz} - \dfrac{df}{dz}\dfrac{d\phi}{dx}}{\dfrac{df}{dz}\dfrac{d\phi}{dy} - \dfrac{df}{dy}\dfrac{d\phi}{dz}},$$

$$\text{'' } \frac{dz}{dx} = \frac{\dfrac{df}{dy}\dfrac{d\phi}{dx} - \dfrac{df}{dx}\dfrac{d\phi}{dy}}{\dfrac{df}{dz}\dfrac{d\phi}{dy} - \dfrac{df}{dy}\dfrac{d\phi}{dz}}.$$

26. Prove that any root of the following equation in y,

$$y^m + xy = 1,$$

satisfies the differential equation

$$y^2 \frac{d^2y}{dx^2} + (m-1)x\frac{dy^3}{dx^3} + (m-3)y\frac{dy^2}{dx^2} = 0.$$

27. How can we ascertain whether an expression such as

$$\phi(x, y) + \sqrt{-1}\,\psi(x, y)$$

admits of being reduced to the form

$$f(x + y\sqrt{-1})?$$

Ans. $\dfrac{d\phi}{dx} = \dfrac{d\psi}{dy},\quad \dfrac{d\phi}{dy} = -\dfrac{d\psi}{dx}.$

28. If $lX + mY + nZ$, $l'X + m'Y + n'Z$, $l''X + m''Y + n''Z$, be substituted for x, y, z, in the quadratic expression of Art. 107; and if a', b', c', d', e', f', be the respective coefficients in the new expression, prove that

$$\begin{vmatrix} a' & f' & e' \\ f' & b' & d' \\ e' & d' & c' \end{vmatrix} = 0, \text{ whenever } \begin{vmatrix} a & f & e \\ f & b & d \\ e & d & c \end{vmatrix} = 0.$$

29. If the transformation be *orthogonal*, i.e. if $x^2 + y^2 + z^2 = X^2 + Y^2 + Z^2$, prove that the preceding determinants are equal to one another.

29. If u be a function of ξ, η, ζ, and $\xi = y + \dfrac{1}{z}$, $\eta = z + \dfrac{1}{x}$, $\zeta = x + \dfrac{1}{y}$,

show that

$$x\frac{du}{dx} + y\frac{du}{dy} + z\frac{du}{dz} + \xi\frac{du}{d\xi} + \eta\frac{du}{d\eta} + \zeta\frac{du}{d\zeta} = 2\left(x\frac{du}{d\zeta} + y\frac{du}{d\xi} + z\frac{du}{d\eta}\right).$$

CHAPTER VI.

SUCCESSIVE DIFFERENTIATION OF FUNCTIONS OF TWO OR MORE VARIABLES.

118. Successive Partial Differentiation.—We have in the preceding chapter considered the manner of determining the partial differential coefficients of the first order in a function of any number of variables.

If u be a function of x, y, z, &c., the expression

$$\frac{du}{dx}, \ \frac{du}{dy}, \ \frac{du}{dz}, \ \&\text{c.},$$

being also functions x, y, z, &c., admit of being differentiated in the same manner as the original function; and the partial differential coefficient of $\dfrac{du}{dx}$, when x *alone* varies, is denoted by

$$\frac{d}{dx}\left(\frac{du}{dx}\right), \ \text{or} \ \frac{d^2u}{dx^2},$$

as in the case of a single variable.

Similarly, the partial differential coefficient of $\dfrac{du}{dx}$, when y *alone* varies, is represented by

$$\frac{d}{dy}\left(\frac{du}{dx}\right), \ \text{or} \ \frac{d^2u}{dydx},$$

and, in general, $\dfrac{d^{m+n}u}{dy^m dx^n}$ denotes that the function u is first differentiated n times in succession, supposing x alone to vary, and the resulting function afterwards differentiated m times in succession, where y alone is supposed to vary; and similarly in all other cases.

We now proceed to show that the values of these partial derived functions are independent of the order in which the variables are supposed to change.

119. If u be a Function of x and y, to prove that

$$\frac{d}{dy}\left(\frac{du}{dx}\right) = \frac{d}{dx}\left(\frac{du}{dy}\right), \text{ or } \frac{d^2u}{dydx} = \frac{d^2u}{dxdy}, \tag{1}$$

where x and y are independent of each other.

Let $u = \phi(x, y)$, then $\dfrac{du}{dx}$ represents the limiting value of

$$\frac{\phi(x+h, y) - \phi(x, y)}{h}$$

when h is infinitely small.

This expression being regarded as a function of y, let y become $y + k$, x remaining constant; then $\dfrac{d}{dy}\left(\dfrac{du}{dx}\right)$ is the limiting value of

$$\frac{\phi(x+h, y+k) - \phi(x, y+k) - \phi(x+h, y) + \phi(x, y)}{hk}$$

when both h and k become infinitely small, or evanescent.

In like manner $\dfrac{du}{dy}$ is the limiting value of

$$\frac{\phi(x, y+k) - \phi(x, y)}{k}$$

when k is infinitely small; hence $\dfrac{d}{dx}\left(\dfrac{du}{dy}\right)$ is the limiting value of

$$\frac{\phi(x+h, y+k) - \phi(x+h, y) - \phi(x, y+k) + \phi(x, y)}{hk}$$

when both h and k are infinitely small.

Since this function is the same as the preceding for all

finite values of h and k, it will continue to be so in the limit; hence we have

$$\frac{d}{dx}\left(\frac{du}{dy}\right) = \frac{d}{dy}\left(\frac{du}{dx}\right).$$

In like manner $\dfrac{d^3u}{dx^2\,dy} = \dfrac{d^3u}{dy\,dx^2}$,

for by the preceding $\dfrac{d^2u}{dx\,dy} = \dfrac{d^2u}{dy\,dx}$;

$$\therefore \frac{d}{dx}\left(\frac{d^2u}{dx\,dy}\right) = \frac{d}{dx}\left|\frac{d^2u}{dy\,dx}\right| = \frac{d}{dx}\cdot\frac{d}{dy}\left|\frac{du}{dx}\right| = \frac{d}{dy}\cdot\frac{d}{dx}\left|\frac{du}{dx}\right| :$$

similarly in all other cases. Hence, in general,

$$\frac{d^{p+q}u}{dx^p\,dy^q} = \frac{d^{p+q}u}{dy^q\,dx^p}.$$

Again, in the case of functions of three or more variables, by similar reasoning it can be proved that

$$\frac{d^3u}{dz\,dx\,dy} = \frac{d^3u}{dx\,dy\,dz}, \&c.$$

Hence we infer that *the order of differentiation is in all cases indifferent*, provided the variables are *independent* of each other.

EXAMPLES FOR VERIFICATION.

1. If $u = \phi\left(\dfrac{x}{y}\right)$, verify that $\dfrac{d^2u}{dy\,dx} = \dfrac{d^2u}{dx\,dy}$.

2. If $u = \tan^{-1}\left(\dfrac{x}{y}\right)$, ,, $\dfrac{d^3u}{dy^2\,dx} = \dfrac{d^3u}{dx\,dy^2}$.

3. If $u = \sin(ax^n + by^n)$, ,, $\dfrac{d^4u}{dx^2\,dy^2} = \dfrac{d^4u}{dy^2\,dx^2}$.

120. Condition that $P\,dx + Q\,dy$ shall be a total Differential.—This implies that $P\,dx + Q\,dy$ should be the exact differential of some function of x and y. Denoting this function by u, then

$$du = P\,dx + Q\,dy,$$

Condition for a Total Differential. 147

and, by (1), Art. 95, we must have

$$P = \frac{du}{dx}, \qquad Q = \frac{du}{dy};$$

$$\therefore \frac{dP}{dy} = \frac{d^2u}{dy\,dx}, \qquad \frac{dQ}{dx} = \frac{d^2u}{dx\,dy}.$$

Hence the required condition is

$$\frac{dP}{dy} = \frac{dQ}{dx}. \tag{2}$$

121. If u be any Function of x and y, to prove that

$$\frac{d}{dy}\left(F(u)\frac{du}{dx}\right) = \frac{d}{dx}\left(F(u)\frac{du}{dy}\right), \tag{3}$$

where x and y are independent variables.

Here each side, on differentiation, becomes

$$F(u)\frac{d^2u}{dx\,dy} + F'(u)\frac{du}{dx}\frac{du}{dy}; \quad \therefore \text{ &c.}$$

122. More generally, to prove that

$$\frac{d}{dy}\left(u\frac{dv}{dx}\right) = \frac{d}{dx}\left(u\frac{dv}{dy}\right), \tag{4}$$

where u and v are both functions of z, and z is a function of x and y.

For $\quad\dfrac{d}{dy}\left(u\dfrac{dv}{dx}\right) = \dfrac{du}{dy}\dfrac{dv}{dx} + u\dfrac{d^2v}{dy\,dx},$

but $\quad\dfrac{du}{dy} = \dfrac{du}{dz}\dfrac{dz}{dy}, \quad \dfrac{dv}{dx} = \dfrac{dv}{dz}\dfrac{dz}{dx};$

$\therefore \quad \dfrac{d}{dy}\left(u\dfrac{dv}{dx}\right) = \dfrac{du}{dz}\dfrac{dv}{dz}\dfrac{dz}{dx}\dfrac{dz}{dy} + u\dfrac{d^2v}{dy\,dx};$

and $\quad\dfrac{d}{dx}\left(u\dfrac{dv}{dy}\right)$ has evidently the same value.

L 2

123. Euler's Theorem of Homogeneous Functions.—In Art. 102 it has been shown that

$$x\frac{du}{dx} + y\frac{du}{dy} = nu,$$

where u is a homogeneous function of the nth degree in x and y.

Moreover, as $\dfrac{du}{dx}$ and $\dfrac{du}{dy}$ are homogeneous functions of the degree $n-1$, we have, by the same theorem,

$$x\frac{d}{dx}\left(\frac{du}{dx}\right) + y\frac{d}{dy}\left(\frac{du}{dx}\right) = (n-1)\frac{du}{dx},$$

$$x\frac{d}{dx}\left(\frac{du}{dy}\right) + y\frac{d}{dy}\left(\frac{du}{dy}\right) = (n-1)\frac{du}{dy}:$$

multiplying the former of these equations by x, and the latter by y, we get, after addition,

$$x^2\frac{d^2u}{dx^2} + 2xy\frac{d^2u}{dxdy} + y^2\frac{d^2u}{dy^2} = (n-1)\left(x\frac{du}{dx} + y\frac{du}{dy}\right)$$

$$= (n-1)nu. \qquad (5)$$

This result can be readily extended to homogeneous functions of any number of independent variables.

A more complete investigation of Euler's Theorems will be found in Chapter VIII.

124. To find the Successive Differential Coefficients with respect to t, of the Function

$$\phi(x + at,\ y + \beta t),$$

where x, y, a, β, are independent of t, and of each other.

By Art. 117 we have in this case, where ϕ stands for the expression $\phi(x + at,\ y + \beta t)$,

$$\frac{d\phi}{dt} = a\frac{d\phi}{dx} + \beta\frac{d\phi}{dy}.$$

Hence
$$\frac{d^2\phi}{dt^2} = a\frac{d}{dt}\left(\frac{d\phi}{dx}\right) + \beta\frac{d}{dt}\left(\frac{d\phi}{dy}\right)$$

$$= a\frac{d}{dx}\left(\frac{d\phi}{dt}\right) + \beta\frac{d}{dy}\left(\frac{d\phi}{dt}\right)$$

$$= a\frac{d}{dx}\left\{a\frac{d\phi}{dx} + \beta\frac{d\phi}{dy}\right\} + \beta\frac{d}{dy}\left\{a\frac{d\phi}{dx} + \beta\frac{d\phi}{dy}\right\}$$

$$= a^2\frac{d^2\phi}{dx^2} + 2a\beta\frac{d^2\phi}{dxdy} + \beta^2\frac{d^2\phi}{dy^2}. \qquad (6)$$

This result can also be written in the form

$$\frac{d^2\phi}{dt^2} = \left\{a\frac{d}{dx} + \beta\frac{d}{dy}\right\}\frac{d\phi}{dt} = \left\{a\frac{d}{dx} + \beta\frac{d}{dy}\right\}^2\phi, \qquad (7)$$

in which $\left(a\dfrac{d}{dx} + \beta\dfrac{d}{dy}\right)^2$ is supposed to be developed in the usual manner, and $\dfrac{d^2\phi}{dx^2}$, &c., substituted for $\left(\dfrac{d}{dx}\right)^2\phi$, &c.

Again, to find $\dfrac{d^3\phi}{dt^3}$.

$$\frac{d^3\phi}{dt^3} = \frac{d}{dt}\left(\frac{d^2\phi}{dt^2}\right) = \frac{d}{dt}\left(a\frac{d}{dx} + \beta\frac{d}{dy}\right)^2\phi$$

$$= \left(a\frac{d}{dx} + \beta\frac{d}{dy}\right)^2\frac{d\phi}{dt} = \left(a\frac{d}{dx} + \beta\frac{d}{dy}\right)^2\left(a\frac{d\phi}{dx} + \beta\frac{d\phi}{dy}\right)$$

$$= \left(a\frac{d}{dx} + \beta\frac{d}{dy}\right)^3\phi.$$

By induction from the preceding it can be readily shown that

$$\frac{d^n\phi}{dt^n} = \left(a\frac{d}{dx} + \beta\frac{d}{dy}\right)^n\phi.$$

This expression, when expanded by the Binomial Theorem, gives the nth differential coefficient of the function in terms of its partial differential coefficients of the nth order in x and y.

Examples.

1. If $u = \sin(x^2 y)$, verify the equation $\dfrac{d^2u}{dxdy} = \dfrac{d^2u}{dydx}$.

2. If $u = \sin(y + ax) + (y - ax)^2$, prove that
$$\frac{d^2u}{dx^2} = a^2 \frac{d^2u}{dy^2}.$$

3. In general, if $u = f(y + ax) + \phi(y - ax)$, prove that
$$\frac{d^2u}{dx^2} = a^2 \frac{d^2u}{dy^2}.$$

4. If $u = y^x$, prove that
$$\frac{d^2u}{dxdy} = y^{x-1}(1 + x \log y) = \frac{d^2u}{dydx}.$$

5. If $u = \dfrac{xyz}{ax + by + cz}$, find the values of
$$\frac{d^2u}{dx^2}, \quad \frac{d^2u}{dy^2}, \quad \text{and} \quad \frac{d^2u}{dz^2}.$$

6. If $u = (x^2 + y^2)^{\frac{1}{2}}$, prove that
$$x^2 \frac{d^2u}{dx^2} + 2xy \frac{d^2u}{dxdy} + y^2 \frac{d^2u}{dy^2} = 0.$$

7. If $u = (x^3 + y^3)^{\frac{1}{2}}$, prove that
$$x^2 \frac{d^2u}{dx^2} + 2xy \frac{d^2u}{dxdy} + y^2 \frac{d^2u}{dy^2} = \frac{3}{4}u.$$

8. If $V = Ay^3 + 3By^2x + 3Cyx^2 + Dx^3$, prove that
$$\frac{d^2V}{dx^2}\frac{dV^2}{dy^2} - 2\frac{d^2V}{dxdy}\frac{dV}{dx}\frac{dV}{dy} + \frac{d^2V}{dy^2}\frac{dV^2}{dx^2} = 54V \begin{vmatrix} x^2, & -xy, & y^2 \\ A, & B, & C \\ B, & C, & D \end{vmatrix},$$

and show that the left-hand side of this equation vanishes when V is a perfect cube.

9. If $u = \dfrac{1}{(x^2 + y^2 + z^2)^{\frac{1}{2}}}$, prove that
$$\frac{d^2u}{dx^2} + \frac{d^2u}{dy^2} + \frac{d^2u}{dz^2} = 0.$$

CHAPTER VII.

LAGRANGE'S THEOREM.

125. Lagrange's Theorem.—Suppose that we are given the equation

$$z = x + y\phi(z), \qquad (1)$$

in which x and y are independent variables, and it is required to expand any function of z in ascending powers of y.

Let the function be denoted by $F(z)$, or by u, and, by Maclaurin's theorem, we have

$$u = u_0 + \frac{y}{1}\left(\frac{du}{dy}\right)_0 + \frac{y^2}{1 \cdot 2}\left(\frac{d^2u}{dy^2}\right)_0 + \ldots + \frac{y^n}{1 \cdot 2 \ldots n}\left(\frac{d^nu}{dy^n}\right)_0 + \&c., \quad (2)$$

where u_0, $\left(\dfrac{du}{dy}\right)_0$, &c., represent the values of u, $\dfrac{du}{dy}$, &c., when zero is substituted for y after differentiation.

It is evident that $u_0 = F(x)$.

To find the other terms, we get by differentiating (1) with respect to x, and also with respect to y,

$$\frac{dz}{dx} = 1 + y\phi'(z)\frac{dz}{dx}, \qquad \frac{dz}{dy} = \phi(z) + y\phi'(z)\frac{dz}{dy},$$

or
$$\frac{dz}{dx}\{1 - y\phi'(z)\} = 1, \qquad \frac{dz}{dy}\{1 - y\phi'(z)\} = \phi(z);$$

hence
$$\frac{dz}{dy} = \phi(z)\frac{dz}{dx}.$$

Also, since u is a function of z, we have

$$\frac{du}{dx} = \frac{du}{dz}\frac{dz}{dx}, \qquad \frac{du}{dy} = \frac{du}{dz}\frac{dz}{dy},$$

hence we obtain

$$\frac{du}{dy} = \phi(z)\frac{du}{dx}. \qquad (3)$$

Again, denoting $\phi(z)$ by Z, we have by Art. 121, since Z is a function of u,

$$\frac{d}{dx}\left(Z\frac{du}{dy}\right) = \frac{d}{dy}\left(Z\frac{du}{dx}\right) = \frac{d^2u}{dy^2}, \text{ from (3)},$$

or

$$\frac{d^2u}{dy^2} = \frac{d}{dx}\left(Z^2\frac{du}{dx}\right). \qquad (4)$$

Hence also

$$\frac{d^3u}{dy^3} = \frac{d^2}{dxdy}\left(Z^2\frac{du}{dx}\right),$$

since x and y are independent variables;

but

$$\frac{d}{dy}\left(Z^2\frac{du}{dx}\right) = \frac{d}{dx}\left(Z^2\frac{du}{dy}\right) = \frac{d}{dx}\left(Z^3\frac{du}{dx}\right), \text{ by (3)},$$

or

$$\frac{d^2}{dxdy}\left(Z^2\frac{du}{dx}\right) = \left(\frac{d}{dx}\right)^2\left(Z^3\frac{du}{dx}\right);$$

hence

$$\frac{d^3u}{dy^3} = \left(\frac{d}{dx}\right)^2\left(Z^3\frac{du}{dx}\right). \qquad (5)$$

To prove that the law here indicated is general, suppose that

$$\frac{d^nu}{dy^n} = \left(\frac{d}{dx}\right)^{n-1}\left(Z^n\frac{du}{dx}\right);$$

then, since

$$\frac{d}{dy}\left(Z^n\frac{du}{dx}\right) = \frac{d}{dx}\left(Z^n\frac{du}{dy}\right) = \frac{d}{dx}\left(Z^{n+1}\frac{du}{dx}\right),$$

we have

$$\frac{d^n}{dx^{n-1}dy}\left(Z^n\frac{du}{dx}\right) = \frac{d^n}{dx^n}\left(Z^{n+1}\frac{du}{dx}\right);$$

and hence

$$\frac{d^{n+1}u}{dy^{n+1}} = \left(\frac{d}{dx}\right)^n\left(Z^{n+1}\frac{du}{dx}\right). \qquad (6)$$

Lagrange's Theorem.

This shows that if the proposed law hold for any integer n, it holds for the integer $n + 1$; but it has been found to hold for $n = 2$ and $n = 3$; accordingly it holds for all integral values of n.

It remains to find the values of $\dfrac{du}{dy}$, $\dfrac{d^2u}{dy^2}$, &c. when we make $y = 0$. Since on this hypothesis Z or $\phi(z)$ becomes $\phi(x)$, and $\dfrac{du}{dx}$ becomes $\dfrac{dF(x)}{dx}$ or $F'(x)$, it is evident from (3), (4), (5), (6), that the values of

$$\frac{du}{dy}, \quad \frac{d^2u}{dy^2}, \quad \frac{d^3u}{dy^3} \cdots \frac{d^{n+1}u}{dy^{n+1}},$$

become at the same time

$$\phi(x)\, F'(x), \quad \frac{d}{dx}\left[\{\phi(x)\}^2 F'(x)\right], \quad \frac{d^2}{dx^2}\left[\{\phi(x)\}^3 F'(x)\right],$$

$$\cdots \frac{d^n}{dx^n}\left[\{\phi(x)\}^{n+1} F'(x)\right].$$

Consequently formula (2) becomes

$$F(z) = F(x) + \frac{y}{1}\phi(x)F'(x) + \frac{y^2}{1 \cdot 2}\frac{d}{dx}\left[\{\phi(x)\}^2 F'(x)\right] + \&c.$$

$$\cdots + \frac{y^{n+1}}{1 \cdot 2 \cdots (n+1)}\frac{d^n}{dx^n}\left[\{\phi(x)\}^{n+1} F'(x)\right] + \&c. \quad (7)$$

This expansion is called Lagrange's Theorem.

If it be merely required to expand z, we get, on making $F(z) = z$,

$$z = x + \frac{y}{1}\phi(x) + \frac{y^2}{1 \cdot 2}\frac{d}{dx}\{\phi(x)\}^2 + \&c.$$

$$+ \frac{y^n}{1 \cdot 2 \cdots n}\frac{d^{n-1}}{dx^{n-1}}\{\phi(x)\}^n + \&c. \quad (8)$$

126. Laplace's Theorem.

More generally, suppose that we are given

$$z = f\{x + y\phi(z)\}, \qquad (9)$$

and that it is required to expand any function $F(z)$ in ascending powers of y.

Let $t = x + y\phi(z)$, then $z = f(t)$, and we have

$$t = x + y\phi\{f(t)\}. \qquad (10)$$

Also $F(z) = F\{f(t)\}$; and the question reduces to the expansion of the function $F\{f(t)\}$ in ascending powers of y by aid of (10); accordingly, formula (7) becomes in this case

$$F(z) = F\{f(t)\} = F\{f(x)\} + \frac{y}{1}\phi\{f(x)\}F'\{f(x)\} + \&c.$$

$$+ \frac{y^{n+1}}{1 \cdot 2 \ldots (n+1)} \frac{d^n}{dx^n}\left\{[\phi\{f(x)\}]^{n+1} F'[f(x)]\right\} + \&c. \qquad (11)$$

This formula is called Laplace's Theorem, and is, as we have seen, an immediate deduction from the Theorem of Lagrange. These theorems evidently only hold when the expansions are *convergent* series.

EXAMPLES.

1. Expand z, being given the equation
$$z = a + bz^3.$$

Here $x = a,\ y = b,\ \phi(z) = z^3,$

and we get, from formula (8),
$$z = a + ba^3 + 3b^2 a^5 + 12b^3 a^7 + \&c.$$

Lagrange has shown that this expansion represents the least root of the proposed cubic, and that a similar principle holds in like cases.

2. Given $z = a + bz^n$, find the expansion of z.

Ans. $z = a + a^n b + 2na^{2n-1}\dfrac{b^2}{1\cdot 2} + 3n(3n-1)a^{3n-2}\dfrac{b^2}{1\cdot 2\cdot 3} + \&c.$

3. Given $z = x + ye^z$, find the expansion of z.

Ans. $z = x + ye^x + y^2 e^{2x} + \dfrac{y^3}{1\cdot 2}3e^{3x} + \dfrac{y^4}{1\cdot 2\cdot 3}4^2 e^{4x} + \&c.$

4. $z = a + e \sin z$, expand (1) z, (2) $\sin z$.

(1). Ans. $z = a + e \sin a + \dfrac{e^2}{1\cdot 2}\dfrac{d}{da}(\sin^2 a) + \dfrac{e^3}{1\cdot 2\cdot 3}\left(\dfrac{d}{da}\right)^2(\sin^3 a) + \&c.$

(2). ,, $\sin z = \sin a + e \sin a \cos a + \dfrac{e^2}{1\cdot 2}\dfrac{d}{da}(\sin^2 a \cos a) + \&c.$

5. If $z = a + \dfrac{x}{2}(z^2 - 1)$, prove that
$$z = a + \dfrac{x}{1}\dfrac{(a^2-1)}{2} + \dfrac{x^2}{1\cdot 2}\dfrac{d}{da}\left(\dfrac{a^2-1}{2}\right)^2 + \cdots$$
$$+ \dfrac{x^n}{1\cdot 2\cdots n}\left(\dfrac{d}{da}\right)^{n-1}\left(\dfrac{a^2-1}{2}\right)^n + \&c.$$

6. Hence prove that
$$(1 - 2ax + x^2)^{-\frac{1}{2}} = 1 + \dfrac{x}{1}\dfrac{d}{da}\left(\dfrac{a^2-1}{2}\right) + \dfrac{x^2}{1\cdot 2}\left(\dfrac{d}{dx}\right)^2\left(\dfrac{a^2-1}{2}\right)^2 + \cdots$$
$$+ \dfrac{x^n}{1\cdot 2\cdots n}\left(\dfrac{d}{da}\right)^n\left(\dfrac{a^2-1}{2}\right)^n + \&c.$$

CHAPTER VIII.

EXTENSION OF TAYLOR'S THEOREM TO FUNCTIONS OF TWO OR MORE VARIABLES.

127. **Expansion of** $\phi(x+h, y+k)$. Suppose u to be a function of two variables x and y, represented by the equation $u = \phi(x, y)$; then substituting $x + h$ for x, we get, by Taylor's Theorem,

$$\phi(x+h, y) = \phi(x, y) + h\frac{d}{dx}\{\phi(x, y)\} + \frac{h^2}{1 \cdot 2}\frac{d^2}{dx^2}\{\phi(x, y)\} + \&c.$$

Again, let y become $y + k$, and we get

$$\phi(x+h, y+k) = \phi(x, y+k) + h\frac{d}{dx}\{\phi(x, y+k)\}$$

$$+ \frac{h^2}{1 \cdot 2}\frac{d^2}{dx^2}\{\phi(x, y+k)\} + \&c. \quad (1)$$

But

$$\phi(x, y+k) = \phi(x, y) + k\frac{d}{dy}\{\phi(x, y)\} + \frac{k^2}{1 \cdot 2}\frac{d^2}{dy^2}\{\phi(x, y)\} + \&c.$$

$$= u + k\frac{du}{dy} + \frac{k^2}{1 \cdot 2}\frac{d^2u}{dy^2} + \&c.$$

Also

$$h\frac{d}{dx}\{\phi(x, y+k)\} = h\frac{du}{dx} + hk\frac{d^2u}{dxdy} + \frac{hk^2}{1 \cdot 2}\frac{d^3u}{dxdy^2} + \&c.,$$

and

$$\frac{h^2}{1 \cdot 2}\frac{d^2}{dx^2}\{\phi(x, y+k)\} = \frac{h^2}{1 \cdot 2}\frac{d^2u}{dx^2} + \frac{h^2k}{1 \cdot 2}\frac{d^3u}{dx^2dy} + \&c.$$

Substituting these values in (1), we get

$$\phi(x+h, y+k)^* = u + h\frac{du}{dx} + k\frac{du}{dy}$$

$$+ \frac{h^2}{1.2}\frac{d^2u}{dx^2} + hk\frac{d^2u}{dxdy} + \frac{k^2}{1.2}\frac{d^2u}{dy^2} + \&c. \quad (2)$$

128. This expansion can also be arrived at otherwise as follows:—Substitute $x + \alpha t$ and $y + \beta t$ for x and y, respectively, in the expression $\phi(x, y)$, then the new function

$$\phi(x + \alpha t, y + \beta t),$$

in which x, y, α, β, are constants with respect to t, may be regarded as a function of t, and represented by $F(t)$; thus

$$\phi(x + \alpha t, y + \beta t) = F(t).$$

The latter function $F(t)$, when expanded by Maclaurin's Theorem, becomes, by Art. 79,

$$F(t) = F(0) + \frac{t}{1}F'(0) + \frac{t^2}{1.2}F''(0) + \ldots$$

$$+ \frac{t^n}{\underline{|n}}F^{(n)}(\theta t), \quad (3)$$

where $F'(0)$ is the value of $F(t)$ when $t=0$, i.e. $F(0) = \phi(x, y) = u$; also $F'(0)$, $F''(0)$, &c. are the values of

$$\frac{d\phi}{dt}, \frac{d^2\phi}{dt^2}, \&c.,$$

when $t = 0$; where ϕ stands for $\phi(x + \alpha t, y + \beta t)$.

Moreover, by Art. 117, we have

$$\frac{d\phi}{dt} = \alpha\frac{d\phi}{dx} + \beta\frac{d\phi}{dy},$$

---

* Since it is indifferent whether we first change x into $x + h$, and afterwards change y into $y + k$, or *vice versâ*; the expansion given above furnishes an independent proof of the results arrived at in Art. 119.

but, when $t = 0$, $\phi(x + \alpha t, y + \beta t)$ becomes u, or $F(0)$, and $\dfrac{d\phi}{dt}$ becomes $\alpha \dfrac{du}{dx} + \beta \dfrac{du}{dy}$ at the same time.

Hence
$$F'(0) = \alpha \frac{du}{dx} + \beta \frac{du}{dy}.$$

Also, by the same Article,
$$\frac{d^2\phi}{dt^2} = \alpha^2 \frac{d^2\phi}{dx^2} + 2\alpha\beta \frac{d^2\phi}{dxdy} + \beta^2 \frac{d^2\phi}{dy^2},$$

which, when $t = 0$, reduces to
$$F''(0) = \alpha^2 \frac{d^2u}{dx^2} + 2\alpha\beta \frac{d^2u}{dxdy} + \beta^2 \frac{d^2u}{dy^2}, \tag{4}$$

&c. &c. &c.

These equations may also be written in the symbolic form
$$F''(0) = \left(\alpha \frac{d}{dx} + \beta \frac{d}{dy}\right)^2 u,$$

$$F'''(0) = \left(\alpha \frac{d}{dx} + \beta \frac{d}{dy}\right)^3 u,$$

. . . . .

$$F^{(n)}(0) = \left(\alpha \frac{d}{dx} + \beta \frac{d}{dy}\right)^n u.$$

Again, $\left(\alpha \dfrac{d}{dx}\right)^r u = \alpha^r \dfrac{d^r u}{dx^r}$, &c., since α, β, are independent of x and y: and hence the general term in the expansion of $F(t)$ can be at once written down by aid of the Binomial Theorem.

Finally, we have, on substituting h for αt, and k for βt,

$$\phi(x+h,\ y+k) = u + h\frac{du}{dx} + k\frac{du}{dy} + \frac{h^2}{1\ .\ 2}\frac{d^2u}{dx^2} + hk\frac{d^2u}{dxdy}$$

$$+ \frac{k^2}{1\ .\ 2}\frac{d^2u}{dy^2} + \ldots + \frac{1}{\underline{|n+1}}\left(h\frac{d}{dx} + k\frac{d}{dy}\right)^{n+1}\phi(x+\theta h,\ y+\theta k). \quad (5)$$

129. **Expansion of $\phi(x+h,\ y+k,\ z+l)$**.—A function of three variables, x, y, z, admits of being treated in a similar manner, and accordingly the expression

$$\phi(x+\alpha t,\ y+\beta t,\ z+\gamma t),$$

when u is substituted for $\phi(x,\ y,\ z)$, becomes

$$\phi(x+\alpha t,\ y+\beta t,\ z+\gamma t) = u + \frac{t}{1}\left(\alpha\frac{d}{dx} + \beta\frac{d}{dy} + \gamma\frac{d}{dz}\right)u$$

$$+ \frac{t^2}{1\ .\ 2}\left(\alpha\frac{d}{dx} + \beta\frac{d}{dy} + \gamma\frac{d}{dz}\right)^2 u + \&c.,$$

or

$$\phi(x+h,\ y+k,\ z+l) = u + \left(h\frac{d}{dx} + k\frac{d}{dy} + l\frac{d}{dz}\right)u$$

$$+ \frac{1}{1\ .\ 2}\left(h\frac{d}{dx} + k\frac{d}{dy} + l\frac{d}{dz}\right)^2 u + \&c.$$

$$= u + h\frac{du}{dx} + k\frac{du}{dy} + l\frac{du}{dz} + \frac{h^2}{1\ .\ 2}\frac{d^2u}{dx^2} + \frac{k^2}{1\ .\ 2}\frac{d^2u}{dy^2} + \frac{l^2}{1\ .\ 2}\frac{d^2u}{dz^2}$$

$$+ hk\frac{d^2u}{dxdy} + lh\frac{d^2u}{dzdx} + kl\frac{d^2u}{dydz} + \&c. \quad (6)$$

The general term in this expansion, and also the remainder after n terms, can be easily written down.

These results admit of obvious generalization for any number of variables.

Also, by making x, y, z each cypher in (6), we have

$$\phi(h, k, l) = (u)_0 + h\left(\frac{du}{dx}\right)_0 + k\left(\frac{du}{dy}\right)_0 + l\left(\frac{du}{dz}\right)_0$$

$$+ \frac{h^2}{1 \cdot 2}\left(\frac{d^2u}{dx^2}\right)_0 + \&c. \ldots$$

where $\left(\dfrac{du}{dx}\right)_0$, $\left(\dfrac{du}{dy}\right)_0$, ... denote the values which the functions $\dfrac{du}{dx}$, $\dfrac{du}{dy}$, ... assume on making $x = 0$, $y = 0$, and $z = 0$.

This result may be regarded as the extension of Maclaurin's Theorem.

130. **Symbolic Expression for preceding Results.**—Since

$$e^{h\frac{d}{dx}+k\frac{d}{dy}} \equiv 1 + \left(h\frac{d}{dx} + k\frac{d}{dy}\right) + \frac{1}{1 \cdot 2}\left(h\frac{d}{dx} + k\frac{d}{dy}\right)^2 + \ldots$$

$$+ \frac{1}{\underline{|n}}\left(h\frac{d}{dx} + k\frac{d}{dy}\right)^n + \&c.,$$

equation (5) may be written in the shape

$$e^{h\frac{d}{dx}+k\frac{d}{dy}}\phi(x, y) = \phi(x + h, y + k). \qquad (7)$$

This is analogous to the form given for Taylor's Theorem in Art. 67, and may be deduced from it as follows:—

We have seen that the operation represented by $e^{h\frac{d}{dx}}$ when applied to any function is equivalent to changing x into $x + h$ throughout in the function.

Accordingly, $e^{h\frac{d}{dx}}\phi(x, y) = \phi(x + h, y)$, since y is independent of x.

In like manner, the operation $e^{k\frac{d}{dy}}$, when applied to any function, changes y into $y+k$;

$$\therefore\ e^{k\frac{d}{dy}} \cdot e^{h\frac{d}{dx}} \phi(x, y) = e^{k\frac{d}{dy}} \phi(x+h, y) = \phi(x+h, y+k),$$

or
$$e^{k\frac{d}{dy}+h\frac{d}{dx}} \phi(x, y) = \phi(x+h, y+k),$$

assuming that the symbols $k\dfrac{d}{dy}$ and $h\dfrac{d}{dx}$ are combined according to the same laws* as ordinary algebraic expressions.

In an analogous manner we obtain the symbolic formula

$$e^{h\frac{d}{dx}+k\frac{d}{dy}+l\frac{d}{dz}} \phi(x, y, z) = \phi(x+h, y+k, z+l). \qquad (8)$$

131. If in the development (2), dx be substituted for h, and dy for k, it becomes

$$\phi(x+dx, y+dy) = \phi + \frac{d\phi}{dx} dx + \frac{d\phi}{dy} dy$$

$$+ \frac{1}{1.2}\left(\frac{d^2\phi}{dx^2} dx^2 + 2\frac{d^2\phi}{dx\,dy} dx\,dy + \frac{d^2\phi}{dy^2} dy^2\right) + \&c. \qquad (9)$$

If the sum of all the terms of the degree n in dx and dy be denoted by $d^n\phi$, the preceding result may be written in the form

$$\phi(x+dx, y+dy) = \phi + \frac{d\phi}{1} + \frac{d^2\phi}{1.2} + \frac{d^3\phi}{1.2.3} + \ldots$$

$$+ \frac{d^n\phi}{\underline{|n}} + \&c.$$

Since dx, dy, are infinitely small quantities of the first

---

* That this is the case appears immediately from the equations $\dfrac{d^2u}{dx\,dy} = \dfrac{d^2u}{dy\,dx}$, $\dfrac{d^3u}{dx^2\,dy} = \dfrac{d^3u}{dy\,dx^2}$, &c.

order, each term in the preceding expansion is infinitely small in comparison with the preceding one.

Hence, since $d^2\phi$ is infinitely small in comparison with $d\phi$, if infinitely small quantities of the second and higher orders be neglected in comparison with those of the first, in accordance with Art. 38, we get

$$d\phi = \phi(x + dx, y + dy) - \phi(x, y) = \frac{d\phi}{dx} dx + \frac{d\phi}{dy} dy,$$

which agrees with the result in Art. 97.

132. Euler's Theorems of Homogeneous Functions.—We now proceed to give another proof of Euler's Theorems in addition to those contained in Arts. 102 and 123.

If we substitute gx for h and gy for k in the expansion (5), it becomes

$$\phi(x + gx, y + gy) = u + g\left(x\frac{du}{dx} + y\frac{du}{dy}\right)$$

$$+ \frac{g^2}{1.2}\left(x^2\frac{d^2u}{dx^2} + 2xy\frac{d^2u}{dxdy} + y^2\frac{d^2u}{dy^2}\right) + \&c.,$$

where u stands for $\phi(x, y)$.

But $\quad \phi(x + gx, y + gy) = \phi\{(1 + g)x, (1 + g)y\};$

and, if $\phi(x, y)$ be a homogeneous function of the nth degree in x and y, it is evident that the result of substituting $(1 + g)x$ for x, and $(1 + g)y$ for y in it, is equivalent to multiplying it by $(1 + g)^n$. Hence, we have for homogeneous functions,

$$\phi(x + gx, y + gy) = (1 + g)^n \phi(x, y) = (1 + g)^n u,$$

or $\quad (1 + g)^n u = u + g\left(x\frac{du}{dx} + y\frac{du}{dy}\right)$

$$+ \frac{g^2}{1.2}\left(x^2\frac{d^2u}{dx^2} + 2xy\frac{d^2u}{dxdy} + y^2\frac{d^2u}{dy^2}\right) + \&c.,$$

where u is a homogeneous function of the nth degree in x and y.

Since the preceding equation holds for *all values* of g, if we expand and equate like powers of g, we obtain

$$x \frac{du}{dx} + y \frac{du}{dy} = nu,$$

$$x^2 \frac{d^2u}{dx^2} + 2xy \frac{d^2u}{dxdy} + y^2 \frac{d^2u}{dy^2} = n(n-1)u,$$

$$x^3 \frac{d^3u}{dx^3} + 3x^2y \frac{d^3u}{dx^2dy} + 3xy^2 \frac{d^3u}{dxdy^2} + y^3 \frac{d^3u}{dy^3} = n(n-1)(n-2)u,$$

&c.   &c.   &c.

The foregoing method of demonstration admits of being easily extended to the case of a homogeneous function of three or more variables.

Thus, substituting gx for h, gy for k, gz for l, in formula (6) Art. 129, and proceeding as before, we get

$$x \frac{du}{dx} + y \frac{du}{dy} + z \frac{du}{dz} = nu,$$

$$x^2 \frac{d^2u}{dx^2} + y^2 \frac{d^2u}{dy^2} + z^2 \frac{d^2u}{dz^2} + 2xy \frac{d^2u}{dxdy} + 2zx \frac{d^2u}{dzdx}$$

$$+ 2yz \frac{d^2u}{dydz} = n(n-1)u,$$

&c.   &c.   &c.

These formulæ are due to Euler, and are of importance in the general theory of curves and surfaces, as well as in other applications of analysis.

The preceding method of proof is taken from Lagrange's *Mécanique Analytique*.

CHAPTER IX.

MAXIMA AND MINIMA OF FUNCTIONS OF A SINGLE VARIABLE.

133. Definition of a Maximum or a Minimum.—If any function increase continuously as the variable on which it depends increases up to a certain value, and diminish for higher values of the variable, then, *in passing from its increasing to its decreasing stage*, the function attains what is called a maximum value.

In like manner, if the function decrease as the variable increases up to a certain value, and increase for higher values of the variable, the function passes through a minimum stage.

Many cases of maxima and minima can be best determined without the aid of the Differential Calculus; we shall commence with a few geometrical and algebraic examples of this class.

134. Geometrical Example.—*To find the area of the greatest triangle which can be inscribed in a given ellipse.* Suppose the ellipse projected orthogonally into a circle; then any triangle inscribed in the ellipse is projected into a triangle inscribed in the circle, and the areas of the triangles are to one another in the ratio of the area of the ellipse to that of the circle (Salmon's Conics, Art. 368). Hence the triangle in the ellipse is a maximum when that in the circle is a maximum; but in the latter case the maximum triangle is evidently equilateral, and it is easily seen that its area is to that of the circle as $\sqrt{27}$ to 4π. Hence the area of the greatest triangle inscribed in the ellipse is

$$\frac{3ab\sqrt{3}}{4},$$

where a, b are the semiaxes.

Moreover, the centre of the ellipse is evidently the point of intersection of the bisectors of the sides of the triangle.

Examples.

1. Prove that the area of the greatest ellipse inscribed in a given triangle is $\dfrac{\pi}{\sqrt{27}}$ (area of the triangle).

2. Find the area of the least ellipse circumscribed to a given triangle.

3. Place a chord of a given length in an ellipse, so that its distance from the centre shall be a maximum.
The lines joining its extremities to the centre must be conjugate diameters.

4. Show that the preceding construction is impossible when the length of the given chord is $> a\sqrt{2}$ or $< b\sqrt{2}$; where a and b are the semiaxes of the ellipse. Prove in this case that if the distance of the chord from the centre be a maximum or a minimum the chord is parallel to an axis of the curve.

5. A chord of an ellipse passes through a given point, find when the triangle formed by joining its extremities to the centre is a maximum.

6. Prove that the area of the maximum polygon of n sides, inscribed in a given ellipse, is represented by $\dfrac{n}{2} ab \sin \dfrac{2\pi}{n}$.

135. Algebraic Examples of Maxima and Minima. —Many cases of maxima and minima can be solved by ordinary algebra. We shall confine our attention to one simple class of examples.

Let $f(x)$ represent the function whose maximum or minimum values are required, and suppose $u = f(x)$, and solve for x; then the values of u for which x changes from *real to imaginary*, are the solutions of the problem. This method is, in general, inapplicable when the equation in x is beyond the second degree. We shall illustrate the process by a few examples:—

Examples.

1. To divide a number into two parts such that their product shall be a maximum.

Let a denote the number, x one of the parts, then $x(a-x)$ is to be a maximum, by hypothesis.

Here $u = x(a-x)$, or $x^2 - ax + u = 0$;

solving for x we get

$$x = \frac{u}{2} \pm \sqrt{\frac{a^2}{4} - u};$$

accordingly, the maximum value of u is $\dfrac{a^2}{4}$, since greater values would make x imaginary.

2. To find the maximum and minimum values of the fraction $\dfrac{x}{x^2+1}$.

Here $u = \dfrac{x}{x^2+1}$, or $x^2 + 1 = \dfrac{x}{u}$; $\therefore x = \dfrac{1}{2u} \pm \dfrac{\sqrt{(1-2u)(1+2u)}}{2u}$.

In this case we infer that the maximum and minimum values of u are $\dfrac{1}{2}$ and $-\dfrac{1}{2}$; and the proposed fraction accordingly lies between the limits $\dfrac{1}{2}$ and $-\dfrac{1}{2}$ for all real values of x.

These results can be also easily established, as follows. We have in all cases

$$(x+y)^2 = (x-y)^2 + 4xy.$$

Accordingly, if $x + y$ be given, xy is greatest when $x - y = 0$, or when $x = y$. Conversely, if xy be given, the least value of $x + y$ is when $x = y$.

Hence, denoting xy by a^2, the minimum value of $x + \dfrac{a^2}{x}$ is $2a$, for positive values of x.

Again, it is evident that when a function attains a maximum value, its inverse becomes a minimum; and *vice versâ*.

Accordingly, the max. value of $\dfrac{x}{x^2+a^2}$ is $\dfrac{1}{2a}$, under the same condition.

3. Find the greatest value of $\dfrac{x}{(a+x)(b+x)}$.

Here $\dfrac{(a+x)(b+x)}{x}$ is to be a minimum, or $\dfrac{ab}{x} + x$ is a min.; $\therefore x = \sqrt{ab}$, and the max. value in question is $\dfrac{1}{(\sqrt{a}+\sqrt{b})^2}$.

4. $\dfrac{(x+a)(x+b)}{x+c}$.

Let $x + c = z$, and the fraction becomes $\dfrac{(z+a-c)(z+b-c)}{z}$.

In order that this should have a real min. value, $(a-c)(b-c)$ must be positive; i.e. the value of c must not lie between those of a and b, &c.

5. Find the least value of $a \tan \theta + b \cot \theta$.   *Ans.* $2\sqrt{ab}$.

6. Prove that the expression $\dfrac{x+a}{x^2+bx+c^2}$ will always lie between two fixed finite limits if $a^2 + c^2 > ab$ and $b^2 < 4c^2$; that there will be two limits between which it cannot lie if $a^2 + c^2 > ab$ and $b^2 > 4c^2$: and that it will be capable of all values if $a^2 + c^2 < ab$.

136. **To find the Maximum and Minimum values of**

$$\dfrac{ax^2 + 2bxy + cy^2}{a'x^2 + 2b'xy + c'y^2}.$$

Let u denote the proposed fraction, and substitute z for $\dfrac{x}{y}$; then we get

$$u = \frac{az^2 + 2bz + c}{a'z^2 + 2b'z + c'}; \qquad (1)$$

or $\quad (a - a'u)z^2 + 2(b - b'u)z + c - c'u = 0.$

Solving for z, this gives

$$(a - a'u)z + b - b'u = \pm \sqrt{(b - b'u)^2 - (a - a'u)(c - c'u)}. \qquad (2)$$

There are three cases, according as the roots of the equation

$$(b'^2 - a'c')u^2 + (ac' + ca' - 2bb')u + b^2 - ac = 0 \qquad (3)$$

are real and unequal, real and equal, or imaginary.

(1). Let the roots be real and unequal, and denoted by α and β (of which β is the greater); then, if $b'^2 - a'c' > 0$, we shall have

$$(a - a'u)z + b - b'u = \pm \sqrt{(b'^2 - a'c')(u - \alpha)(u - \beta)}.$$

Here, so long as u is not greater than α, z is real; but when $u > \alpha$ and $< \beta$, z becomes imaginary; consequently, the lesser* root (α) is a maximum value of u. In like manner, it can be easily seen that the greater root (β) is a minimum.

Accordingly, when the roots of the denominator, $a'x^2 + 2b'x + c' = 0$, are real and unequal, the fraction admits of all possible, positive, or negative values, with the exception of those which lie between α and β.

If either $a' = 0$, or $c' = 0$, the radical becomes

$$b' \sqrt{(u - \alpha)(u - \beta)},$$

and, as before, the greater root is a minimum, and the lesser a maximum, value of u.

---

* In general, in seeking the maximum or minimum values of y from the equation, $y = \phi(x)$, if for all values of y between the limits α and β, the corresponding values of x are imaginary, while x is real when $y = \alpha$, or $y = \beta$; then it is evident that the lesser of the quantities, α, β, is a maximum, and the greater a minimum, value of y. This result also admits of a simple geometrical proof, by considering the curve whose equation is $y = \phi(x)$.

(2.) When $a = \beta$, the expression under the radical sign is positive for all values of u, and consequently u does not admit of either a maximum or a minimum value.

(3.) When the roots a and β are imaginary, the expression under the radical sign is necessarily positive, and u in this case also does not admit of either a maximum or a minimum value.

Hence, in the two latter cases, the fraction admits of all possible values between $+\infty$ and $-\infty$.

In the preceding, the roots of the denominator are supposed real; if they be imaginary, i.e. if $b'^2 - a'c' < 0$, we have

$$(a - a'u)z + b - b'u = \pm\sqrt{(a'c' - b'^2)(u - a)(\beta - u)}.$$

It is easily seen that z is imaginary for all values of u except those lying between a and β. Accordingly, the greater root is a maximum, and the lesser a minimum, value of u.

Hence, in this case, the fraction represented by u lies between the limits a and β for all real values of x and y.

137. Quadratic for determining z.—Again, the value of z, corresponding to a maximum or a minimum value of u, must satisfy the equation

$$(a - a'u)z + b - b'u = 0.$$

Substituting for u in (1) its value derived from this latter equation, we obtain the following quadratic in z:

$$(ab' - ba')z^2 + z(ac' - ca') + bc' - cb' = 0. \qquad (4)$$

This equation determines the values of z which correspond to the maximum and minimum values of u. It can be easily seen that if the roots of equation (3) are real so also are those of (4); and *vice versâ*.

The student will observe in the preceding investigation that when u attains a maximum or a minimum value, the corresponding equation in z, obtained from (2), has equal roots. This is, as will be seen more fully in the next Article, the essential criterion of a maximum or a minimum value, in general.

Find the maximum or minimum values of u in the following cases:—

EXAMPLES.

1. $u = \dfrac{x^2 + 2x + 11}{x^2 + 4x + 10}$. *Ans.* $u = 2$, a max., $u = \dfrac{5}{6}$ a min.

2. $u = \dfrac{x^2 - x + 1}{x^2 + x - 1} = 1 + \dfrac{2 - 2x}{x^2 + x - 1}$.

$\dfrac{1-x}{x^2+x-1}$ is a max. or a min. according as $\dfrac{x^2+x-1}{1-x}$ is a min. or a max., i.e.

as $\dfrac{1}{1-x} - x$ is a maximum or a minimum.

∴ $x = 0$, or $x = 2$; the former gives a maximum, the latter a minimum solution.

We now proceed to a general investigation of the conditions for a maximum and minimum, by aid of the principles of the Differential Calculus.

138. Condition for a Maximum or Minimum.—If the increment of a variable, x, be positive, then the corresponding increment of any function, $f(x)$, has the same sign as that of $f'(x)$, by Art. 6; hence, as x increases, $f(x)$ increases or diminishes according as $f'(x)$ is positive or negative.

Consequently, *when $f(x)$ changes from an increasing to a decreasing state, or vice versâ, its derived function $f'(x)$ must change its sign.* Let a be a value of x corresponding to a maximum or a minimum value of $f(x)$; then, in the case of a maximum we must have for small values of h,

$$f(a) > f(a+h), \text{ and } f(a) > f(a-h);$$

and, for a minimum,

$$f(a) < f(a+h), \text{ and } f(a) < f(a-h).$$

Accordingly, in either case the expressions

$$f(a+h) - f(a), \text{ and } f(a-h) - f(a),$$

have both the same sign.

Again, by formulæ* (29), Art. 75, we have

$$f(a + h) - f(a) = hf'(a) + \frac{h^2}{1 \cdot 2} f''(a + \theta h),$$

$$f(a - h) - f(a) = -hf'(a) + \frac{h^2}{1 \cdot 2} f''(a - \theta_1 h).$$

Now, when h is very small, and $f''(a)$ finite, the second term in the right-hand side in each of these equations is very small in comparison with the first, and hence $f(a + h) - f(a)$ and $f(a - h) - f(a)$ cannot have the same sign unless $f'(a) = 0$.

Hence, *the values of x which render $f(x)$ a maximum or a minimum are in general roots of the derived equation $f'(x) = 0$.*

This result can also be arrived at from geometrical considerations; for, let $y = f(x)$ be the equation of a curve, then, at a point from which the ordinate y attains a maximum or a minimum value, the tangent to the curve is evidently parallel to the axis of x; and, consequently $f'(x) = 0$, by Art. 10.

Moreover, if x be eliminated between the equations $f(x) = u$ and $f'(x) = 0$, the roots of the resulting equation in u are, in general, the maximum and minimum values of $f(x)$.

This is the extension of the principle arrived at in Art. 134.

Again, since $f'(a) = 0$, we have

$$\left. \begin{array}{l} f(a + h) - f(a) = \dfrac{h^2}{1 \cdot 2} f''(a + \theta h), \\[2mm] f(a - h) - f(a) = \dfrac{h^2}{1 \cdot 2} f''(a - \theta_1 h) \end{array} \right\}. \qquad (5)$$

---

* In the investigation of maxima and minima given above, Lagrange's form of Taylor's Theorem has been employed. For students who are unacquainted with this form of the Theorem, it may be observed that the conditions for a maximum or minimum can be readily established from the form of Taylor's Series given in Art. 54, viz.,

$$f(a + h) - f(a) = hf'(a) + \frac{h^2}{1 \cdot 2} f''(a) + \frac{h^3}{1 \cdot 2 \cdot 3} f'''(a) + \&c.;$$

for when h is very small and the coefficients $f'(a), f''(a)$, &c. finite, it is evident that the sign of the series at the right-hand side depends on that of its first term, and hence all the results arrived at in the above and the subsequent Articles can be readily established.

Condition for a Maximum or a Minimum.

But the expressions at the left-hand side in these equations are both positive for small values of h when $f''(a)$ is positive; and negative, when $f''(a)$ is negative; therefore $f(a)$ is a maximum or a minimum according as $f''(a)$ is negative or positive.

If, however, $f''(a)$ vanish along with $f'(a)$, we have, by Art. 75,

$$f(a+h) - f(a) = \frac{h^3}{1.2.3} f'''(a) + \frac{h^4}{1.2.3.4} f^{\text{iv}}(a + \theta h),$$

$$f(a-h) - f(a) = \frac{-h^3}{1.2.3} f'''(a) + \frac{h^4}{1.2.3.4} f^{\text{iv}}(a - \theta_1 h).$$

Hence it follows that in this case, $f(a)$ is neither a maximum nor a minimum unless $f'''(a)$ also vanish; but if $f'''(a) = 0$, then $f(a)$ is a maximum when $f^{\text{iv}}(a)$ is negative, and a minimum when $f^{\text{iv}}(a)$ is positive.

In general, let $f^{(n)}(a)$ be the first derived function that does not vanish; then, if n be odd, $f(a)$ is neither a maximum nor a minimum; if n be even, $f(a)$ is a maximum or a minimum according as $f^{(n+1)}(a)$ is negative or positive.

The student who is acquainted with the elements of the theory of plane curves will find no difficulty in giving the geometrical interpretation of the results arrived at in this and the subsequent Articles.

Examples.

1. $u = a \sin x + b \cos x$.

Here the maximum and minimum values are given by the equation

$$\frac{du}{dx} = a \cos x - b \sin x = 0, \text{ or } \tan x = \frac{a}{b}.$$

Hence, the max. value of u is $\sqrt{a^2 + b^2}$, and the min. is $-\sqrt{a^2 + b^2}$. This is also evident independently, since u may be written in the form

$$\sqrt{a^2 + b^2} \sin(x + a),$$

where $\tan a = \frac{b}{a}$.

2. $u = x - \sin x$.

In this case $\quad \dfrac{du}{dx} = 1 - \cos x, \quad \dfrac{d^2 u}{dx^2} = \sin x, \quad \dfrac{d^3 u}{dx^3} = \cos x.$

Accordingly, if $\frac{du}{dx} = 0$, we have $\frac{d^2u}{dx^2} = 0$, and $\frac{d^3u}{dx^3} = 1$.

Consequently, the function $x - \sin x$ does not admit of either a maximum or a minimum value.

This result can also be easily seen from geometrical considerations.

3. $u = a \cos x + b \cos 2x$, a and b being both positive.

Here
$$\frac{du}{dx} = -a \sin x - 2b \sin 2x,$$

$$\frac{d^2u}{dx^2} = -a \cos x - 4b \cos 2x.$$

The maximum and minimum values are given by the equation $a \sin x + 2b \sin 2x = 0$:

$\therefore$ we have, (1), $\sin x = 0$; or (2), $\cos x = \dfrac{-a}{4b}$.

The simplest solution of (1) is $x = 0$, in which case

$$u = a + b, \quad \frac{d^2u}{dx^2} = -a - 4b;$$

consequently this gives a maximum solution.

Again, let $x = \pi$, and we have $u = b - a$, $\frac{d^2u}{dx^2} = a - 4b$; consequently this gives a maximum or a minimum solution, according as a is $<$ or $> 4b$.

If $a = 4b$, we get when $x = \pi$, $\frac{d^2u}{dx^2} = 0$.

On proceeding to the next differentiation we have

$$\frac{d^3u}{dx^3} = a(\sin x + 2 \sin 2x), = 0 \text{ when } x = \pi.$$

Again, $\frac{d^4u}{dx^4} = a(\cos x + 4 \cos 2x) = 3a$. Consequently the solution is a minimum in this case.

Again, the solution (2) is impossible unless a be less than $4b$. In this case, i.e. when $a < 4b$, we easily find $\frac{d^2u}{dx^2}$ positive, and accordingly this gives a min. value of u, viz. $-\dfrac{a^2}{8b} - b$.

4. Find the value of x for which $\sec x - x$ is a maximum or a minimum.

Ans. $\sin x = \dfrac{\sqrt{5} - 1}{2}$.

139. Application to Rational Algebraic Expressions.—Suppose $f(x)$ a rational function containing no fractional power of x, and let the real roots of $f'(x) = 0$, arranged in order of magnitude, be a, β, γ, &c.; no two of which are supposed equal.

Then $\qquad f'(x) = (x - a)(x - \beta)(x - \gamma) \ldots$

and $\qquad f''(a) = (a - \beta)(a - \gamma) \ldots$

But by hypothesis, $a - \beta$, $a - \gamma$, &c. are all positive; hence $f''(a)$ is also positive, and consequently a corresponds to a minimum value of $f(x)$.

Again, $\qquad f''(\beta) = (\beta - a)(\beta - \gamma) \ldots$

here $\beta - a$ is negative, and the remaining factors are positive; hence $f''(\beta)$ is negative, and $f(\beta)$ a maximum.

Similarly, $f(\gamma)$ is a minimum, &c.

140. Maxima and Minima Values occur alternately.—We have seen that this principle holds in the case just considered.

A general proof can easily be given as follows :—Suppose $f(x)$ a maximum when $x = a$, and also when $x = b$, where b is the greater; then when $x = a + h$, the function is decreasing, and when $x = b - h$, it is increasing (where h is a small increment); but in passing from a decreasing to an increasing state it must pass through a minimum value; hence between two maxima one minimum at least must exist.

In like manner it can be shown that between two minima one maximum must exist.

141. Case of Equal Roots.—Again, if the equation $f'(x) = 0$ has two roots each equal to a, it must be of the form

$$f'(x) = (x - a)^2 \psi(x).$$

In this case $f''(a) = 0$, $f'''(a) = 2\psi(a)$, and accordingly, from Art. 138, a corresponds to neither a maximum nor a minimum value of the function $f(x)$.

In general, if $f'(x)$ have n roots equal to a, then

$$f'(x) = (x - a)^n \psi(x).$$

Here, when n is even, $f(a)$ is neither a maximum nor a minimum solution: and when n is odd, $f(a)$ is a maximum or a minimum according as $\psi(a)$ is negative or positive.

174 *Maxima and Minima of Functions of a Single Variable.*

142. Case where $f'(x) = \infty$. The investigation in Art. 138 shows that a function *in general changes its sign in passing through zero*.

In like manner it can be shown that a function changes its sign, in general, in *passing through an infinite value*; i.e. if $\phi(a) = \infty$, $\phi(a-h)$ and $\phi(a+h)$ have in general *opposite* signs, for small values of h.

For, if u and $\dfrac{1}{u}$ represent any function and its reciprocal, they have necessarily the same sign; because if u be positive, $\dfrac{1}{u}$ is positive, and if negative, negative.

Suppose u_1, u_2, u_3, three successive values of u, and $\dfrac{1}{u_1}$, $\dfrac{1}{u_2}$, $\dfrac{1}{u_3}$, the corresponding reciprocals.

Then, if $u_2 = 0$, by Art. 138, u_1 and u_3 have *in general* opposite signs.

Hence, if $\dfrac{1}{u_2} = \infty$, $\dfrac{1}{u_1}$ and $\dfrac{1}{u_3}$ have also opposite signs; and we infer that the values of x which satisfy the equation $f'(x) = \infty$ may furnish maxima and minima values of $f(x)$.

143. We now return to the equation

$$f'(x) = (x-a)^n \psi(x),$$

in which n is supposed to have any real value, positive, negative, integral, or fractional.

In this case, when $x = a$, $f'(x)$ is zero or infinity according as n is positive or negative.

To determine whether the corresponding value of $f(x)$ is a real maximum or minimum, we shall investigate whether $f'(x)$ changes its sign or not as x passes through a.

When $\qquad x = a+h, \quad f'(a+h) = h^n \psi(a+h)$,

„ $\qquad x = a-h, \quad f'(a-h) = (-h)^n \psi(a-h)$:

now, when h is infinitely small, $\psi(a+h)$ and $\psi(a-h)$ become each ultimately equal to $\psi(a)$: and therefore $f'(a+h)$ and $f'(a-h)$ have the same or opposite signs according as $(-1)^n$ is positive or negative.

(1). If n be an *even integer*, positive or negative, $f'(x)$ does not change sign in passing through a, and accordingly a corresponds to neither a maximum nor a minimum solution.

(2). If n be an *odd integer*, positive or negative, $f'(a + h)$ and $f'(a - h)$ have opposite signs, and a corresponds to a real maximum or minimum.

(3). If n be a fraction of the form $\pm\dfrac{2r}{p}$, then $(-1)^{\pm\frac{2r}{p}}$ $= 1^{\pm\frac{r}{p}} = 1$, and a corresponds to *neither* a maximum nor a minimum.

(4). If n be of the form $\pm\dfrac{(2r+1)}{p}$, then $(-1)^{\pm\frac{2r+1}{p}} = (-1)^{\pm\frac{1}{p}}$;

this is *imaginary* if p be even, but has a real value (-1) when p is odd. In the former case, $f'(a - h)$ becomes *imaginary*; in the latter, $f'(a + h)$ and $f'(a - h)$ have opposite signs, and $f(a)$ is a real maximum or minimum.

Thus in all cases of real maximum and minimum values the index n must be the quotient of two odd numbers.

Examples.

1. $\qquad f(x) = ax^2 + 2bx + c.$

Here $\qquad f'(x) = 2(ax + b) = 0;\qquad$ hence $x = -\dfrac{b}{a}$,

$\qquad f''(x) = 2a.$

And $\dfrac{ac - b^2}{a}$ is a maximum or a minimum value of $ax^2 + 2bx + c$, according as a is negative or positive.

2. $\qquad f(x) = 2x^3 - 15x^2 + 36x + 10.$

Here $\qquad f'(x) = 6(x^2 - 5x + 6) = 6(x - 2)(x - 3).$

(1.) Let $x = 2$; then $f''(x)$ is negative; hence $f(2)$ or 38 is a maximum.

(2.) Let $x = 3$; then $f''(x)$ is positive; hence $f(3)$ or 37 is a minimum.

It is evident that neither of these values is an absolute maximum or minimum; for when $x = \infty$, $f(x) = \infty$, and when $x = -\infty$, $f(x) = -\infty$; accordingly, the proposed function admits of all possible values, positive or negative.

Again, neither $+\infty$ nor $-\infty$ is a proper maximum or minimum value, because for large values of x, $f(x)$ *constantly* increases in one case, and constantly diminishes in the other.

It is easily seen that as x increases from $-\infty$ to $+2$, $f(x)$ increases from $-\infty$ to 38; as x increases from 2 to 3, $f(x)$ diminishes from 38 to 37; and as x increases from 3 to ∞, $f(x)$ increases from 37 to ∞. When considered geometrically, the preceding investigation shows that in the curve represented by the equation

$$y = 2x^3 - 15x^2 + 36x + 10,$$

the tangent is parallel to the axis of x at the points $x = 2$, $y = 38$; and $x = 3$, $y = 37$; and that the ordinate is a maximum in the former, and a minimum in the latter case, &c.

3. $f(x) = a + b(x - c)^{\frac{2}{3}}$.   *Ans.* $x = c$.   Neither a max. nor a min.

4. $f(x) = b + c(x - a)^{\frac{2}{3}} + d(x - a)^{\frac{4}{3}}$.

Substitute $a + h$ for x, and the equation becomes

$$f(a + h) = b + ch^{\frac{2}{3}} + dh^{\frac{4}{3}};$$

also

$$f(a - h) = b + ch^{\frac{2}{3}} + dh^{\frac{4}{3}};$$

but when h is very small $h^{\frac{4}{3}}$ is very small in comparison with $h^{\frac{2}{3}}$, and accordingly b is a minimum or a maximum value of $f(x)$ according as c is positive or negative.

5. $f(x) = 5x^6 + 12x^5 - 15x^4 - 40x^3 + 15x^2 + 60x + 17$.

Ans. $x = \pm 1$ gives neither a max. nor a min.; $x = -2$ gives a min.

6. $\dfrac{(x - 1)(x - 6)}{x - 10}$.   Let $x - 10 = z$, and the fraction becomes

$$\dfrac{(z + 9)(z + 4)}{z}, \text{ or } z + 13 + \dfrac{36}{z}.$$

The maximum and minimum values are given by the equation $1 - \dfrac{36}{z^2} = 0$;

$\therefore z = \pm 6$, and hence $x = 16$ or 4; the former gives a minimum, the latter a maximum value of the fraction.

7. $$f(x) = \dfrac{(x - 1)^3}{(x + 1)^2}.$$

Hence $$f'(x) = \dfrac{(x - 1)^2}{(x + 1)^3}(x + 5).$$

If $x = 1$, $f(x)$ is neither a maximum nor a minimum; if $x = -5$, $f(x)$ is a maximum.

Max. and Min. of $\dfrac{ax^2 + 2bxy + cy^2}{a'x^2 + 2b'xy + c'y^2}$.

Again, the reciprocal function $\dfrac{(x+1)^2}{(x-1)^3}$ is evidently a max. when $x = -1$; for if we substitute for x, $-1 + h$, and $-1 - h$, successively, the resulting values are *both negative*; and consequently the proposed function is a minimum in this case.

This furnishes an example of a solution corresponding to $f'(x) = \infty$. See Art. 142.

144. We shall now return to the fraction

$$\dfrac{ax^2 + 2bxy + cy^2}{a'x^2 + 2b'xy + c'y^2},$$

the maximum and minimum values of which have been already considered in Art. 136.

Write as before the equation in the form

$$z^2(a - a'u) + 2z(b - b'u) + (c - c'u) = 0,$$

where $z = \dfrac{x}{y}$.

Differentiate with respect to z, and, as $\dfrac{du}{dz} = 0$ for a maximum or a minimum, we have

$$z(a - a'u) + (b - b'u) = 0.$$

Multiply this latter equation by z, and subtract from the former, when we get

$$z(b - b'u) + (c - c'u) = 0.$$

Hence, eliminating z between these equations, we obtain

$$(a - a'u)(c - c'u) = (b - b'u)^2,$$

or $\quad u^2(a'c' - b'^2) - u(ac' + ca' - 2bb') + (ac - b^2) = 0;\quad$ (3)

the same equation (3) as before.

The quadratic for z,

$$z^2(ab' - ba') + z(ac' - ca') + bc' - cb' = 0, \quad (4)$$

is obtained by eliminating u from the two preceding linear equations.

N

This equation can also be written in a determinant form, as follows:—

$$\begin{vmatrix} 1 & -z & z^2 \\ a & b & c \\ a' & b' & c' \end{vmatrix} = 0.$$

It may be observed that the coefficients in (3) are *invariants* of the quadratic expressions in the numerator and denominator of the proposed fraction, as is evident from the principle that its maximum and minimum values cannot be altered by linear transformations.

This result can also be proved as follows:—

Let
$$u = \frac{aX^2 + 2bXY + cY^2}{a'X^2 + 2b'XY + c'Y^2},$$

where X, Y denote any functions of x and y; then in seeking the maximum and minimum values of u we may substitute z for $\dfrac{X}{Y}$, when it becomes

$$u = \frac{az^2 + 2bz + c}{a'z^2 + 2b'z + c'},$$

and we obviously get the *same* maximum and minimum values for u, whether we regard it as determined from the original fraction or from the equivalent fraction in z.

Again, let X, Y be linear functions of x and y, i.e.

$$X = lx + my, \quad Y = l'x + m'y,$$

then u becomes of the form

$$\frac{Ax^2 + 2Bxy + Cy^2}{A'x^2 + 2B'xy + C'y^2},$$

where A, B, C, A', B', C', denote the coefficients in the transformed expressions; hence, since the quadratics which determine the maximum and minimum values of u must have the same roots in both cases, we have

$$AC - B^2 = \lambda(ac - b^2), \quad AC' + CA' - 2BB' = \lambda(ac' + ca' - 2bb'),$$
$$A'C' - B'^2 = \lambda(a'c' - b'^2). \qquad \text{Q.E.D.}$$

It can be seen without difficulty that

$$\lambda = (lm' - ml')^2.$$

We shall illustrate the use of the equations (3) and (4) by applying them to the following question, which occurs in the determination of the principal radii of curvature at any point on a curved surface.

145. To find the Maxima and Minima Values of

$$r \cos^2 a + 2s \cos a \cos \beta + t \cos^2 \beta,$$

where $\cos a$ and $\cos \beta$ are connected by the equation

$$(1 + p^2) \cos^2 a + 2pq \cos a \cos \beta + (1 + q^2) \cos^2 \beta = 1,$$

and p, q, r, s, t are independent of a and β.

Denoting the proposed expression by u, and substituting z for $\dfrac{\cos a}{\cos \beta}$, we get

$$u = \frac{rz^2 + 2sz + t}{(1 + p^2)z^2 + 2pqz + (1 + q^2)}.$$

The maximum and minimum values of this fraction, by the preceding Article, are given by the quadratic

$$u^2\{1 + p^2 + q^2\} - u\{(1 + q^2)r - 2pqs + (1 + p^2)t\} + rt - s^2 = 0; \quad (6)$$

while the corresponding values of z or $\dfrac{\cos a}{\cos \beta}$ are given by

$$z^2\{(1 + p^2)s - pqr\} + z\{(1 + p^2)t - (1 + q^2)r\}$$
$$+ \{pqt - (1 + q^2)s\} = 0.^* \quad (7)$$

The student will observe that the roots of the denominator in the proposed fraction are imaginary, and, consequently, the values of the fraction lie between the roots of the quadratic (6), in accordance with Art. 136.

---

* Lacroix, *Dif. Cal.*, pp. 575, 576.

146. To find the Maximum and Minimum Radius Vector of the Ellipse

$$ax^2 + 2bxy + cy^2 = 1.$$

(1). Suppose the axes rectangular; then

$r^2 = x^2 + y^2$ is to be a maximum or a minimum.

Let $\dfrac{x}{y} = z$, and we get

$$r^2 = \frac{z^2 + 1}{az^2 + 2bz + c}.$$

Hence the quadratic which determines the maximum and minimum distances from the centre is

$$r^4(ac - b^2) - r^2(a + c) + 1 = 0.$$

The other quadratic, viz.

$$bx^2 - (a - c)xy - by^2 = 0,$$

gives the *directions* of the axes of the curve.

(2.) If the axes of co-ordinates be inclined at an angle ω, then

$$r^2 = x^2 + y^2 + 2xy \cos \omega$$

$$= \frac{z^2 + 2z \cos \omega + 1}{az^2 + 2bz + c};$$

and the quadratic becomes in this case

$$r^4(ac - b^2) - r^2(a + c - 2b \cos \omega) + \sin^2 \omega = 0,$$

the coefficients in which are the *invariants* of the quadratic expressions forming the numerator and denominator in the expression for r^2.

The equation which determines the directions of the axes ? the conic can also be easily written down in this case.

147. To investigate the Maximum and Minimum Values of

$$\frac{ax^3 + 3bx^2y + 3cxy^2 + dy^3}{a'x^3 + 3b'x^2y + 3c'xy^2 + d'y^3}.$$

Substituting z for $\dfrac{x}{y}$, and denoting the fraction by u, we have

$$u = \frac{az^3 + 3bz^2 + 3cz + d}{a'z^3 + 3b'z^2 + 3c'z + d'}.$$

Proceeding, as in Art. 144, we find that the values of u and z are given by aid of the two quadratics

$$az^2 + 2bz + c = (a'z^2 + 2b'z + c')u,$$
$$bz^2 + 2cz + d = (b'z^2 + 2c'z + d')u.$$

Eliminating u between these equations, we get the following biquadratic in z:—

$$z^4(ab' - ba') + 2z^3(ac' - ca') + z^2\{ad' - a'd + 3(bc' - cb')\}$$
$$+ 2z(bd' - db') + (cd' - c'd) = 0. \qquad (8)$$

Eliminating z between the same equations, we obtain a biquadratic in u, whose roots are the maxima and minima values of the proposed fraction. Again, as in Art. 144, it can easily be shown that the coefficients in the equation in u are *invariants* of the cubics in the numerator and denominator of the fraction.

148. To cut the Maximum and Minimum Ellipse from a Right Cone which stands on a given circular base.—Let AD represent the axis of the cone, and suppose BP to be the axis major of the required section; O its centre; a, b, its semi-axes. Through O and P draw LM and PR parallel to BC. Then $BP = 2a$, $b^2 = LO \cdot OM$ (Euclid, Book III., Pr. 35); but $LO = \dfrac{PR}{2}$, $OM = \dfrac{BC}{2}$; $\therefore b^2 = \dfrac{1}{4} \cdot BC \cdot PR$. Hence $BP^2 \cdot PR$ is to be a maximum or a minimum.

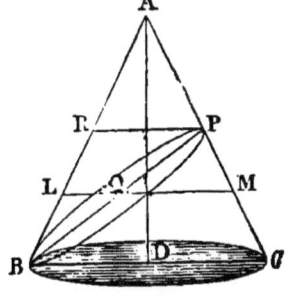

Fig. 7.

Let $\angle BAD = a$, $PBC = \theta$, $BC = c$.

Then
$$BP = BC \frac{\sin BCP}{\sin BPC} = \frac{c \cos a}{\cos(\theta - a)}.$$

$$PR = BP \frac{\sin PBR}{\sin PRB} = \frac{c \cos(\theta + a)}{\cos(\theta - a)};$$

$$\therefore u = \frac{\cos(\theta + a)}{\cos^3(\theta - a)} \text{ is a maximum or a minimum.}$$

Hence $\dfrac{du}{d\theta} = \dfrac{\sin 2\theta - 2 \sin 2a}{\cos^4(\theta - a)} = 0$; $\therefore \sin 2\theta = 2 \sin 2a$.

The solution becomes impossible when $2 \sin 2a > 1$; i.e. if the vertical angle of the cone be $> 30°$.

The problem admits of two solutions when a is less than $15°$. For, if θ_1 be the least value of θ derived from the equation $\sin 2\theta = 2 \sin 2a$; then the value $\dfrac{\pi}{2} - \theta_1$ evidently gives a second solution.

Again, by differentiation, we get

$$\frac{d^2u}{d\theta^2} = \frac{2 \cos 2\theta}{\cos^4(\theta - a)} \text{ (when } \sin 2\theta = 2 \sin 2a\text{).}$$

This is positive or negative according as $\cos 2\theta$ is positive or negative. Hence the *greater* value of θ corresponds to a maximum section, and the *lesser* to a minimum.

In the limiting case, when $a = 15°$, the two solutions coincide. However, it is easily shown that the corresponding section gives neither a maximum nor a minimum solution of the problem. For, we have in this case $\theta = 45°$; which value gives $\dfrac{d^2u}{d\theta^2} = 0$. On proceeding to the next differentiation, we find, when $\theta = 45°$,

$$\frac{d^3u}{d\theta^3} = \frac{-4}{\cos^4(45° - a)} = -\frac{64}{9}.$$

Hence the solution is neither a maximum nor a minimum.
When $a > 15°$, both solutions are *impossible*.

149. The principle, that when a function is a maximum or a minimum its reciprocal is at the same time a minimum or a maximum, is of frequent use in finding such solutions.

There are other considerations by which the determination of maxima and minima values is often facilitated.

Thus, whenever u is a maximum or a minimum, so also is $\log(u)$, unless u vanishes along with $\dfrac{du}{dx}$.

Again, any constant may be added or subtracted, i.e. if $f(x)$ be a maximum, so also is $f(x) \pm c$.

Also, if any function, u, be a maximum, so will be any positive power of u, in general.

150. Again, if $z = f(u)$, then $dz = f'(u)\,du$, and consequently z is a maximum or a minimum; either (1) when $du = 0$, i.e. *when u is a maximum or a minimum*; or (2) *when $f'(u) = 0$*.

In many questions the values of u are restricted, by the conditions of the problem,* to lie between given limits; accordingly, in such cases, any root of $f'(u) = 0$ does not furnish a real maximum or minimum solution unless it lies between the given limiting values of u.

We shall illustrate this by one or two geometrical examples.

(1). *In an ellipse, to find when the rectangle under a pair of conjugate diameters is a maximum or a minimum.* Let r be any semi-diameter of the ellipse, then the square of the conjugate semi-diameter is represented by $a^2 + b^2 - r^2$, and we have

$$u = r^2(a^2 + b^2 - r^2) \text{ a maximum or a minimum.}$$

Here
$$\frac{du}{dr} = 2(a^2 + b^2 - 2r^2)r.$$

Accordingly the maximum and minimum values are, (1) those for which r is a maximum or a minimum; i.e. $r = a$, or $r = b$; and, (2) those given by the equation

$$r(a^2 + b^2 - 2r^2) = 0;$$

---

* See *Cambridge Mathematical Journal*, vol. iii. p. 237.

or $\quad r = 0$, and $r = \sqrt{\dfrac{a^2 + b^2}{2}}$.

The solution $r = 0$ is inadmissible, since r must lie between the limits a and b: the other solution corresponds to the equiconjugate diameters. It is easily seen that the solution in (2) is the maximum, and that in (1) the minimum value of the rectangle in question.

151. As another example, we shall consider the following problem*:—

Given in a plane triangle two sides (a, b) to find the maximum and minimum values of

$$\frac{1}{c} \cdot \cos \frac{A}{2},$$

where A and c have the usual significations.

Squaring the expression in question, and substituting x for c, we easily find for the quantity whose maximum and minimum values are required the following expression:

$$\frac{1}{x} + \frac{2b}{x^2} - \frac{a^2 - b^2}{x^3},$$

neglecting a constant multiplier.

Accordingly, the solutions of the problem are—(1) the maximum and minimum values of x, i.e. $a + b$ and $a - b$. (2) the solutions of the equation $\dfrac{du}{dx}$, i.e. of

$$\frac{1}{x^2} + \frac{4b}{x^3} - \frac{3(a^2 - b^2)}{x^4} = 0,$$

or $\quad x^2 + 4bx - 3(a^2 - b^2) = 0;$

whence we get $\quad x = \sqrt{3a^2 + b^2} - 2b,$

neglecting the negative root, which is inadmissible.

Again, if $b > a$, $\sqrt{3a^2 + b^2} - 2b$ is negative, and accordingly in this case the solution given by (2) is inadmissible.

---

* This problem occurs in Astronomy, in finding when a planet appears brightest, the orbits being supposed circular.

If $a > b$, it remains to see whether $\sqrt{3a^2 + b^2} - 2b$ lies between the limits $a + b$ and $a - b$. It is easily seen that $\sqrt{3a^2 + b^2} - 2b$ is $> a - b$: the remaining condition requires

$$a + b > \sqrt{3a^2 + b^2} - 2b,$$

or
$$a + 3b > \sqrt{3a^2 + b^2},$$

or
$$a^2 + 6ab + 9b^2 > 3a^2 + b^2,$$

i.e.
$$4b^2 + 3ab > a^2,$$

or
$$4b^2 + 3ab + \frac{9a^2}{16} > \frac{25a^2}{16}; \therefore 2b + \frac{3a}{4} > \frac{5a}{4};$$

or, finally,
$$b > \frac{a}{4}.$$

We see accordingly that this gives no real solution unless the lesser of the given sides exceeds one-fourth of the greater.

When this condition is fulfilled, it is easily seen that the corresponding solution is a maximum, and that the solutions corresponding to $x = a + b$, and $x = a - b$, are both minima solutions.

152. Maxima and Minima Values of an Implicit Function.—Suppose it be required to find the maxima or minima values of y from the equation

$$f(x, y) = 0.$$

Differentiating, we get

$$\frac{du}{dx} + \frac{du}{dy}\frac{dy}{dx} = 0,$$

where u represents $f(x, y)$. But the maxima and minima values of y must satisfy the equation $\frac{dy}{dx} = 0$: accordingly the maximum and minimum values are got by combining* the equations $\dfrac{du}{dx} = 0$, and $u = 0$.

153. Maximum and Minimum in case of a Function of two dependent Variables.—To determine the maximum or minimum values of a function of two variables, x and y, which are connected by a relation of the form

$$f(x, y) = 0.$$

Let the proposed function, $\phi(x, y)$ be represented by u; then, by Art. 101, we have

$$\frac{du}{dx} = \frac{\dfrac{d\phi}{dx}\dfrac{df}{dy} - \dfrac{d\phi}{dy}\dfrac{df}{dx}}{\dfrac{df}{dy}}.$$

But the maxima and minima values of u satisfy the equation $\dfrac{du}{dx} = 0$, hence the values of x and y derived from the equations $f(x, y) = 0$, and

$$\frac{d\phi}{dx}\frac{df}{dy} - \frac{d\phi}{dy}\frac{df}{dx} = 0,$$

furnish the solutions required. To determine whether the solution so determined is a maximum or a minimum, it is necessary to investigate the sign of $\dfrac{d^2u}{dx^2}$. We add an example for illustration.

154. *Given the four sides of a quadrilateral, to find when its area is a maximum.*

Let a, b, c, d be the lengths of the sides, ϕ the angle between a and b, ψ that between c and d. Then $ab \sin \phi + cd \sin \psi$ is a maximum; also

$$a^2 + b^2 - 2ab \cos \phi = c^2 + d^2 - 2cd \cos \psi$$

being each equal to the square of the diagonal.

---

* This result is evident also from geometrical considerations.

Hence $\quad ab\cos\phi + cd\cos\psi\dfrac{d\psi}{d\phi} = 0$

for a maximum or a minimum; also,

$$ab\sin\phi = cd\sin\psi\dfrac{d\psi}{d\phi};$$

$\therefore \tan\phi + \tan\psi = 0$, or $\phi + \psi = 180°$.

Hence the quadrilateral is inscribable in a circle.

That the solution arrived at is a *maximum* is evident from geometrical considerations; it can also be proved to be so by aid of the preceding principles.

For, substitute $\dfrac{ab\sin\phi}{cd\sin\psi}$ instead of $\dfrac{d\psi}{d\phi}$, and we get

$$\dfrac{du}{d\phi} = \dfrac{ab\sin(\phi+\psi)}{\sin\psi}.$$

Hence $\dfrac{d^2u}{d\phi^2} = \dfrac{ab\cos(\phi+\psi)}{\sin\psi}\left(1 + \dfrac{d\psi}{d\phi}\right) +$ a term which vanishes when $\phi + \psi = 180°$; and the value of $\dfrac{d^2u}{d\phi^2}$ becomes in this case

$$-\dfrac{ab}{\sin\psi}\left(1 + \dfrac{ab}{cd}\right),$$

which being negative, the solution is a maximum.

Examples.

1. Prove that $a \sec \theta + b \csc \theta$ is a minimum when $\tan \theta = \sqrt[3]{\dfrac{b}{a}}$.

2. Find when $4x^3 - 15x^2 + 12x - 1$ is a maximum or a minimum.

    Ans. $x = \frac{1}{2}$, a max.; $x = 2$, a min.

3. If a and b be such that $f(a) = f(b)$, show that $f(x)$ has, in general, a maximum or a minimum value for some value of x between a and b.

4. Find the value of x which makes
$$\frac{\sin x \cdot \cos x}{\cos^2(60° - x)}$$
a maximum. *Ans.* $x = 30°$.

5. If $\dfrac{f(x) + \phi(x)}{f(x) - \phi(x)}$ be a maximum, show immediately that $\dfrac{f(x)}{\phi(x)}$ is a minimum.

6. Find the value of $\cos x$ when $\dfrac{\sin^2 x}{\sqrt{5 - 4\cos x}}$ is a maximum.

    Ans. $\cos x = \dfrac{5 - \sqrt{13}}{6}$.

7. Find when $\dfrac{1 + 3x}{\sqrt{4 + 5x^2}}$ is a maximum. ,, $x = \dfrac{12}{5}$.

8. Apply the method of Ex. 5 to the expression $\dfrac{x^2 + ax + b}{x^2 - ax + b}$.

9. What are the values of x which make the expression
$$2x^3 - 21x^2 + 36x - 20$$
a maximum or a minimum? and (2) what are the maximum and minimum values of the expression? *Ans.* $x = 1$, a max.; $x = 6$, a min.

10. $u = x^m(a - x)^n$. *Ans.* $x = \dfrac{ma}{m + n}$, a maximum.

11. Given the angle C of a triangle; prove that $\sin^2 A + \sin^2 B$ is a maximum, and $\cos^2 A + \cos^2 B$ a minimum, when $A = B$.

12. Find the least value of $ae^{kx} + be^{-kx}$. *Ans.* $2\sqrt{ab}$.

13. $\dfrac{(a + x)(b + x)}{(a - x)(b - x)}$. ,, $x = \pm\sqrt{ab}$.

14. Show that $b + c(x-a)^{\frac{2}{3}}$, when $x = a$, is a minimum or a maximum according as c is positive or negative.

15. $u = x \cos x$. *Ans.* $x = \cot x$.

16. Prove that $x^{\frac{1}{x}}$ is a maximum when $x = e$.

17. $\tan^m x \cdot \tan^n(a-x)$ is a maximum when $\tan(a - 2x) = \dfrac{n-m}{n+m} \tan a$?

18. Prove that $\dfrac{x}{\log x}$ is a minimum when $x = e$.

19. Given the vertical angle of a triangle and its area, find when its base is a minimum.

20. Given one angle A of a right-angled spherical triangle, find when the difference betweeen the sides which contain it is a maximum.

Here $\tan c \cos A = \tan b$; and since $c - b$ is a maximum, $\dfrac{dc}{db} = 1$.

Hence we find $\tan b = \sqrt{\cos A}$.

This question admits of another easy solution; for, as in Art. 112, we have

$$\frac{\sin(c-b)}{\sin(c+b)} = \tan^2 \frac{A}{2};$$

consequently $\sin(c - b)$ becomes a maximum along with $\sin(c + b)$, since A is constant; and hence $c - b$ is a maximum when $c + b = 90°$.

This problem occurs in Astronomy, in finding when the part of the equation of time which arises from the obliquity of the ecliptic is a maximum.

21. Prove that the problem, to describe a circle with its centre on the circumference of a given circle, so that the length of the arc intercepted within the given circle shall be a maximum, is reducible to the solution of the equation $\theta = \cot \theta$.

22. A perpendicular is let fall from the centre on a tangent to an ellipse, find when the intercept between the point of contact and the foot of the perpendicular is a maximum. Prove that $p = \sqrt{ab}$, and intercept $= a - b$.

23. A semicircle is described on the axis-major of an ellipse; draw a line from one extremity of the axis so that the portion intercepted between the circle and the ellipse shall be a maximum.

24. Draw two conjugate diameters of an ellipse, so that the sum of the perpendiculars from their extremities on the axis-major shall be a maximum.

25. Through a point O on the produced diameter AB of a semicircle draw a secant ORR', so that the quadrilateral $ABRR'$ inscribed in the semicircle shall be a maximum.

Prove that, in this case, the projection of RR' on AB is equal in length to the radius of the circle.

26. If $\sin \phi = k \sin \psi$, and $\psi + \psi' = a$, where a and k are constants, prove that $\cos \psi' \cos \phi$ is a maximum when $\tan^2 \phi = \tan \psi \tan \psi'$.

27. Find the area of the ellipse

$$ax^2 + 2hxy + by^2 = c$$

in terms of the coefficients in its equation, by the method of Art. 146.

(1) for rectangular axes. *Ans.* $\dfrac{\pi c}{\sqrt{ab - h^2}}$.

(2) for oblique. ,, $\dfrac{\pi c \sin \omega}{\sqrt{ab - h^2}}$.

28. A triangle inscribed in a given circle has its base parallel to a given line, and its vertex at a given point; find an expression for the cosine of its vertical angle when the area is a maximum.

29. Find when the base of a triangle is a minimum, being given the vertical angle and the ratio of one side to the difference between the other and a fixed line.

30. Of all spherical triangles of equal area, that of the least perimeter is equilateral?

31. Let $u^3 + x^3 - 3axu = 0$; determine whether the value $x = 0$ gives u a maximum or minimum.   *Ans.* Neither.

32. Show that the maximum and minimum values of the cubic expression

$$ax^3 + 3bx^2 + 3cx + d$$

are the roots of the quadratic

$$a^2 z^2 - 2Gz - \Delta = 0;$$

where $G = a^2 d - 3abc + 2b^3$, and $\Delta = a^2 d^2 + 4ac^3 + 4db^3 - 3b^2 c^2 - 6abcd$.

33. Through a fixed point within a given angle draw a line so that the triangle formed shall be a minimum.

The line is bisected in the given point.

34. Prove in general that the chord drawn through a given point so as to cut off the minimum area from a given curve is bisected at that point.

35. If the portion, AB, of the tangent to a given curve intercepted by two fixed lines OA, OB, be a minimum, prove that $PA = NB$, where P is the point of contact of the tangent, and N the foot of the perpendicular let fall on the tangent from O.

36. The portion of the tangent to an ellipse intercepted between the axes is a minimum: find its length.   *Ans.* $a + b$.

37. Prove that the maximum and minimum values of the expression, Art. 147, are roots of the biquadratic

$$(a - ua')^2 (d - ud')^2 + 4(a - ua')(c - uc')^3 + 4(d - ud')(b - ub')^3$$
$$- 3(b - ub')^2 (c - uc')^2 - 6(a - ua')(b - ub')(c - uc')(d - ud') = 0.$$

CHAPTER X.

MAXIMA AND MINIMA OF FUNCTIONS OF TWO OR MORE INDEPENDENT VARIABLES.

155. **Maxima and Minima for Two Variables.**—In accordance with the principles established in the preceding chapter, if $\phi(x, y)$ be a maximum for the particular values x_0 and y_0, of the independent variables x and y, then for all small positive or negative values of h and k, $\phi(x_0, y_0)$ must be greater than $\phi(x_0 + h, y_0 + k)$; and for a minimum it must be less.

Again, since x and y are independent, we may suppose either of them to vary, the other remaining constant; accordingly, as in Art. 138, it is necessary for a maximum or minimum value that

$$\frac{du}{dx} = 0, \text{ and } \frac{du}{dy} = 0; \qquad (1)$$

omitting the case where either of these functions becomes infinite.

156. **Lagrange's Condition.**—We now proceed to consider whether the values found by this process correspond to real maxima or minima, or not.

Suppose x_0, y_0 to be values of x and y which satisfy the equations

$$\frac{du}{dx} = 0, \text{ and } \frac{du}{dy} = 0,$$

and let A, B, C be the values which $\dfrac{d^2u}{dx^2}$, $\dfrac{d^2u}{dxdy}$, $\dfrac{d^2u}{dy^2}$ assume when x_0 and y_0 are substituted for x and y; then we shall have

$$\phi(x_0 + h, y_0 + k) - \phi(x_0, y_0) = \frac{1}{1 \cdot 2}(Ah^2 + 2Bhk + Ck^2) + \&c. \quad (2)$$

But when h and k are very small, the remainder of the expansion becomes in general very small in comparison with the quantity $Ah^2 + 2Bhk + Ck^2$; accordingly the sign of $\phi(x_0 + h, y_0 + k) - \phi(x_0, y_0)$ depends on that of

$$Ah^2 + 2Bhk + Ck^2, \text{ i.e. of } \frac{(Ah + Bk)^2 + k^2(AC - B^2)}{A}.$$

Now, in order that this expression should be either always positive or always negative for all small values of h and k, it is necessary that $AC - B^2$ should not be negative; as, if it be negative, the numerator in the preceding expression would be positive when $k = 0$, and negative when $Ah + Bk = 0$.

Hence, the condition for a real maximum or minimum is that AC should not be less than B^2, or

$$\frac{d^2u}{dx^2}\frac{d^2u}{dy^2} > \text{ or } = \left(\frac{d^2u}{dxdy}\right)^2;$$

and, when this condition is satisfied, the solution is a maximum or a minimum value of the function according as the sign of A is negative or positive.

If B^2 be $> AC$ the solution is neither a maximum nor a minimum.

The necessity of the preceding condition was first established by Lagrange;* by whom also the corresponding conditions in the case of a function of any number of variables were first discussed.

Again, if $A = 0$, $B = 0$, $C = 0$, then for a real maximum or minimum it is necessary that all the terms of the third degree in h and k in expansion (2) should vanish at the same time, while the quantity of the fourth degree in h and k should preserve the same sign for all values of these quantities. See Art. 138.

The spirit of the method, as well as the processes employed in its application, will be illustrated by the following examples.

157. To find the position of the point the sum of the squares of whose distances from n given points situated in the same plane shall be a minimum.

---

* *Théorie des Fonctions.* Deuxième Partie. Ch. onzième.

Let the co-ordinates of the given points referred to rectangular axes be

$$(a_1, b_1), (a_2, b_2), (a_3, b_3) \ldots (a_n, b_n), \text{ respectively};$$

(x, y) those of the point required; then we have

$$u = (x - a_1)^2 + (y - b_1)^2 + (x - a_2)^2 + (y - b_2)^2 + \ldots$$
$$+ (x - a_n)^2 + (y - b_n)^2$$

a minimum;

$$\therefore \frac{du}{dx} = x - a_1 + x - a_2 + \ldots + x - a_n = nx - (a_1 + a_2 + \ldots + a_n) = 0;$$

$$\frac{du}{dy} = y - b_1 + y - b_2 + \ldots + y - b_n = ny - (b_1 + b_2 + \ldots + b_n) = 0.$$

Hence $\quad x = \dfrac{a_1 + a_2 + \ldots + a_n}{n}, \quad y = \dfrac{b_1 + b_2 + \ldots + b_n}{n};$

and the point required is the centre of mean position of the n given points.

From the nature of the problem it is evident that this result corresponds to a minimum.

This can also be established by aid of Lagrange's condition, for we have

$$A = \frac{d^2u}{dx^2} = n, \quad B = \frac{d^2u}{dx\,dy} = 0, \quad C = \frac{d^2u}{dy^2} = n.$$

In this case $AC - B^2$ is positive, and A also positive; and accordingly the result is a minimum.

158. To find the Maximum or Minimum Value of the expression

$$ax^2 + by^2 + 2hxy + 2gx + 2fy + c.$$

Denoting the expression by u, we have

$$\frac{1}{2}\frac{du}{dx} = ax + hy + g = 0,$$

$$\frac{1}{2}\frac{du}{dy} = hx + by + f = 0.$$

Multiplying the first equation by x, the second by y, and subtracting their sum from the given expression, we get

$$u = gx + fy + c;$$

whence, eliminating x and y between the three equations, we obtain

$$u(ab - h^2) = \begin{vmatrix} a & h & g \\ h & b & f \\ g & f & c \end{vmatrix}. \qquad (3)$$

This result may also be written in the form

$$u\frac{d\Delta}{dc} = \Delta,$$

where Δ denotes the discriminant of the proposed expression.

Again, $\dfrac{d^2u}{dx^2} = 2a$, $\dfrac{d^2u}{dy^2} = 2b$, $\dfrac{d^2u}{dxdy} = 2h$.

Hence, if $ab - h^2$ be positive, the foregoing value of u is a maximum or a minimum according as the sign of a is negative or positive.

If $h^2 > ab$, the solution is neither a maximum nor a minimum.

The geometrical interpretation of the preceding result is evident; viz., if the co-ordinates of the centre be substituted for x and y in the equation of a conic, $u = 0$, the resulting value of u is either a maximum or a minimum if the curve be an ellipse, but is neither a maximum nor a minimum for a hyperbola; as is also evident from other considerations.

159. To find the Maxima and Minima Values of the Fraction

$$\frac{ax^2 + by^2 + 2hxy + 2gx + 2fy + c}{a'x^2 + b'y^2 + 2h'xy + 2g'x + 2f'y + c'}.$$

Let the numerator and denominator be represented by ϕ_1 and ϕ_2; then, denoting the fraction by u, we get

$$\phi_1 = u\phi_2. \qquad (a)$$

Examples for Two Variables.

Differentiate with respect to x and y separately, then

$$\frac{d\phi_1}{dx} = \frac{du}{dx}\phi_2 + u\frac{d\phi_2}{dx}, \quad \frac{d\phi_1}{dy} = \frac{du}{dy}\phi_2 + u\frac{d\phi_2}{dy};$$

but for a maximum or a minimum we must have

$$\frac{du}{dx} = 0, \quad \frac{du}{dy} = 0;$$

hence, the required solutions are given by the equations

$$ax + hy + g = u(a'x + h'y + g'),$$
$$hx + by + f = u(h'x + b'y + f').$$

Multiplying the former by x, the latter by y, and subtracting the sum from the equation (a), we get

$$gx + fy + c = u(g'x + f'y + c').$$

These equations may be written

$$(a - a'u)x + (h - h'u)y + g - g'u = 0,$$
$$(h - h'u)x + (b - b'u)y + f - f'u = 0,$$
$$(g - g'u)x + (f - f'u)y + c - c'u = 0.$$

Eliminating x and y, we get the determinant

$$\begin{vmatrix} a - a'u & h - h'u & g - g'u \\ h - h'u & b - b'u & f - f'u \\ g - g'u & f - f'u & c - c'u \end{vmatrix} = 0. \qquad (4)$$

The roots of this cubic equation in u are the **maxima and minima** required.

This cubic is the same as that which gives the three systems of right lines that pass through the points of intersection of the conics $\phi_1 = 0$, $\phi_2 = 0$.[*]

---

[*] Salmon's *Conic Sections*, Art. 370.

The cubic is written by Dr. Salmon in the form

$$\Delta' u^3 + \Theta' u^2 + \Theta u + \Delta = 0, \qquad (5)$$

where Δ, Δ' denote the *discriminants* of the expressions ϕ_1 and ϕ_2, and Θ, Θ' are their two other invariants.

On the proof of the property that the coefficients are invariants compare Art. 144.

The cubic reduces to a quadratic if either the numerator or the denominator be resolvable into linear factors; for in this case either $\Delta = 0$, or $\Delta' = 0$.

If both the numerator and denominator be resolvable into factors, the cubic reduces to the linear equation

$$\Theta' u + \Theta = 0,$$

and has but one solution, as is evident also geometrically.

160. To find the Maxima or Minima Values of $x^2 + y^2 + z^2$, **where**

$$ax^2 + by^2 + cz^2 + 2hxy + 2gxz + 2fzy = 1.$$

Let $u = x^2 + y^2 + z^2$; substitute x' and y' for $\dfrac{x}{z}$ and $\dfrac{y}{z}$, and we have

$$u = \frac{x'^2 + y'^2 + 1}{ax'^2 + by'^2 + c + 2hx'y' + 2gx' + 2fy'}.$$

Accordingly the cubic of formula (4) becomes in this case

$$\begin{vmatrix} a - u^{-1} & h & g \\ h & b - u^{-1} & f \\ g & f & c - u^{-1} \end{vmatrix} = 0. \qquad (6)$$

This is the well-known cubic* for determining the axes of a surface of the second degree in terms of the coefficients in its equation: when expanded it becomes

$$u^{-3} - (a + b + c)u^{-2} + (ab + bc + ac - f^2 - g^2 - h^2)u^{-1}$$
$$+ (af^2 + bg^2 + ch^2 - abc - 2fgh) = 0.$$

---

* See Salmon's *Geometry of Three Dimensions*, 3rd ed., Art. 82.

161. Application of Lagrange's Condition.

—In applying this condition to the general case of Art. 159, we write the equation in the form

$$\phi_1 = u\phi_2,$$

from which we get, on making $\dfrac{du}{dx} = 0$, and $\dfrac{du}{dy} = 0$,

$$\frac{d^2\phi_1}{dx^2} = u\frac{d^2\phi_2}{dx^2} + \phi_2\frac{d^2u}{dx^2},$$

$$\frac{d^2\phi_1}{dxdy} = u\frac{d^2\phi_2}{dxdy} + \phi_2\frac{d^2u}{dxdy},$$

$$\frac{d^2\phi_1}{dy^2} = u\frac{d^2\phi_2}{dy^2} + \phi_2\frac{d^2u}{dy^2},$$

but $\quad \dfrac{d^2\phi_1}{dx^2} = 2a, \quad \dfrac{d^2\phi_2}{dx^2} = 2a', \quad \dfrac{d^2\phi_1}{dxdy} = 2h,$ &c.

Hence

$$\phi_2^2\left\{\frac{d^2u}{dx^2}\frac{d^2u}{dy^2} - \left(\frac{d^2u}{dxdy}\right)^2\right\} = 4\{(a - a'u)(b - b'u) - (h - h'u)^2\}.$$

Accordingly, the sign of $AC - B^2$ is the same as that of the quadratic expression

$$(ab - h^2) - (ab' + ba' - 2hh')u + (a'b' - h'^2)u^2, \qquad (7)$$

where u is a *root* of the cubic (4) or (5).

If Δ_2 represent the determinant in (4), the preceding quadratic expression may be written in the form $\dfrac{d\Delta_2}{dc}$.

Again, u_1, u_2, u_3 representing the roots of the cubic (4); a, β, those of the quadratic (7); if u_1 be a real maximum or minimum value of u, we must have $(u_1 - a)(u_1 - \beta)(a'b' - h'^2)$ a positive quantity.

Accordingly, if $a'b' - h'^2$ be positive, u_1 must not lie between the values a and β. Similarly for the other roots.

If all the roots of the cubic lie outside the limits α and β, they correspond to real maxima or minima, but any root which lies between α and β gives no maximum or minimum.

In the particular case discussed in Art. 160 the roots of the cubic (6) are all real, and those of the quadratic

$$\begin{vmatrix} a - u^{-1}, & h \\ h, & b - u^{-1} \end{vmatrix} = 0$$

are interposed between the roots of the cubic. (See Salmon's *Higher Algebra*, Art. 44). Accordingly, in this case the two extreme roots furnish real maxima and minima solutions, while the intermediate root gives *neither*. This agrees with what might have been anticipated from the properties of the Ellipsoid; viz., the axes a and c are real maximum and minimum distances from the centre to the surface, while the mean axis b is neither.

It would be unsuited to the elementary nature of this treatise to enter into further details on the subject here.

162. Maxima or Minima of Functions of three Variables.—Next, let $u = \phi(x, y, z)$, and suppose x_0, y_0, z_0 to be values of x, y, z, which render u a maximum or a minimum; then, if x, y, z be *independent* of each other, by the same reasoning as before, it is obvious that x_0, y_0, z_0 must satisfy the three equations

$$\frac{du}{dx} = 0, \quad \frac{du}{dy} = 0, \quad \frac{du}{dz} = 0;$$

omitting the case of infinite values.

Accordingly we must have

$$\phi(x_0 + h, y_0 + k, z_0 + l) - \phi(x_0, y_0, z_0) = A\frac{h^2}{1 \cdot 2} + B\frac{k^2}{1 \cdot 2} + C\frac{l^2}{1 \cdot 2}$$

$$+ Fkl + Ghl + Hhk + \&c.$$

where A, B, C, F, G, H, are the values that

$$\frac{d^2u}{dx^2}, \quad \frac{d^2u}{dy^2}, \quad \frac{d^2u}{dz^2}, \quad \frac{d^2u}{dydz}, \quad \frac{d^2u}{dxdz}, \quad \frac{d^2u}{dxdy}$$

respectively assume when x_0, y_0, z_0 are substituted for x, y, z in them.

Now, in this, as in the case of two independent variables, it is necessary for a real maximum or minimum value that the preceding quadratic function should be either always positive or always negative for all small real values of h, k, and l.

Substituting al for h, and βl for k, and suppressing the positive factor l^2, the expression becomes

$$A a^2 + B \beta^2 + C + 2F\beta + 2Ga + 2Ha\beta, \qquad (8)$$

or $\quad A\left[a^2 + 2a\dfrac{(H\beta + G)}{A}\right] + B\beta^2 + 2F\beta + C.$

Completing the square in the first term, and multiplying by A, we get

$$(Aa + H\beta + G)^2 + (AB - H^2)\beta^2 + 2(AF - GH)\beta + (AC - G^2).$$

Moreover, since the first term is a perfect square, in order that the expression should preserve the same sign, it is necessary that the quadratic

$$(AB - H^2)\beta^2 + 2(AF - CH)\beta + AC - G^2$$

should be positive for all values of β: hence we must have

$$AB - H^2 > 0, \qquad (9)$$

and $\quad (AB - H^2)(AC - G^2) > (AF - GH)^2,$

or $\quad A(ABC + 2FGH - AF^2 - BG^2 - CH^2) > 0, \qquad (10)$

i.e. A and Δ must have the same sign, Δ denoting the *discriminant* of the quadratic expression (8), as before.

Accordingly, the conditions (9) and (10) are necessary that x_0, y_0, z_0 should correspond to a real maximum or minimum value of the function u.

When these conditions are fulfilled, if the sign of A be positive, the function in (8) is also positive, and the solution is a minimum; if A be negative, the solution is a maximum.

163. Maxima and Minima for any number of Variables.—The preceding theory admits of easy extension to functions of any number of independent variables. The values which give maxima and minima in that case are got by equating to zero the partial derived functions for each variable separately, and the quadratic function in the expansion must preserve the same sign for all values; i.e. it must be equivalent to a number of squares, multiplied by constant coefficients, having each the same sign.

The number of *independent* conditions to be fulfilled in the case of n independent variables is simply $n-1$, and not 2^n-1, as stated by some writers on the Differential Calculus. A simple and general investigation of these conditions will be given in a note at the end of the Book.

164. To investigate the Maximum or Minimum Value of the Expression

$$ax^2 + by^2 + cz^2 + 2hxy + 2gzx + 2fyz + 2px + 2qy + 2rz + d.$$

Let u denote the function in question, then for its maximum or minimum value we have

$$\frac{du}{dx} = 2(ax + hy + gz + p) = 0,$$

$$\frac{du}{dy} = 2(hx + by + fz + q) = 0,$$

$$\frac{du}{dz} = 2(gx + fy + cz + r) = 0;$$

hence, adopting the method of Art. 158, we get

$$u = px + qy + rz + d.$$

Eliminating x, y, z between these four equations, we obtain

$$\begin{vmatrix} a & h & g & p \\ h & b & f & q \\ g & f & c & r \\ p & q & r & d \end{vmatrix} = u \begin{vmatrix} a & h & g \\ h & b & f \\ g & f & c \end{vmatrix}.$$

Again, since $\dfrac{d^2u}{dx^2} = 2a$, $\dfrac{d^2u}{dy^2} = 2b$, &c.,

the result is neither a maximum nor a minimum unless

$$\begin{vmatrix} a & h \\ h & b \end{vmatrix} \text{ is positive, and } \begin{vmatrix} a & h & g \\ h & b & f \\ g & f & c \end{vmatrix} \text{ has the same sign as } a.$$

The student who is acquainted with the theory of surfaces of the second degree will find no difficulty in giving the geometrical interpretation of the preceding result.

165. To find a point such that the sum of the squares of its distances from n given points shall be a Minimum.—Let (a, b, c), (a', b', c'), &c., be the co-ordinates of the given points referred to rectangular axes; x, y, z, the co-ordinates of the required point; then

$$(x - a)^2 + (y - b)^2 + (z - c)^2$$

is equal to the square of the distance between the points (a, b, c), and (x, y, z).

Hence

$$u = (x - a)^2 + (y - b)^2 + (z - c)^2 + (x - a')^2 + (y - b')^2 + (z - c)'^2$$
$$+ \&c. = \Sigma(x - a)^2 + \Sigma(y - b)^2 + \Sigma(z - c)^2,$$

where the summation is extended to each of the n points. For the maximum or minimum value, we have

$$\frac{du}{dx} = 2\Sigma(x - a) = 2nx - 2\Sigma a = 0,$$

$$\frac{du}{dy} = 2\Sigma(y - b) = 2ny - 2\Sigma b = 0,$$

$$\frac{du}{dz} = 2\Sigma(z - c) = 2nz - 2\Sigma c = 0;$$

$$\therefore x_0 = \frac{\Sigma a}{n}, \quad y_0 = \frac{\Sigma b}{n}, \quad z_0 = \frac{\Sigma c}{n};$$

i.e. x_0, y_0, z_0 are the co-ordinates of the centre of mean position of the given points. This is an extension of the result established in Art. 157.

Again $\quad \dfrac{d^2u}{dx^2} = 2n, \quad \dfrac{d^2u}{dy^2} = 2n, \quad \dfrac{d^2u}{dz^2} = 2n, \quad \dfrac{d^2u}{dxdy} = 0,$ &c.

The expressions (10) and (11) are both positive in this case, and hence the solution is a minimum.

It may be observed with reference to examples of maxima and minima, that in most cases the circumstances of the problem indicate whether the solution is a maximum, a minimum, or neither, and accordingly enable us to dispense with the labour of investigating Lagrange's conditions.

EXAMPLES.

Find the maximum and minimum values, if any such exist, of

1. $$\frac{ax + by + c}{x^2 + y^2 + 1}.$$ Ans. $\dfrac{c \pm \sqrt{a^2 + b^2 + c^2}}{2}.$

2. $$\frac{ax + by + c}{\sqrt{x^2 + y^2 + 1}}.$$,, $\pm \sqrt{a^2 + b^2 + c^2}.$

3. $x^4 + y^4 - x^2 + xy - y^2.$

    (α). $x = 0, \; y = 0,$ a maximum.

    (β). $x = y = \pm \dfrac{1}{2},$ a minimum.

    (γ). $x = -y = \pm \dfrac{\sqrt{3}}{2},$ a minimum.

4. $ax^2 + bxy + dz^2 + lxz + myz.$

    $x = y = z = 0,$ neither a maximum nor a minimum.

5. If $u = ax^3y^2 - x^4y^2 - x^3y^3$, prove that $x = \dfrac{a}{2}, \; y = \dfrac{a}{3}$ makes u a maximum.

6. Prove that the value of the minimum found in Art. 165 is the $\dfrac{1}{n}$-th part of the sum of the squares of the mutual distances between the n points, taken two and two.

7. Find the maximum value of

$$(ax + by + cz) e^{-a^2x^2 - \beta^2 y^2 - \gamma^2 z^2}. \quad \text{Ans. } \sqrt{\frac{1}{2e}\left(\frac{a^2}{\alpha^2} + \frac{b^2}{\beta^2} + \frac{c^2}{\gamma^2}\right)}.$$

8. Find the values of x and y for which the expression

$$(a_1 x + b_1 y + c_1)^2 + (a_2 x + b_2 y + c_2)^2 + \ldots + (a_n x + b_n y + c_n)^2$$

becomes a minimum.

CHAPTER XI.

METHOD OF UNDETERMINED MULTIPLIERS APPLIED TO THE INVESTIGATION OF MAXIMA AND MINIMA IN IMPLICIT FUNCTIONS.

166. Method of Undetermined Multipliers.—In many cases of maxima and minima the variables which enter into the function are not independent of one another, but are connected by certain equations of condition.

The most convenient process to adopt in such cases is what is styled the method of undetermined* multipliers. We shall illustrate this process by considering the case of a function of four variables which are connected by two equations of condition.

Thus, let
$$u = \phi(x_1, x_2, x_3, x_4),$$
where x_1, x_2, x_3, x_4 are connected by the equations

$$F_1(x_1, x_2, x_3, x_4) = 0, \quad F_2(x_1, x_2, x_3, x_4) = 0. \tag{1}$$

The condition for a maximum or a minimum value of u evidently requires the equation

$$\frac{d\phi}{dx_1} dx_1 + \frac{d\phi}{dx_2} dx_2 + \frac{d\phi}{dx_3} dx_3 + \frac{d\phi}{dx_4} dx_4 = 0.$$

Moreover, the differentials are also connected by the relations

$$\frac{dF_1}{dx_1} dx_1 + \frac{dF_1}{dx_2} dx_2 + \frac{dF_1}{dx_3} dx_3 + \frac{dF_1}{dx_4} dx_4 = 0,$$

$$\frac{dF_2}{dx_1} dx_1 + \frac{dF_2}{dx_2} dx_2 + \frac{dF_2}{dx_3} dx_3 + \frac{dF_2}{dx_4} dx_4 = 0.$$

Multiplying the first of the two latter equations by the arbitrary

---

* This method is also due to Lagrange. See *Méc. Anal.*, tome 1., p. 74.

Method of Undetermined Multipliers. 205

quantity λ_1, the other by λ_2, and adding their sum to the preceding equation, we get

$$\left(\frac{d\phi}{dx_1} + \lambda_1 \frac{dF_1}{dx_1} + \lambda_2 \frac{dF_2}{dx_1}\right)dx_1 + \left(\frac{d\phi}{dx_2} + \lambda_1 \frac{dF_1}{dx_2} + \lambda_2 \frac{dF_2}{dx_2}\right)dx_2$$

$$+ \left(\frac{d\phi}{dx_3} + \lambda_1 \frac{dF_1}{dx_3} + \lambda_2 \frac{dF_2}{dx_3}\right)dx_3 + \left(\frac{d\phi}{dx_4} + \lambda_1 \frac{dF_1}{dx_4} + \lambda_2 \frac{dF_2}{dx_4}\right)dx_4 = 0.$$

As λ_1, λ_2 are completely at our disposal, we may suppose them determined so as to make the coefficients of dx_1 and dx_2 vanish. Then we shall have

$$\left(\frac{d\phi}{dx_3} + \lambda_1 \frac{dF_1}{dx_3} + \lambda_2 \frac{dF_2}{dx_3}\right)dx_3 + \left(\frac{d\phi}{dx_4} + \lambda_1 \frac{dF_1}{dx_4} + \lambda_2 \frac{dF_2}{dx_4}\right)dx_4 = 0.$$

Again, since we may regard x_3, x_4 as *independent variables*, and x_1, x_2 as dependent on them in consequence of the equations (1), it follows that the coefficients of dx_3 and dx_4 in the last equation must be separately zero, for a maximum or a minimum; consequently, we must have

$$\frac{d\phi}{dx_3} + \lambda_1 \frac{dF_1}{dx_3} + \lambda_2 \frac{dF_2}{dx_3} = 0,$$

$$\frac{d\phi}{dx_4} + \lambda_1 \frac{dF_1}{dx_4} + \lambda_2 \frac{dF_2}{dx_4} = 0.$$

These, along with equations (1) and

$$\frac{d\phi}{dx_1} + \lambda_1 \frac{dF_1}{dx_1} + \lambda_2 \frac{dF_2}{dx_1} = 0,$$

$$\frac{d\phi}{dx_2} + \lambda_1 \frac{dF_1}{dx_2} + \lambda_2 \frac{dF_2}{dx_2} = 0,$$

are theoretically sufficient to determine the six unknown quantities, x_1, x_2, x_3, x_4, λ_1, λ_2; and thus to furnish a solution of the problem in general.

This method is especially applicable when the functions F_1, F_2, &c., are homogeneous; for if we multiply the preceding differential equations by x_1, x_2, x_3, x_4, respectively, and add, we can often find the result with facility by aid of Euler's Theorem of Art. 103.

There is no difficulty in extending the method of undetermined multipliers to a function of n variables, x_1, x_2, x_3, ... x_n, the variables being connected by m equations of condition.

$$F_1 = 0, \; F_2 = 0, \; F_3 = 0, \; \ldots \; F_m = 0,$$

m being less than n; for if we differentiate as before, and multiply the differentials of the equations of condition by the arbitrary multipliers, λ_1, λ_2, ... λ_m respectively; by the same method of reasoning as that given above, we shall have the n following equations,

$$\frac{d\phi}{dx_1} + \lambda_1 \frac{dF_1}{dx_1} + \ldots + \lambda_m \frac{dF_m}{dx_1} = 0,$$

$$\frac{d\phi}{dx_2} + \lambda_1 \frac{dF_1}{dx_2} + \ldots + \lambda_m \frac{dF_m}{dx_2} = 0,$$

$$\cdot \quad \cdot \quad \cdot \quad \cdot \quad \cdot \quad \cdot$$

$$\frac{d\phi}{dx_n} + \lambda_1 \frac{dF_1}{dx_n} + \ldots + \lambda_m \frac{dF_m}{dx_n} = 0.$$

These, combined with the m equations of condition, are theoretically sufficient for the determination of the $m + n$ unknown quantities

$$x_1, x_2, \ldots x_n, \lambda_1, \lambda_2, \ldots \lambda_m.$$

Examples.

1. To find the triangle of maximum area inscribed in a given circle.

Let R denote the radius of the circle, A, B, C, the angles of an inscribed triangle, u its area; then

$$u = \frac{abc}{4R} = 2R^2 \sin A \sin B \sin C.$$

Also, $\quad A + B + C = 180°; \; \therefore \; dA + dB + dC = 0;$

and, taking logarithmic differentials, we get

$$\cot A \, dA + \cot B \, dB + \cot C \, dC = 0,$$

and consequently

$$\tan A = \tan B = \tan C; \text{ hence } A = B = C = 60°;$$

and therefore the triangle is equilateral.

2. Find a point such that the sum of the squares of the perpendiculars drawn from it to the sides of a given triangle shall be a minimum.

Let x, y, z denote the perpendiculars: a, b, c the sides of the triangle; then

$$u = x^2 + y^2 + z^2 \text{ is to be a minimum};$$

also $\quad ax + by + cz =$ double the area of a triangle $= 2\Delta$ (suppose);

$$\therefore xdx + ydy + zdz = 0, \; adx + bdy + cdz = 0,$$

$\therefore x = \lambda a, \; y = \lambda b, \; z = \lambda c$: multiplying these equations by a, b, c, respectively, and adding, we obtain

$$ax + by + cz = \lambda(a^2 + b^2 + c^2), \text{ or } \lambda = \frac{2\Delta}{a^2 + b^2 + c^2};$$

$$\therefore x = \frac{2\Delta a}{a^2 + b^2 + c^2}, \quad y = \frac{2\Delta b}{a^2 + b^2 + c^2}, \quad z = \frac{2\Delta c}{a^2 + b^2 + c^2},$$

which determine the position of the point. The minimum sum is obviously

$$\frac{4\Delta^2}{a^2 + b^2 + c^2}.$$

3. Similarly, to find a point such that the sum of the squares of its distances from four given planes shall be a minimum. Suppose A, B, C, D to represent the areas of the faces of the tetrahedron formed by the four planes; x, y, z, w, the perpendiculars on these faces respectively; then, as in the preceding example, we have

$$Ax + By + Cz + Dw = \text{three times the volume of the tetrahedron} = 3V \text{ (suppose)},$$

and $\quad u = x^2 + y^2 + z^2 + w^2$, a minimum;

$$\therefore xdx + ydy + zdz + wdw = 0,$$

$$Adx + Bdy + Cdz + Ddw = 0;$$

hence $\quad x = \lambda A, \; y = \lambda B, \; z = \lambda C, \; w = \lambda D;$

and proceeding as before, we get $u = \dfrac{9V^2}{A^2 + B^2 + C^2 + D^2}.$

4. To prove that of all rectangular parallelepipeds of the same volume the cube has the least surface.

Let x, y, z represent the lengths of the edges of the parallelepiped; then, if A denote the given volume, we have

$$xyz = A, \text{ and } xy + xz + yz \text{ a minimum};$$

$$\therefore yzdx + xzdy + xydz = 0,$$

$$(y + z)dx + (x + z)dy + (x + y)dz = 0;$$

hence $\quad yz = \lambda(y + z), \; xz = \lambda(x + z), \; xy = \lambda(x + y):$

from which it appears immediately that $x = y = z$.

167. To find the Maximum and Minimum Values of

$$ax^2 + by^2 + cz^2 + 2hxy + 2gzx + 2fyz,$$

where the variables are connected by the equations

$$Lx + My + Nz = 0, \text{ and } x^2 + y^2 + z^2 = 1.$$

In this case we get the following equations:

$$ax + hy + gz + \lambda_1 L + \lambda_2 x = 0,$$

$$hx + by + fz + \lambda_1 M + \lambda_2 y = 0,$$

$$gx + fy + cz + \lambda_1 N + \lambda_2 z = 0.$$

Multiply the first by x, the second by y, the third by z, and add; then

$$u + \lambda_2 = 0, \text{ or } \lambda_2 = -u.$$

Hence
$$(a - u)x + hy + gz + \lambda_1 L = 0,$$

$$hx + (b - u)y + fz + \lambda_1 M = 0,$$

$$gx + fy + (c - u)z + \lambda_1 N = 0,$$

$$Lx + My + Nz \qquad\qquad = 0:$$

eliminating x, y, z and λ_1, we get the determinant equation

$$\begin{vmatrix} a - u, & h, & g, & L \\ h, & b - u, & f, & M \\ g, & f, & c - u, & N \\ L, & M, & N, & 0 \end{vmatrix} = 0. \qquad (2)$$

The roots of this quadratic determine the maximum and minimum values of u.

The preceding result enables us to determine the principal radii of curvature at a given point on a surface whose equation is given in rectangular co-ordinates.

Again, the term independent of u in this determinant is evidently

$$\begin{vmatrix} a, & h, & g, & L \\ h, & b, & f, & M \\ g, & f, & c, & N \\ L, & M, & N, & 0 \end{vmatrix},$$

and the coefficient of u^2 is $L^2 + M^2 + N^2$. Accordingly, the product of the roots of the quadratic (2) is equal to the fraction whose numerator is the latter determinant, and denominator $L^2 + M^2 + N^2$. From this can be immediately deduced an expression for the *measure of curvature** at any point on a surface.

---

* Salmon's *Geometry of Three Dimensions*, Art. 295.

Examples.

1. Find the minimum value of
$$x_1^2 + x_2^2 + x_3^2 + \ldots + x_n^2,$$
where $x_1, x_2, \ldots x_n$ are subject to the condition
$$a_1 x_1 + a_2 x_2 + \ldots + a_n x_n = k. \qquad Ans. \quad \frac{k^2}{a_1^2 + a_2^2 \ldots + a_n^2}.$$

2. Find the maximum value of
$$x^p y^q z^r,$$
where the variables are subject to the condition
$$ax + by + cz = p + q + r. \qquad Ans. \quad \left(\frac{p}{a}\right)^p \left(\frac{q}{b}\right)^q \left(\frac{r}{c}\right)^r.$$

3. If $\tan \frac{\theta}{2} \tan \frac{\phi}{2} = m$, find when $\sin \theta - m \sin \phi$ is a maximum.

4. Find the maximum value of $(x + 1)(y + 1)(z + 1)$ where $a^x b^y c^z = A$.
$$Ans. \quad \frac{\{\log(Aabc)\}^3}{27 \log a \cdot \log b \cdot \log c}.$$

5. Find the volume of the greatest rectangular parallelepiped inscribed in the ellipsoid whose equation is
$$\frac{x^2}{a^2} + \frac{y^2}{b^2} + \frac{z^2}{c^2} = 1. \qquad Ans. \quad \frac{8\,abc}{3\sqrt{3}}.$$

6. Find the maximum or the minimum values of u, being given that
$$u = a^2 x^2 + b^2 y^2 + c^2 z^2, \quad x^2 + y^2 + z^2 = 1, \quad \text{and } lx + my + nz = 0.$$
Proceeding by the method of Art. 167, we get
$$a^2 x + \lambda x + \mu l = 0, \quad b^2 y + \lambda y + \mu m = 0, \quad c^2 z + \lambda z + \mu n = 0.$$
Again, multiplying by x, y, z, respectively, and adding, we get $\lambda = -u$.
$$\therefore (u - a^2) x = \mu l, \quad (u - b^2) y = \mu m, \quad (u - c^2) z = \mu n.$$
Hence, the required values of u are the roots of the quadratic
$$\frac{l^2}{u - a^2} + \frac{m^2}{u - b^2} + \frac{n^2}{u - c^2} = 0.$$

Examples.

7. Given $\dfrac{x^2}{a^2} + \dfrac{y^2}{b^2} + \dfrac{z^2}{c^2} = 1$, and $lx + my + nz = 0$, find when $x^2 + y^2 + z^2$ is a maximum or minimum. Proceeding, as in the last example, we get the quadratic

$$\frac{a^2 l^2}{u - a^2} + \frac{b^2 m^2}{u - b^2} + \frac{c^2 n^2}{u - c^2} = 0.$$

This question can be at once reduced to the last by substituting in our equations ax, by, and cz, instead of x, y, z.

8. Of all triangular pyramids having a given triangle for base, and a given altitude above that base, find that whose surface is least.

Ans. Value of minimum surface is $\dfrac{(a + b + c)}{2} \sqrt{r^2 + p^2}$, where a, b, c represent the sides of the triangular base; r, the radius of its inscribed circle; and p, the given altitude.

9. Divide the quadrant of a circle into three parts, such that the sum of the products of the sines of every two shall be a maximum or a minimum; and determine which it is.

10. Of all polygons of a given number of sides circumscribed to a circle, the regular polygon is of minimum area? For, let $\phi_1, \phi_2, \ldots \phi_n$ be the external angles of the polygon, then the area can be easily seen to be in general

$$r^2 \left(\tan \frac{\phi_1}{2} + \tan \frac{\phi_2}{2} + \ldots + \tan \frac{\phi_n}{2} \right),$$

where $\phi_1 + \phi_2 \ldots + \phi_n = 2\pi.$

Hence, for a minimum, $\phi_1 = \phi_2 = \phi_3 = \ldots = \phi_n.$

11. Of all polygons of a given number of sides circumscribed to any closed oval curve which has no singular points, that which has the minimum area touches the curve at the middle point of each of the sides.

12. Given the ratio $\sin \phi : \sin \psi$, and the angle θ, find when the ratio $\sin (\phi + \theta) : \sin (\psi + \theta)$ is a maximum or a minimum. *Ans.* $\phi + \psi = \theta.$

13. Required the dimensions of an open cylindrical vessel of given capacity, so that the smallest possible quantity of material shall be employed in its construction, the thickness of the base and sides being given.
Ans. Its altitude must be equal to the radius of its base.

14. Show how to determine the maximum and minimum values of $x^2 + y^2 + z^2$ subject to the conditions

$$(x^2 + y^2 + z^2)^2 = a^2 x^2 + b^2 y^2 + c^2 z^2,$$

$$lx + my + nz = 0.$$

CHAPTER XII.

TANGENTS AND NORMALS TO CURVES.

168. Equation of the Tangent.—If (x, y), (x_1, y_1), be the co-ordinates of any two points, P, Q, taken on a curve, and if (X, Y) be any point on the line which joins P and Q; then the equation of the line PQ is

$$Y - y = (X - x)\frac{y_1 - y}{x_1 - x},$$

in which X and Y represent the current co-ordinates.

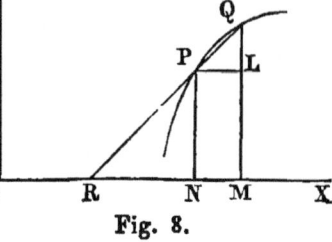

Fig. 8.

If now the point Q be taken infinitely near to P, the line PQ becomes the tangent at the point P, and, as in Art. 10, we have for its equation

$$Y - y = (X - x)\frac{dy}{dx}, \qquad (1)$$

where X, Y are the co-ordinates of any point on the line, and x, y those of its point of contact.

For example, to find the equation of the tangent to the curve

$$x^n y^m = a^{n+m}.$$

Taking the logarithmic differentials of both sides, we get

$$\frac{n}{x} + \frac{m}{y}\frac{dy}{dx} = 0; \quad \therefore \frac{dy}{dx} = -\frac{ny}{mx},$$

and the equation of the tangent becomes

$$\frac{nX}{x} + \frac{mY}{y} = m + n.$$

If we make $X = 0$, and $Y = 0$, separately, we get $\dfrac{m+n}{n}x$ and $\dfrac{m+n}{n}y$ for the lengths of the intercepts made by the tangent on the axes of x and y, respectively. This result furnishes an easy geometrical method of drawing the tangent at any point on a curve of this class.

If $m = 1$, $n = 1$, the preceding equation represents a hyperbola; if $m = 2$, and $n = -1$, it represents a parabola.

169. If the equation of the curve be of the form $f(x, y) = 0$, and if $f(x, y)$ be denoted by u, we have from Art. 100,

$$\frac{dy}{dx} = -\frac{\dfrac{du}{dx}}{\dfrac{du}{dy}},$$

and hence the equation of the tangent becomes

$$(X - x)\frac{du}{dx} + (Y - y)\frac{du}{dy} = 0. \qquad (2)$$

The points on the curve at which the tangents are parallel to the axis of x must satisfy the equation $\dfrac{du}{dx} = 0$; they are accordingly given by the intersection of the curve, $u = 0$, with the curve whose equation is $\dfrac{du}{dx} = 0$. The y co-ordinates at such points are evidently in general either maxima or minima.

Similar remarks apply to the points at which the tangents are parallel to the axis of y.

To find the tangents parallel to the line $y = mx + n$. The points of contact must evidently satisfy

$$\frac{du}{dx} + m\frac{du}{dy} = 0.$$

The points of intersection of the curve represented by

this equation with the given curve are the points of contact of the system of parallel tangents in question.

The results in this and the preceding Article evidently apply to oblique as well as to rectangular axes.

Examples.

1. To find the equation of the tangent to the ellipse

$$\frac{x^2}{a^2} + \frac{y^2}{b^2} = 1.$$

Here
$$\frac{du}{dx} = \frac{2x}{a^2}, \quad \frac{du}{dy} = \frac{2y}{b^2},$$

and the required equation is

$$\frac{x}{a^2}(X - x) + \frac{y}{b^2}(Y - y) = 0,$$

or
$$\frac{xX}{a^2} + \frac{yY}{b^2} = \frac{x^2}{a^2} + \frac{y^2}{b^2} = 1.$$

2. Find the equation of the tangent at any point on the curve

$$\frac{x^m}{a^m} + \frac{y^m}{b^m} = 1. \qquad Ans. \quad \frac{Xx^{m-1}}{a^m} + \frac{Yy^{m-1}}{b^m} = 1.$$

3. If two curves, whose equations are denoted by $u = 0$, $u' = 0$, intersect in a point (x, y), and if ω be their angle of intersection, prove that

$$\tan \omega = \frac{\dfrac{du}{dx}\dfrac{du'}{dy} - \dfrac{du'}{dx}\dfrac{du}{dy}}{\dfrac{du}{dx}\dfrac{du'}{dx} + \dfrac{du}{dy}\dfrac{du'}{dy}}.$$

4. Hence, if the curves intersect at right angles, we must have

$$\frac{du}{dx}\frac{du'}{dx} + \frac{du}{dy}\frac{du'}{dy} = 0.$$

5. Apply this to find the condition that the curves

$$\frac{x^2}{a^2} + \frac{y^2}{b^2} = 1, \quad \frac{x^2}{a'^2} + \frac{y^2}{b'^2} = 1$$

should intersect at right angles. $\qquad Ans.\ a^2 - b^2 = a'^2 - b'^2.$

170. Equation of Normal.—Since the normal at any point on a curve is perpendicular to the tangent, its equation, when the co-ordinate axes are rectangular, is

$$(Y - y)\frac{dy}{dx} + X - x = 0,$$

or
$$\frac{du}{dx}(Y - y) = \frac{du}{dy}(X - x). \qquad (3)$$

The points at which normals are parallel to the line $y = mx + n$ are given by aid of the equation of the curve $u = 0$ along with the equation

$$\frac{du}{dy} = m\frac{du}{dx}.$$

EXAMPLES.

1. Find the equation of the normal at any point (x, y) on the ellipse

$$\frac{x^2}{a^2} + \frac{y^2}{b^2} = 1. \qquad \text{Ans. } \frac{a^2 X}{x} - \frac{b^2 Y}{y} = a^2 - b^2.$$

2. Find the equation of the normal at any point on the curve

$$y^m = ax^n. \qquad \text{Ans. } nYy + mXx = ny^2 + mx^2.$$

171. Subtangent and Subnormal.—In the accompanying figure, let PT represent the tangent at the point P, PN the normal; OM, PM the co-ordinates at P; then the lines TM and MN are called the subtangent and subnormal corresponding to the point P.

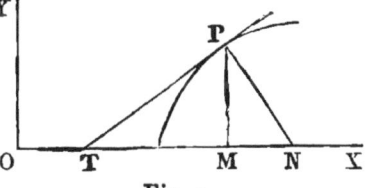

Fig. 9.

To find the expressions for their lengths, let $\phi = \angle PTM$, then

$$\frac{PM}{TM} = \tan\phi = \frac{dy}{dx}; \quad \therefore TM = \frac{y}{\frac{dy}{dx}},$$

$$\frac{MN}{PM} = \tan\phi = \frac{dy}{dx}, \qquad MN = y\frac{dy}{dx}.$$

The lengths of PT and PN are sometimes called the lengths of the tangent and the normal at P: it is easily seen that

$$PN = y\sqrt{1 + \left(\frac{dy}{dx}\right)^2}, \quad PT = \frac{y\sqrt{1 + \left(\frac{dy}{dx}\right)^2}}{\frac{dy}{dx}}.$$

EXAMPLES.

1. To find the length of the subnormal in the ellipse

$$\frac{x^2}{a^2} + \frac{y^2}{b^2} = 1.$$

Here
$$y\frac{dy}{dx} = -\frac{b^2}{a^2}x;$$

the negative sign signifies that MN is measured from M in the negative direction along the axis of x, i.e. the point N lies between M and the centre O; as is also evident from the shape of the curve.

2. Prove that the subtangent in the logarithmic curve, $y = a^x$, is of constant length.

3. Prove that the subnormal in the parabola, $y^2 = 2mx$, is equal to m.

4. Find the length of the part of the normal to the catenary

$$y = \frac{a}{2}\left(e^{\frac{x}{a}} + e^{-\frac{x}{a}}\right),$$

intercepted by the axis of x. *Ans.* $\frac{y^2}{a}$.

5. Find at what point the subtangent to the curve whose equation is

$$xy^2 = a^2(a - x)$$

is a maximum. *Ans.* $x = \frac{a}{2}$, $y = a$.

172. Perpendicular on Tangent.—Let p be the length of the perpendicular from the origin on the tangent at any point on the curve

$$F(x, y) = c,$$

then the equation of the tangent may be written

$$X \cos \omega + Y \sin \omega = p,$$

where ω is the angle which the perpendicular makes with the axis of x.

Denoting $F(x, y)$ by u, and comparing this form of the equation with that in (2), and representing the common value of the fraction by λ,

we get
$$\frac{\frac{du}{dx}}{\cos \omega} = \frac{\frac{du}{dy}}{\sin \omega} = \frac{x \frac{du}{dx} + y \frac{du}{dy}}{p} = \lambda.$$

Hence
$$\lambda^2 = \left(\frac{du}{dx}\right)^2 + \left(\frac{du}{dy}\right)^2,$$

and
$$p = \frac{x \frac{du}{dx} + y \frac{du}{dy}}{\sqrt{\left(\frac{du}{dx}\right)^2 + \left(\frac{du}{dy}\right)^2}}. \tag{4}$$

Cor. If $F(x, y)$ be a homogeneous expression of the nth degree in x and y, then by Euler's formula, Art. 102, we have

$$x \frac{du}{dx} + y \frac{du}{dy} = nu = nc,$$

and the expression for the length of the perpendicular becomes in this case

$$\frac{nc}{\sqrt{\left(\frac{du}{dx}\right)^2 + \left(\frac{du}{dy}\right)^2}}.$$

173. **In the curve**

$$\frac{x^m}{a^m} + \frac{y^m}{b^m} = 1$$

to prove that

$$p^{\frac{m}{m-1}} = (a \cos \omega)^{\frac{m}{m-1}} + (b \sin \omega)^{\frac{m}{m-1}}. \tag{5}$$

By Ex. 2, Art. 169, the equation of the tangent is

$$\frac{Xx^{m-1}}{a^m} + \frac{Yy^{m-1}}{b^m} = 1;$$

comparing this with the form

$$X \cos \omega + Y \sin \omega = p,$$

we get $\quad \dfrac{\cos \omega}{p} = \dfrac{x^{m-1}}{a^m}, \qquad \dfrac{\sin \omega}{p} = \dfrac{y^{m-1}}{b^m},$

or $\quad \left(\dfrac{a \cos \omega}{p}\right)^{\frac{1}{m-1}} = \dfrac{x}{a}, \qquad \left(\dfrac{b \sin \omega}{p}\right)^{\frac{1}{m-1}} = \dfrac{y}{b}.$

Hence, substituting in the equation of the curve, we obtain the result required.

174. **Locus of Foot of Perpendicular for the same Curve.**—Let X, Y be the co-ordinates of the point in question, and we have, evidently, $\cos \omega = \dfrac{X}{p}$, $\sin \omega = \dfrac{Y}{p}$: substituting these values for $\cos \omega$ and $\sin \omega$ in (5), it becomes

$$(X^2 + Y^2)^{\frac{m}{m-1}} = (aX)^{\frac{m}{m-1}} + (bY)^{\frac{m}{m-1}},$$

since $p^2 = X^2 + Y^2$.

175. **Another Form of the Equation to a Tangent.**—If the equation of a curve of the nth degree be written in the form

$$\phi(x, y) = u_n + u_{n-1} + u_{n-2} + \ldots + u_2 + u_1 + u_0 = 0,$$

where u_n denotes the homogeneous part of the nth degree in the equation, u_{n-1} that of the $(n-1)^{th}$, &c.; then, by Cor. Art. 103, we have

$$x\frac{d\phi}{dx} + y\frac{d\phi}{dy} = -\{u_{n-1} + 2u_{n-2} + \&c. \ldots + nu_0\}.$$

Hence the equation of the tangent in Art. 169 becomes

$$X\frac{d\phi}{dx} + Y\frac{d\phi}{dy} + u_{n-1} + 2u_{n-2} + \ldots + nu_0 = 0; \qquad (6)$$

an equation of the $(n-1)^{th}$ degree in x and y.

176. Number of Tangents from an External Point.—To find the number of tangents which can be drawn to a curve of the nth degree from a point (a, β), we substitute a for X, and β for Y in (6), and it becomes

$$a\frac{d\phi}{dx} + \beta\frac{d\phi}{dy} + u_{n-1} + 2u_{n-2} + \ldots + nu_0 = 0. \qquad (7)$$

This represents a curve of the $(n-1)^{th}$ degree in x and y, and the points of its intersection with the given curve are the points of contact of all the tangents which can be drawn from the point (a, β) to the curve. Moreover, as two curves of the degrees n and $n-1$ intersect in general in $n(n-1)$ points, real or imaginary (Salmon's *Conic Sections*, Art. 214), it follows that there can *in general* be $n(n-1)$ real or imaginary tangents drawn from an external point to a curve of the nth degree.

If the curve be of the second degree, equation (7) becomes

$$a\frac{d\phi}{dx} + \beta\frac{d\phi}{dy} + u_1 + 2u_0 = 0,$$

an equation of the first degree, which evidently represents the polar of (a, β) with respect to the conic.

In the curve of the third degree

$$u_3 + u_2 + u_1 + u_0 = 0,$$

equation (7) becomes

$$a\frac{d\phi}{dx} + \beta\frac{d\phi}{dy} + u_2 + 2u_1 + 3u_0 = 0,$$

which represents a conic that passes through the points of contact of the tangents to the curve from the point (a, β).

This conic is called the *polar conic* of the point. For the origin it becomes

$$u_2 + 2u_1 + 3u_0 = 0.$$

177. Number of Normals which pass through a Given Point.—If a normal pass through the point (a, β), we must have from (3),

$$(a - x) \frac{du}{dy} = (\beta - y) \frac{du}{dx}.$$

This represents a curve of the nth degree, which intersects the given curve *in general* in n^2 points, real or imaginary, the normals at which all pass through the point (a, β).

For example, the points on the ellipse

$$\frac{x^2}{a^2} + \frac{y^2}{b^2} = 1,$$

at which the normals pass through a given point (a, β), are determined by the intersection of the ellipse with the hyperbola

$$xy(a^2 - b^2) = a^2 ay - b^2 \beta x.$$

For the modification in the results of this and the preceding article arising from the existence of *singular* points on the curve, the student is referred to Salmon's *Higher Plane Curves*, Arts. 66, 67, 111.

178. Differential of the Arc of a Plane Curve. Direction of the Tangent.—If the length of the arc of a curve, measured from a fixed point A on it, be denoted by s, then an infinitely small portion of it is represented by ds. Again, if ϕ' represent the angle QPL (fig. 8), we have

$$\cos \phi' = \frac{PL}{PQ}, \text{ and } \sin \phi' = \frac{QL}{PQ};$$

but in the limit, $PL = dx$, $QL = dy$, and $PQ = ds$,* and also ϕ' becomes PTX, or ϕ (fig. 9).

---

* In Art. 37 it has been proved that the difference between the length of an infinitely small arc and its chord is an infinitely small quantity of the *second* order in comparison with the length of the chord; i.e. $\dfrac{\text{arc } PQ - PQ}{PQ}$ is infinitely small of the second order, and therefore this fraction vanishes in the limit. Hence $\dfrac{\text{arc } PQ}{\text{ord } PQ} = 1$, ultimately.

Hence
$$\cos\phi = \frac{dx}{ds}, \quad \sin\phi = \frac{dy}{ds}; \qquad (8)$$

squaring and adding, we get
$$\left(\frac{dx}{ds}\right)^2 + \left(\frac{dy}{ds}\right)^2 = 1. \qquad (9)$$

Hence, also, we have
$$ds^2 = dx^2 + dy^2,$$
and therefore
$$ds = \sqrt{1 + \frac{dy^2}{dx^2}}\, dx. \qquad (10)$$

On account of the importance of these results, we shall give another proof, as follows :—

Let, as before, PR be the tangent to the curve at the point P,

$OM = x$, $PM = y$,

$MN = PL = \Delta x$, $QL = \Delta y$.

$\angle PTX = \phi$, arc $PQ = \Delta s$,

Then, if the curvature of the elementary portion PQ of the curve be continuous, we have evidently the line

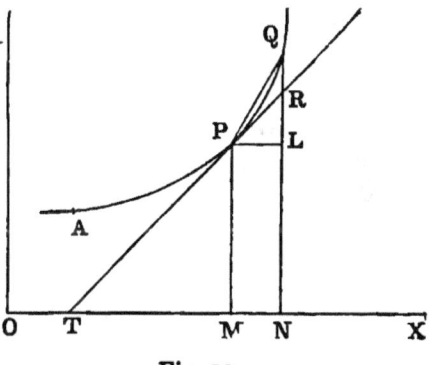

$PQ < \text{arc } PQ < PR + QR$;

Fig. 10.

or $\quad \sqrt{\Delta x^2 + \Delta y^2} < \Delta s < \Delta x \sec\phi + \Delta y - \Delta x \tan\phi$;

$$\therefore \sqrt{1 + \left(\frac{\Delta y}{\Delta x}\right)^2} < \frac{\Delta s}{\Delta x} < \sec\phi + \frac{\Delta y}{\Delta x} - \tan\phi.$$

Again, in the limit $\dfrac{\Delta y}{\Delta x} = \dfrac{dy}{dx} = \tan\phi$, and $\sqrt{1 + \left(\dfrac{\Delta y}{\Delta x}\right)^2}$

becomes $\sqrt{1 + \left(\dfrac{dy}{dx}\right)^2}$ or $\sec \phi$; accordingly each of the preceding expressions converges to the same limiting value, and we have $\dfrac{ds}{dx} = \sqrt{1 + \left(\dfrac{dy}{dx}\right)^2}$; which establishes the required result.

179. **Polar Co-ordinates.**—The position of any point in a plane is determined when its distance from a fixed point called a *pole*, and the angle which that distance makes with a fixed line, are known; these are called the polar co-ordinates of the point, and are usually denoted by the letters r and θ. The fixed line is called the *prime vector*, and r is called the *radius vector* of the point.

The equation of a curve referred to polar co-ordinates is generally written in one or other of the forms,

$$r = f(\theta), \text{ or } F(r, \theta) = 0,$$

according as r is given *explicitly or implicitly* in terms of θ. Also, if θ be positive when measured *above* the prime vector, it must be regarded as *negative* when measured *below* it.

180. **Angle between Tangent and Radius Vector.** Let O be the pole, P and Q two near points on the curve, PM a perpendicular on OQ, $OP = r$, $POX = \theta$, and ψ the angle between the tangent and radius vector. Then

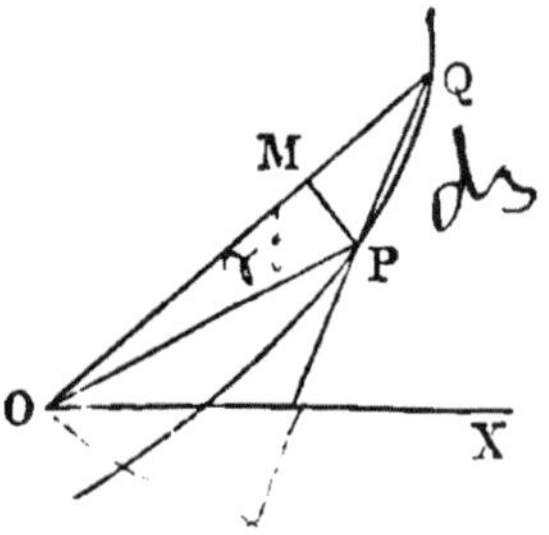

$$\tan OQP = \frac{PM}{QM}, \quad \sin OQP = \frac{PM}{PQ},$$

$\cos OQP = \dfrac{QM}{QP}$: but in the limit when Q and P coincide, the angle OQP becomes equal to ψ, and*

$$\frac{QM}{PQ} = \frac{dr}{ds}, \quad \frac{PM}{PQ} = \frac{rd\theta}{ds}, \text{ at the same time;}$$

or $\quad \cos \psi = \dfrac{dr}{ds}, \quad \sin \psi = \dfrac{rd\theta}{ds}, \quad \tan \psi = \dfrac{rd\theta}{dr}.$ \hfill (11)

Fig. 11.

---

* These results can be easily established from Art. 37.

Also,
$$\left(\frac{rd\theta}{ds}\right)^2 + \left(\frac{dr}{ds}\right)^2 = 1. \quad (12)$$

Hence, also, we can determine an expression for the differential of an arc in polar co-ordinates; for, since

$$\frac{PQ^2}{QM^2} = 1 + \frac{PM^2}{QM^2},$$

we get, on proceeding to the limit,

$$\frac{ds}{dr} = \sqrt{1 + \frac{r^2 d\theta^2}{dr^2}};$$

or
$$ds = \sqrt{1 + \frac{r^2 d\theta^2}{dr^2}}\, dr. \quad (13)$$

These results are of importance in the general theory of curves.

181. **Application to the Logarithmic Spiral.**—The curve whose equation is $r = a^\theta$ is called the logarithmic spiral. In this curve we have

$$\tan\psi = \frac{rd\theta}{dr} = \frac{1}{\log a}.$$

Accordingly, the angle between the radius vector and the tangent is constant. On account of this property the curve is also called the *equiangular spiral*.

182. **Polar Subtangent and Subnormal.**—Through the origin O let ST be drawn perpendicular to OP, meeting the tangent in T, and the normal in S. The lines OT and OS are called the polar subtangent and subnormal, for the point P. To find their values, we have

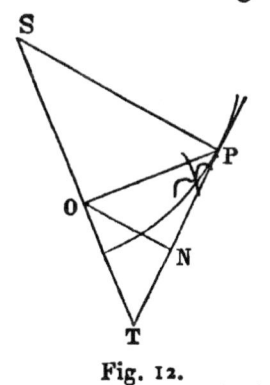

$$OT = OP \tan OPT = r\tan\psi = \frac{r^2 d\theta}{dr}.$$

$$\left. \begin{array}{l} OS = OP \tan OPS = r\cot\psi = \dfrac{dr}{d\theta}. \\[1em] \text{Also, if} \quad u = \dfrac{1}{r}, \quad OT = -\dfrac{d\theta}{du}. \end{array} \right\} \quad (14)$$

Fig. 12.

Again, if ON be drawn perpendicular to PT, we have

$$PN = OP \cos \psi = r \frac{dr}{ds}. \tag{15}$$

183. Expression for Perpendicular on Tangent.— As before, let $p = ON$, then

$$p = r \sin \psi = \frac{r^2 d\theta}{ds};$$

hence $$\frac{1}{p^2} = \frac{ds^2}{r^4 d\theta^2} = \frac{dr^2 + r^2 d\theta^2}{r^4 d\theta^2} = \frac{dr^2}{r^4 d\theta^2} + \frac{1}{r^2},$$

or $$\frac{1}{p^2} = u^2 + \left(\frac{du}{d\theta}\right)^2. \tag{16}$$

The equations in polar co-ordinates of the tangent and the normal at any point on a curve can be found without difficulty: they have, however, been omitted here, as they are of little or no practical advantage.

Examples.

1. To find the length of the perpendicular from a focus on the tangent to an ellipse.

The focal equation of the curve is

$$r = \frac{a(1 - e^2)}{1 - e \cos \theta}, \text{ or } u = \frac{1 - e \cos \theta}{a(1 - e^2)};$$

hence $$\frac{du}{d\theta} = \frac{e \sin \theta}{a(1 - e^2)};$$

$$\therefore \frac{1}{p^2} = \frac{1 + e^2 - 2e \cos \theta}{a^2(1 - e^2)^2} = \frac{1}{a^2(1 - e^2)}\left(\frac{2a}{r} - 1\right).$$

2. Prove that the polar subnormal is constant, in the curve $r = a\theta$; and the polar subtangent, in the curve $r\theta = a$.

184. Inverse Curves.—If on any radius vector OP, drawn from a fixed origin O, a point P' be taken such that the rectangle $OP \cdot OP'$ is constant, the point P' is called the inverse of the point P; and if P describe any curve, P' describes another curve called the inverse of the former.

The polar equation of the inverse is obtained immediately from that of the original curve by substituting $\dfrac{k^2}{r}$ instead of r in its equation; where k^2 is equal to the constant $OP \cdot OP'$.

Again, let P, Q be two points, and P', Q' the inverse points; then since $OP \cdot OP' = OQ \cdot OQ'$, the four points P, Q, Q', P', lie on a circle, and hence the triangles OQP and $OP'Q'$ are equiangular;

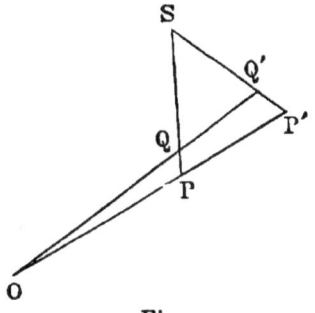

Fig. 13.

$$\therefore \frac{PQ}{P'Q'} = \frac{OP}{OQ'} = \frac{OP \cdot OQ}{OQ \cdot OQ'} = \frac{OP \cdot OQ}{k^2}. \qquad (17)$$

Again, if P, Q be infinitely near points, denoting the lengths of the corresponding elements of the curve and of its inverse by ds and ds', the preceding result becomes

$$ds = \frac{r^2}{k^2} ds'. \qquad (18)$$

185. Direction of the Tangent to an Inverse Curve.—Let the points P, Q belong to one curve, and P', Q' to its inverse; then when P and Q coincide, the lines PQ, $P'Q'$ become the tangents at the inverse points P and P': again, since the angle SPP' = the angle $SQ'Q$, it follows that the tangents at P and P' form an isosceles triangle with the line PP'.

By aid of this property the tangent at any point on a curve can be drawn, whenever that at the corresponding point of the inverse curve is known.

It follows immediately from the preceding result, that if *two curves intersect at any angle, their inverse curves intersect at the same angle.*

186. Equation to the Inverse of a Given Curve.—

Suppose the curve referred to rectangular axes drawn through the pole O, and that x and y are the co-ordinates of a point P on the curve, X and Y those of the inverse point P'; then

$$\frac{x}{X} = \frac{OP}{OP'} = \frac{OP \cdot OP'}{OP'^2} = \frac{k^2}{X^2 + Y^2}; \text{ similarly } \frac{y}{Y} = \frac{k^2}{X^2 + Y^2};$$

hence the equation of the inverse is got by substituting

$$\frac{k^2 x}{x^2 + y^2} \text{ and } \frac{k^2 y}{x^2 + y^2}$$

instead of x and y in the equation of the original curve

Again, let the equation of the original curve, as in Art. 174, be

$$u_n + u_{n-1} + u_{n-2} + \ldots + u_2 + u_1 + u_0 = 0.$$

When $\dfrac{k^2 x}{x^2 + y^2}$ and $\dfrac{k^2 y}{x^2 + y^2}$ are substituted for x and y, u_n becomes evidently $\dfrac{k^{2n} u_n}{(x^2 + y^2)^n}$.

Accordingly, the equation of the inverse curve is

$$k^{2n} u_n + k^{2n-2} u_{n-1} (x^2 + y^2) + k^{2n-4} u_{n-2} (x^2 + y^2)^2 + \ldots$$
$$+ u_0 (x^2 + y^2)^n = 0. \qquad (19)$$

For instance, the equation of any right line is of the form

$$u_1 + u_0 = 0;$$

hence that of its inverse with respect to the origin is

$$k^2 u_1 + u_0 (x^2 + y^2) = 0.$$

This represents a circle passing through the pole, as is well known, except when $u_0 = 0$; i.e. when the line passes through the pole O.

Again, the equation of the inverse of the circle

$$x^2 + y^2 + u_1 + u_0 = 0,$$

with respect to the origin, is

$$(k^4 + k^2 u_1 + u_0 (x^2 + y^2)) (x^2 + y^2) = 0,$$

which represents another circle, along with the two imaginary right lines $x^2 + y^2 = 0$.

Again, the general equation of a conic is of the form

$$u_2 + u_1 + u_0 = 0;$$

hence that of its inverse with respect to the origin is

$$k^4 u_2 + k^2 u_1 (x^2 + y^2) + u_0 (x^2 + y^2)^2 = 0,$$

which represents a curve of the fourth degree of the class called "bicircular quartics."

If the origin be on the conic the absolute term u_0 vanishes, and the inverse is the curve of the third degree represented by

$$k^2 u_2 + u_1 (x^2 + y^2) = 0.$$

This curve is called a "circular cubic."

If the focus be the origin of inversion, the inverse is a curve called the Limaçon of Pascal. The form of this curve will be given in a subsequent Chapter.

187. **Pedal Curves.**—If from any point as origin a perpendicular be drawn to the tangent to a given curve, the locus of the foot of the perpendicular is called the *pedal* of the curve with respect to the assumed origin.

In like manner, if perpendiculars be drawn to the tangents to the pedal, we get a new curve called the *second pedal* of the original, and so on. With respect to its pedal, the original curve is styled the *first negative* pedal, &c.

188. **Tangent at any Point to the Pedal of a given Curve.**—Let ON, ON' be the perpendiculars from the origin O on the tangents drawn at two points P and Q on the given curve, and T the intersection of these tangents; join NN'; then since the angles ONT and $ON'T$ are right angles, the quadrilateral $ONN'T$ is inscribable in a circle,

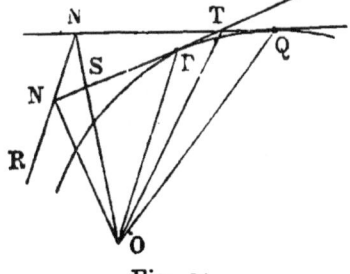

Fig. 14.

$$\therefore \angle ON'N = \angle OTN.$$

In the limit when P and Q coincide, $\angle OTN = \angle OPN$, and NN' becomes the tangent to the locus of N; hence the

Q 2

latter tangent makes the same angle with ON that the tangent at P makes with OP. This property enables us to draw the tangent at any point N on the pedal locus in question.

Again, if p' represent the perpendicular on the tangent at N to the first pedal, from similar triangles we evidently have
$$r = \frac{p^2}{p'}.$$

Hence, if the equation of a curve be given in the form $r = f(p)$, that of its first *pedal* is of the form $\frac{p^2}{p'} = f(p)$, in which p and p' are respectively analogous to r and p in the original curve. In like manner the equation of the next pedal can be determined, and so on.

189. **Reciprocal Polars.**—If on the perpendicular ON a point P' be taken, such that $OP' \cdot ON$ is constant (k^2 suppose), the point P' is evidently the pole of the line PN with respect to the circle of radius k and centre O; and if all the tangents to the curve be taken, the locus of their poles is a new curve. We shall denote these curves by the letters A and B, respectively. Again, by elementary geometry, the point of intersection of any two lines is the pole of the line joining the poles of the lines.* Now, if the lines be taken as two infinitely near tangents to the curve A, the line joining their poles becomes a tangent to B; accordingly, the tangent to the curve B has its pole on the curve A. Hence A is the locus of the poles of the tangents to B.

In consequence of this reciprocal relation, the curves A and B are called *reciprocal polars* of each other with respect to the circle whose radius is k.

Since to every tangent to a curve corresponds a point on its reciprocal polar, it follows that to a number of points *in directum* on one curve correspond a number of tangents to its reciprocal polar, which pass through a common point.

Again, it is evident that the *reciprocal polar to any curve is the inverse to its pedal with respect to the origin.*

We have seen in Art. 180 that the greatest number of tangents from a point to a curve of the nth degree is $n(n-1)$;

---

* Townsend's *Modern Geometry*, vol. i., p. 219.

hence the greatest number of points in which its reciprocal polar can be cut by a line is $n(n-1)$, or the degree of the reciprocal polar is $n(n-1)$. For the modification in this result, arising from singular points in the original curve, as well as for the complete discussion of reciprocal polars, the student is referred to Salmon's *Higher Plane Curves*.

As an example of reciprocal polars we shall take the curve considered in Art. 173.

If r denote the radius vector of the reciprocal polar corresponding to the perpendicular p in the proposed curve, we have

$$p = \frac{k^2}{r}.$$

Substituting this value for p in equation (5), we get

$$\left(\frac{k^2}{r}\right)^{\frac{m}{m-1}} = (a \cos \omega)^{\frac{m}{m-1}} + (b \sin \omega)^{\frac{m}{m-1}},$$

or
$$k^{\frac{2m}{m-1}} = (ax)^{\frac{m}{m-1}} + (by)^{\frac{m}{m-1}},$$

which is the equation of the reciprocal polar of the curve represented by the equation

$$\frac{x^m}{a^m} + \frac{y^m}{b^m} = 1.$$

In the particular case of the ellipse,

$$\frac{x^2}{a^2} + \frac{y^2}{b^2} = 1,$$

the reciprocal polar has for its equation

$$k^4 = a^2 x^2 + b^2 y^2.$$

The theory of reciprocal polars indicated above admits of easy generalization. Thus, if we take the poles with respect to any conic section (U) of all the tangents to a given curve A, we shall get a new curve B; and it can be easily seen, as before, that the poles of the tangents to B are situated on the curve A. Hence the curves are said to be *reciprocal polars* with respect to the conic U.

It may be added, that if two curves have a common point,

their reciprocal polars have a common tangent; and if the curves touch, their reciprocal polars also touch.

For illustrations of the great importance of this "principle of duality," and of reciprocal polars as a method of investigation, the student is referred to Salmon's *Conics*, ch. xv.

We next proceed to illustrate the preceding by discussing a few elementary properties of the curves which are comprised under the equation $r^m = a^m \cos m\theta$.

190. **Pedal and Reciprocal Polar of** $r^m = a^m \cos m\theta$. We shall commence by finding the angle between the radius vector and the perpendicular on the tangent.

In the accompanying figure we have $\tan PON = \cot OPN = -\dfrac{dr}{rd\theta}$.

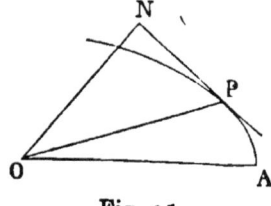

Fig. 15.

But $\qquad m \log r = m \log a + \log (\cos m\theta)$;

hence $\qquad \dfrac{dr}{rd\theta} = -\tan m\theta$,

and accordingly, $\qquad \angle PON = m\theta.$ \hfill (20)

Again, $\qquad p = ON = r \cos m\theta = \dfrac{r^{m+1}}{a^m}$,

or $\qquad r^{m+1} = a^m p.$ \hfill (21)

The equation of the *pedal*, with respect to O, can be immediately found.

For, let $\angle AON = \omega$, and we have

$$\omega = (m+1)\theta.$$

Also, from (21), $\qquad \left(\dfrac{r}{a}\right)^m = \left(\dfrac{p}{a}\right)^{\frac{m}{m+1}}.$

Hence, the equation of the pedal is

$$p^{\frac{m}{m+1}} = a^{\frac{m}{m+1}} \cos\left(\dfrac{m\omega}{m+1}\right). \qquad (22)$$

Consequently, the equation of the pedal is got by substituting $\dfrac{m}{m+1}$ instead of m in the equation of the curve.

By a like substitution the equation of the *second pedal* is easily seen to be

$$r^{\frac{m}{2m+1}} = a^{\frac{m}{2m+1}} \cos \frac{m\theta}{2m+1};$$

and that of the nth pedal

$$r^{\frac{m}{mn+1}} = a^{\frac{m}{mn+1}} \cos \frac{m\theta}{mn+1}. \qquad (23)$$

Again, from Art. 184, it is plain that the inverse to the curve $r^m = a^m \cos m\theta$, with respect to a circle of radius a, is the curve $r^m \cos m\theta = a^m$.

Again, the *reciprocal polar* of the proposed, with respect to the same circle, being the inverse of its pedal, is the curve

$$r^{\frac{m}{m+1}} \cos \frac{m\theta}{m+1} = a^{\frac{m}{m+1}}. \qquad (24)$$

It may be observed that this equation is got by substituting $\dfrac{-m}{m+1}$ for m in the original equation.

Accordingly we see that the pedals, inverse curves, and reciprocal polars of the proposed, are all curves whose equations are of the same form as that of the proposed.

In a subsequent chapter the student will find an additional discussion of this class of curves, along with illustrations of their shape for a few particular values of m.

Examples.

1. The equation of a parabola referred to its focus as pole is
$$r(1 + \cos\theta) = 2a,$$
to find the relation between r and p.

Here $r^{\frac{1}{2}} \cos \dfrac{\theta}{2} = a^{\frac{1}{2}}$, and consequently $p^2 = ar$,

a well-known elementary property of the curve.

2. The equation $r^2 \cos 2\theta = a^2$ represents an equilateral hyperbola; prove that $pr = a^2$.

3. The equation $r^2 = a^2 \cos 2\theta$ represents a Lemniscate of Bernoulli; find the equation connecting p and r in this case. *Ans.* $r^3 = a^2 p$.

4. Find the equation connecting the radius vector and the perpendicular on the tangent in the Cardioid whose equation is

$$r = a(1 + \cos\theta). \qquad Ans.\ r^3 = 2ap^2.$$

It is evident that the Cardioid is the inverse of a parabola with respect to its focus; and the Lemniscate that of an equilateral hyperbola with respect to its centre. Accordingly, we can easily draw the tangents at any point on either of these curves by aid of the Theorem of Art. 185.

5. Show, by the method of Art. 188, that the pedal of the parabola, $p^2 = ar$, with respect to its focus, is the right line $p = a$.

6. Show that the pedal of the equilateral hyperbola $pr = a^2$ is a Lemniscate.

7. Find the pedal of the circle $r^2 = 2ap$. *Ans.* A Cardioid, $r^3 = 2ap^2$.

191. Expression for PN.—To find the value of the intercept between the point of contact P and the foot N of the perpendicular from the origin on the tangent at P.

Let $p = ON$, $\omega = \angle NOA$, $PN = t$; then $\angle NTN' = \angle NON' = \Delta\omega$, also $SN' = TS \sin STN'$;

$\therefore TS = \dfrac{SN'}{\sin NON'}$; but in the

limit, when PQ is infinitely small, $\dfrac{SN'}{\sin NON'}$ becomes $\dfrac{dp}{d\omega}$, and TS becomes PN or t;

$$\therefore t = \frac{dp}{d\omega}. \qquad (25)$$

Also $\qquad OP^2 = ON^2 + PN^2$;

$$\therefore r^2 = p^2 + \left(\frac{dp}{d\omega}\right)^2. \qquad (26)$$

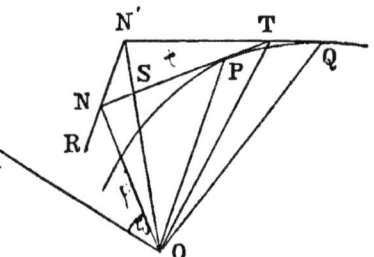

Fig. 16.

192. To prove that

$$\frac{ds}{d\omega} = p + \frac{dt}{d\omega}. \qquad (27)$$

On reference to the last figure we have

$$\frac{ds}{d\omega} = \text{limit of } \frac{PT + TQ}{\Delta\omega}, \quad \frac{dt}{d\omega} = \text{limit of } \frac{QN' - PN}{\Delta\omega};$$

but $\qquad PT + TQ - QN' + PN = TN - TN'$;

hence $\dfrac{ds}{d\omega} - \dfrac{dt}{d\omega} = \text{limit of } \dfrac{TN - TN'}{\Delta\omega} = \text{limit of} \dfrac{SN}{\Delta\omega} = ON = p$;

$$\therefore \frac{ds}{d\omega} = p + \frac{dt}{d\omega}.$$

This result, which is due to Legendre, is of importance in the Integral Calculus, in connexion with the *rectification* of curves.

If $\dfrac{dp}{d\omega}$ be substituted for t, the preceding formula becomes

$$\frac{ds}{d\omega} = p + \frac{d^2p}{d\omega^2}. \tag{28}$$

This shape of the result is of use in connexion with curvature, as will be seen in a subsequent chapter.

193. **Direction of Normal in Vectorial Co-ordinates.**—In some cases the equation of a curve can be expressed in terms of the distances from two or more fixed points or foci. Such distances are called vectorial co-ordinates. For instance, if r_1, r_2 denote the distances from two fixed points, the equation $r_1 + r_2 = const.$ represents an ellipse, and $r_1 - r_2 = const.$, a hyperbola.

Again, the equation

$$r_1 + mr_2 = const.$$

represents a curve called a Cartesian* oval.

Also, the equation

$$r_1 r_2 = const.$$

represents an oval of Cassini, and so on.

The direction of the normal at any point of a curve, in such cases, can be readily obtained by a geometrical construction.

---

* A discussion of the principal properties of Cartesian ovals will be found in Chapter XX.

For, let
$$F(r_1, r_2) = \text{const.}$$
be the equation of the curve, where
$$F_1P = r_1, \quad F_2P = r_2,$$
then we have

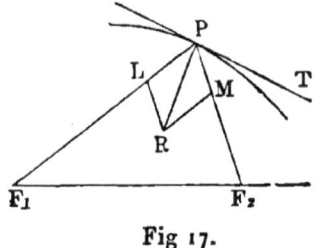

Fig 17.

$$\frac{dF}{dr_1}\frac{dr_1}{ds} + \frac{dF}{dr_2}\frac{dr_2}{ds} = 0.$$

Now, if PT be the tangent at P, then, by Art. 180, we have

$$\frac{dr_1}{ds} = \cos\psi_1, \quad \frac{dr_2}{ds} = \cos\psi_2, \quad \text{where } \psi_1 = \angle TPF_1, \quad \psi_2 = \angle TPF_2.$$

Hence
$$\frac{dF}{dr_1}\cos\psi_1 + \frac{dF}{dr_2}\cos\psi_2 = 0. \tag{29}$$

Again, from any point R on the normal draw RL and RM respectively parallel to F_2P and F_1P, and we have

$$PL : LR = \sin RPM : \sin RPL = \cos\psi_2 : -\cos\psi_1$$

$$= \frac{dF}{dr_1} : \frac{dF}{dr_2}.$$

Accordingly, if we measure on PF_1 and PF_2 lengths PL and PM, which are in the proportion of $\dfrac{dF}{dr_1}$ to $\dfrac{dF}{dr_2}$, then the diagonal of the parallelogram thus formed is the normal required.

This result admits of the following generalization:

Let the equation of the curve* be represented by

$$F(r_1, r_2, r_3, \ldots r_n) = \text{const.},$$

---

* The theorem given above is taken from Poinsot's *Elements de Statique*, Neuvième Edition, p. 435. The principle on which it was founded was, however, given by Leibnitz (*Journal des Savans*, 1693), and was deduced from mechanical considerations. The term *resultant* is borrowed from Mechanics, and is obtained by the same construction as that for the resultant of a number of forces acting at the same point. Thus, to find the resultant of a number of lines $Pa, Pb, Pc, Pd, \ldots$ issuing from a point P, we draw through a a right line aB, equal and parallel to Pb, and in the same direction; through B, a right line BC, equal and parallel to Pc, and so on, whatever be the number of lines: then the line PR, which closes the polygon, is the resultant in question.

where $r_1, r_2, \ldots r_n$ denote the distances from n fixed points. To draw the normal at any point, we connect the point with the n fixed points, and on the joining lines measure off lengths proportional to

$$\frac{dF}{dr_1}, \frac{dF}{dr_2}, \frac{dF}{dr_3}, \ldots \frac{dF}{dr_n}, \text{ respectively};$$

then the direction of the normal is the *resultant* of the lines thus determined.

For, as before, we have

$$\frac{dF}{dr_1}\frac{dr_1}{ds} + \frac{dF}{dr_2}\frac{dr_2}{ds} + \ldots \frac{dF}{dr_n}\frac{dr_n}{ds} = 0.$$

Hence $\quad \dfrac{dF}{dr_1}\cos\psi_1 + \dfrac{dF}{dr_2}\cos\psi_2 + \ldots \dfrac{dF}{dr_n}\cos\psi_n = 0.\quad (30)$

Now, $\quad \dfrac{dF}{dr_1}\cos\psi_1, \ \dfrac{dF}{dr_2}\cos\psi_2, \ \ldots \ \dfrac{dF}{dr_n}\cos\psi_n,$

are evidently proportional to the projections on the tangent of the segments measured off in our construction. Moreover, in any polygon, the projection of one side on any right line is manifestly equal to the sum of the projections of all the other sides on the same line, taken with their proper signs. Consequently, from (30), the projection of the resultant on the tangent is zero; and, accordingly, the resultant is normal to the curve, which establishes the theorem.

It can be shown without difficulty that the normal at any point of a surface whose equation is given in terms of the distances from fixed points can be determined by the same construction.

Examples.

1. A Cartesian oval is the locus of a point, P, such that its distances, PM, PM', from the circumferences of two given circles are to each other in a constant ratio; prove geometrically that the tangents to the oval at P, and to the circles at M and M', meet in the same point.

2. The equation of an ellipse of Cassini is $rr' = ab$, where r and r' are the distances of any point P on the curve, from two fixed points, A and B. If O be the middle point of AB, and PN the normal at P, prove that $\angle APO = \angle BPN$.

3. In the curve represented by the equation $r_1^3 + r_2^3 = a^3$, prove that the normal divides the distance between the foci in the ratio of r_2 to r_1.

194. In like manner, if the equation of a curve be given in terms of the angles $\theta_1, \theta_2, \ldots \theta_n$, which the vectors drawn to fixed points make respectively with a fixed right line, the direction of the tangent at any point is obtained by an analogous construction.

For, let the equation be represented by

$$F(\theta_1, \theta_2, \ldots \theta_n) = const.$$

Then, by differentiation, we have

$$\frac{dF}{d\theta_1}\frac{d\theta_1}{ds} + \frac{dF}{d\theta_2}\frac{d\theta_2}{ds} + \ldots \frac{dF}{d\theta_n}\frac{d\theta_n}{ds} = 0.$$

Hence, as before, from Art. 180, we get

$$\frac{1}{r_1}\frac{dF}{d\theta_1}\sin\psi_1 + \frac{1}{r_2}\frac{dF}{d\theta_1}\sin\psi_2 + \ldots + \frac{1}{r_n}\frac{dF}{d\theta_n}\sin\psi_n = 0. \qquad (31)$$

Accordingly, if we measure on the lines drawn to the fixed points segments proportional to

$$\frac{1}{r_1}\frac{dF}{d\theta_1}, \quad \frac{1}{r_2}\frac{dF}{d\theta_2}, \quad \ldots \quad \frac{1}{r_n}\frac{dF}{d\theta_n},$$

and construct the *resultant* line as before, then this line will be the tangent required. The proof is identical with that of last Article.

195. **Curves Symmetrical with respect to a Line, and Centres of Curves.**—It may be observed here, that if the equation of a curve be unaltered when y is changed into $-y$, then to every value of x correspond equal and opposite values of y; and, when the co-ordinate axes are rectangular, the curve is symmetrical with respect to the axis of x.

In like manner, a curve is symmetrical with respect to the axis of y, if its equation remains unaltered when the sign of x is changed.

Again, if, when we change x and y into $-x$ and $-y$, respectively, the equation of a curve remains unaltered, then every right line drawn through the origin and terminated by the curve is divided into equal parts at the origin. This takes place for a curve of an even degree when the sum of the indices of x and y in each term is even; and for a curve of an odd degree when the like sum is odd. Such a point is called *the centre** *of the curve*. For instance, in conics, when the equation is of the form

$$ax^2 + 2hxy + by^2 = c,$$

the origin is a centre. Also, if the equation of a cubic† be reducible to the form

$$u_3 + u_1 = 0,$$

the origin is a centre, and every line drawn through it is bisected at that point.

Thus we see that when a cubic has a centre, that point lies on the curve. This property holds for all curves of an odd degree.

It should be observed that curves of higher degrees than the second cannot generally have a centre, for it is evidently impossible by transformation of co-ordinates to eliminate the requisite number of terms from the equation of the curve. For instance, to seek whether a cubic has a centre, we substitute $X + a$ for x, and $Y + \beta$ for y, in its equation, and equate to zero the coefficients of X^2, XY and Y^2, as well as the absolute term, in the new equation: as we have but two arbitrary constants (a and β) to satisfy four equations, there will be two equations of condition among its constants in order that the cubic should have a centre. The number of conditions is obviously greater for curves of higher degrees.

---

* For a general meaning of the word "centre," as applied to curves of higher degrees, see Chasles's *Apercu Historique*, p. 233, note.

† This name has been given to curves of the third degree by Dr. Salmon, in his *Higher Plane Curves*, and has been generally adopted by subsequent writers on the subject.

EXAMPLES.

1. Find the lengths of the subtangent and subnormal at any point of the curve
$$y^n = a^{n-1}x.$$
Ans. $nx, \dfrac{y^2}{nx}$.

2. Find the subtangent to the curve
$$x^m y^n = a^{m+n}.$$
Ans. $-\dfrac{nx}{m}$.

3. Find the equation of the tangent to the curve
$$x^5 = a^3 y^2.$$
Ans. $\dfrac{5X}{x} - \dfrac{2Y}{y} = 3$.

4. Show that the points of contact of tangents from a point (a, β) to the curve
$$x^m y^n = a^{m+n}$$
are situated on the hyperbola $(m + n) xy = n\beta x + m\alpha y$.

5. In the same curve prove that the portion of the tangent intercepted between the axes is divided at its point of contact into segments which are to each other in a constant ratio.

6. Find the equation of the tangent at any point to the hypocycloid, $x^{\frac{2}{3}} + y^{\frac{2}{3}} = a^{\frac{2}{3}}$; and prove that the portion of the tangent intercepted between the axes is of constant length.

7. In the curve $x^n + y^n = a^n$, find the length of the perpendicular drawn from the origin to the tangent at any point, and find also the intercept made by the axes on the tangent.
$$\text{Ans. } p = \dfrac{a^n}{\sqrt{x^{2n-2} + y^{2n-2}}}\ ;\ \text{intercept} = \dfrac{a^{2n}}{p x^{n-1} y^{n-1}}.$$

8. If the co-ordinates of every point on a curve satisfy the equations
$$x = c \sin 2\theta (1 + \cos 2\theta), \quad y = c \cos 2\theta (1 - \cos 2\theta),$$
prove that the tangent at any point makes the angle θ with the axis of x.

9. The co-ordinates of any point in the cycloid satisfy the equations
$$x = a(\theta - \sin \theta), \quad y = a(1 - \cos \theta):$$
prove that the angle which the tangent at the point makes with the axis of y is $\dfrac{\theta}{2}$.

Here
$$\frac{dy}{dx} = \frac{\frac{dy}{d\theta}}{\frac{dx}{d\theta}} = \cot\frac{\theta}{2}.$$

10. Prove that the locus of the foot of the perpendicular from the pole on the tangent to an equiangular spiral is the same curve turned through an angle.

11. Prove that the reciprocal polar, with respect to the origin, of an equiangular spiral is another spiral equal to the original one.

12. An equiangular spiral touches two given lines at two given points; prove that the locus of its pole is a circle.

13. Find the equation of the reciprocal polar of the curve

$$r^{\frac{1}{3}} \cos\frac{\theta}{3} = a^{\frac{1}{3}},$$

with respect to the origin. *Ans.* The Cardioid $r^{\frac{1}{2}} = a^{\frac{1}{2}} \cos\frac{\theta}{2}$.

14. Find the equation of the inverse of a conic, the focus being the pole of inversion.

15. Apply Art. 184, to prove that the equation of the inverse of an ellipse with respect to any origin O is of the form

$$2a\rho = OF_1 \cdot \rho_1 + OF_2 \cdot \rho_2,$$

where F_1 and F_2 are the foci, and ρ, ρ_1, ρ_2 represent the distances of any point on the curve from the points O, f_1 and f_2, respectively; f_1 and f_2 being the points inverse to the foci, F_1 and F_2.

16. The equation of a Cartesian oval is of the form

$$r + kr' = a,$$

where r and r' are the distances of any point on the curve from two fixed points, and a, k are constants. Prove that the equation of its inverse, with respect to any origin, is of the form

$$\alpha\rho_1 + \beta\rho_2 + \gamma\rho_3 = 0,$$

where ρ_1, ρ_2, ρ_3 are the distances of any point on the curve from three fixed points, and α, β, γ are constants.

17. In general prove that the inverse of the curve

$$\alpha\rho_1 + \beta\rho_2 + \gamma\rho_3 = 0,$$

with respect to any origin, is another curve whose equation is of similar form.

18. If the radius vector, OP, drawn from the origin to any point P on a

curve be produced to P_1, until PP_1 be a constant length; prove that the normal at P_1 to the locus of P_1, the normal at P to the original curve, and the perpendicular at the origin to the line OP, all pass through the same point.

This follows immediately from the value of the polar subnormal given in Art. 182.

19. If a constant length measured from the curve be taken on the normals along a given curve, prove that these lines are also normals to the new curve which is the locus of their extremities.

20. In the ellipse $\dfrac{x^2}{a^2} + \dfrac{y^2}{b^2} = 1$, if $x = a \sin \phi$, prove that
$$\frac{ds}{d\phi} = a\sqrt{1 - e^2 \sin^2 \phi}.$$

21. If ds be the element of the arc of the inverse of an ellipse with respect to its centre, prove that
$$ds = k^2 \frac{a}{b^2} \frac{\sqrt{1 - e^2 \sin^2 \phi}}{1 + n \sin^2 \phi} d\phi, \quad \text{where } n = \frac{a^2 - b^2}{b^2}.$$

22. If ω be the angle which the normal at any point on the ellipse $\dfrac{x^2}{a^2} + \dfrac{y^2}{b^2} = 1$ makes with the axis-major, prove that
$$ds = \frac{b^2}{a} \frac{d\omega}{(1 - e^2 \sin^2 \omega)^{\frac{3}{2}}}.$$

23. Express the differential of an elliptic arc in terms of the semi-axis major, μ, of the confocal hyperbola which passes through the point.

$$\text{Ans. } \sqrt{\frac{a^2 - \mu^2}{a^2 e^2 - \mu^2}}\, d\mu.$$

24. In the curve $r^m = a^m \cos m\theta$, prove that
$$\frac{ds}{d\theta} = a \sec^{\frac{m-1}{m}} m\theta.$$

25. If $F(x, y) = 0$ be the equation to any plane curve, and ϕ the angle between the perpendicular from the origin on the tangent and the radius vector to the point of contact, prove that
$$\tan \phi = \frac{y\dfrac{dF}{dx} - x\dfrac{dF}{dy}}{x\dfrac{dF}{dx} + y\dfrac{dF}{dy}}.$$

CHAPTER XIII.

ASYMPTOTES.

196. Intersection of a Curve and a Right Line.—Before entering on the subject of this chapter it will be necessary to consider briefly the general question of the intersection of a right line with a curve of the nth degree.

Let the equation of the right line be $y = \mu x + \nu$, and substitute $\mu x + \nu$ instead of y in the equation of the curve; then the roots of the resulting equation in x represent the abscissæ of the points of section of the line and curve.

Moreover, as this equation is always of the nth degree, it follows that *every right line meets a curve of the nth degree in n points, real or imaginary*, and cannot meet it in more.

If two roots in the resulting equation be equal, two of the points of section become coincident, and the line becomes a tangent to the curve.

Again, suppose the equation of the curve written in the form of Art. 175, viz. :

$$u_n + u_{n-1} + u_{n-2} + \ldots u_2 + u_1 + u_0 = 0;$$

then, since u_n is a homogeneous function of the nth degree in x and y, it can be written in the form $x^n f_0\left(\dfrac{y}{x}\right)$; similarly

$$u_{n-1} = x^{n-1} f_1\left(\dfrac{y}{x}\right), \quad u_{n-2} = x^{n-2} f_2\left(\dfrac{y}{x}\right), \text{ &c.}$$

And accordingly, the equation of the curve may be written,

$$x^n f_0\left(\dfrac{y}{x}\right) + x^{n-1} f_1\left(\dfrac{y}{x}\right) + x^{n-2} f_2\left(\dfrac{y}{x}\right) + \text{&c.} = 0. \qquad (1)$$

Substituting $\mu + \dfrac{\nu}{x}$ for $\dfrac{y}{x}$ in this, it becomes

$$x^n f_0\left(\mu + \dfrac{\nu}{x}\right) + x^{n-1} f_1\left(\mu + \dfrac{\nu}{x}\right) + x^{n-2} f_2\left(\mu + \dfrac{\nu}{x}\right) + \text{&c.} = 0.$$

242   *Asymptotes.*

Or, expanding by Taylor's Theorem,

$$x^n f_0(\mu) + x^{n-1}\{\nu f'_0(\mu) + f_1(\mu)\} + x^{n-2}\{\tfrac{1}{1.2}\nu^2 f''_0(\mu) + \nu f'_1(\mu) + f_2(\mu)\}$$
$$+ \&c. = 0. \qquad (2)$$

The roots of this equation determine the points of section in question.

We add a few obvious conclusions from the results arrived at above:—

1°. Every right line must intersect a curve of an odd degree in at least one real point; for every equation of an odd degree has one real root.

2°. A tangent to a curve of the nth degree cannot meet it in more than $n-2$ points besides its points of contact.

3°. Every tangent to a curve of an odd degree must meet it in one other real point besides its point of contact.

4°. Every tangent to a curve of the third degree meets the curve in one other real point.

197. **Definition of an Asymptote.**—An asymptote is a tangent to a curve in the limiting position when its point of contact is situated at an infinite distance.

1°. No asymptote to a curve of the nth degree can meet it in more than $n-2$ points distinct from that at infinity.

2°. Each asymptote to a curve of the third degree intersects the curve in one point besides that at infinity.

198. **Method of finding the Asymptotes to a Curve of the nth Degree.**—If one of the points of section of the line $y = \mu x + \nu$ with the curve be at an infinite distance, one root of equation (2) must be infinite, and accordingly we have in that* case

$$f_0(\mu) = 0. \qquad (3)$$

Again, if two of the roots be infinite, we have in addition

$$\nu f'_0(\mu) + f_1(\mu) = 0. \qquad (4)$$

---

* This can be easily established by aid of the reciprocal equation; for if we substitute $\dfrac{1}{z}$ for x in equation (2), the resulting equation in z will have one root zero when its absolute term vanishes, i.e., when $f_0(\mu) = 0$; it has two roots zero when we have in addition $\nu f'_0(\mu) + f_1(\mu) = 0$; and so on.

Accordingly, when the values of μ and ν are determined so as to satisfy the two preceding equations, the corresponding line

$$y = \mu x + \nu$$

meets the curve in two points in infinity, and consequently is an asymptote. (Salmon's *Conic Sections*, Art. 154.)

Hence, if μ_1 be a root of the equation $f_0(\mu) = 0$, the line

$$y = \mu_1 x - \frac{f_1(\mu_1)}{f_0'(\mu_1)} \tag{5}$$

is in general an asymptote to the curve.

If $f_1(\mu) = 0$ and $f_0(\mu) = 0$ have a common root (μ_1 suppose), the corresponding asymptote in general passes through the origin, and is represented by the equation

$$y = \mu_1 x.$$

In this case u_n and u_{n-1} evidently have a common factor.

The exceptional case when $f_0'(\mu)$ vanishes at the same time will be considered in a subsequent Article.

To each root of $f_0(\mu) = 0$ corresponds an asymptote, and accordingly,* *every curve of the nth degree has in general n asymptotes, real or imaginary.*

From the preceding it follows that every line parallel to an asymptote meets the curve in one point at infinity. This also is immediately apparent from the geometrical property that a system of parallel lines may be considered as meeting in the same point at infinity—a principle introduced by Desargues in the beginning of the seventeenth century, and which must be regarded as one of the first important steps in the progress of modern geometry.

COR. No line parallel to an asymptote can meet a curve of the nth degree in more than $(n-1)$ points besides that at infinity.

Since every equation of an odd degree has one real root, it follows that a curve of an odd degree has one real

---

* Since $f_0(\mu)$ is of the nth degree in μ, unless its highest coefficient vanishes, in which case, as we shall see, there is an additional asymptote parallel to the axis of y.

asymptote, at least, and has accordingly an infinite branch or branches. Hence, *no curve of an odd degree can be a closed curve.*

For instance, no curve of the third degree can be a finite or closed curve.

The equation $f_0(\mu) = 0$, when multiplied by x^n, becomes $u_n = 0$; consequently the n right lines, real or imaginary, represented by this equation, are, in general, parallel to the asymptotes of the curve under consideration.

In the preceding investigation we have not considered the case in which a root of $f_0(\mu) = 0$ either vanishes or is infinite; i.e., where the asymptotes are parallel to either co-ordinate axis. This case will be treated of separately in a subsequent Article.

If all the roots of $f_0(\mu) = 0$ be imaginary the curve has no real asymptote, and consists of one or more closed branches.

EXAMPLES.

To find the asymptotes to the following curves:—

1. $$y^3 = ax^2 + x^3.$$

Substituting $\mu x + \nu$ for y, and equating to zero the coefficients of x^3 and x^2, separately, in the resulting equation, we obtain

$$\mu^3 - 1 = 0, \quad \text{and } 3\mu^2\nu = a;$$

$$\therefore \mu = 1, \quad \nu = \frac{a}{3};$$

hence the curve has but one real asymptote, viz.,

$$y = x + \frac{a}{3}.$$

2. $$y^4 - x^4 + 2ax^2y = b^2x^2.$$

Here the equations for determining the asymptotes are

$$\mu^4 - 1 = 0, \quad \text{and } 4\mu^3\nu + 2a\mu = 0;$$

accordingly, the two real asymptotes are

$$y = x - \frac{a}{2}, \text{ and } y + x + \frac{a}{2} = 0.$$

3. $$x^3 + 3x^2y - xy^2 - 3y^3 + x^2 - 2xy + 3y^2 + 4x + 5 = 0.$$

$$\text{Ans. } y + \frac{x}{3} + \frac{3}{4} = 0, \quad y = x + \frac{1}{4}, \quad y + x = \frac{3}{2}.$$

Asymptotes Parallel to Co-ordinate Axes.

199. Case in which $u_n = 0$ represents the n Asymptotes.—If the equation of the curve contain no terms of the $(n-1)^{th}$ degree, that is, if it be of the form

$$u_n + u_{n-2} + u_{n-3} + \&c. \ldots + u_1 + u_0 = 0,$$

the equations for determining the asymptotes become

$$f_0(\mu) = 0, \text{ and } \nu f_0'(\mu) = 0.$$

The latter equation gives $\nu = 0$, unless $f_0'(\mu)$ vanishes along with $f_0(\mu)$, i.e., unless $f_0(\mu)$ has equal roots.

Hence, in curves whose equations are of the above form, the n right lines represented by the equation $u_n = 0$ are the n asymptotes, unless two of these lines are coincident.

This exceptional case will be considered in Art. 202.

The simplest example of the preceding is that of the hyperbola

$$ax^2 + 2hxy + by^2 = c,$$

in which the terms of the second degree represent the asymptotes (Salmon's *Conic Sections*, Art. 195).

EXAMPLES.

Find the real asymptotes to the curves

1. $xy^2 - x^2y = a^2(x + y) + b^3$.   *Ans.* $x = 0, \ y = 0, \ x - y = 0$.
2. $y^3 - x^3 = a^2x$.   ,, $y - x = 0$.
3. $x^4 - y^4 = a^2xy + b^2y^2$.   ,, $x + y = 0, \ x - y = 0$.

200. Asymptotes parallel to the Co-ordinate Axes.—Suppose the equation of the curve arranged according to powers of x, thus

$$a_0 x^n + (a_1 y + b) x^{n-1} + \&c. = 0;$$

then, if $a_0 = 0$ and $a_1 y + b = 0$, or $y = -\dfrac{b}{a_1}$, two of the roots of the equation in x become infinite; and consequently the line $a_1 y + b = 0$ is an asymptote.

In other words, whenever the highest power of x is wanting in the equation of a curve, the coefficient of the next highest power equated to zero represents an asymptote parallel to the axis of x.

If $a_0 = 0$, and $b = 0$, the axis of x is itself an asymptote.

If x^n and x^{n-1} be both wanting, the coefficient of x^{n-2} represents a pair of asymptotes, real or imaginary, parallel to the axis of x; and so on.

In like manner, the asymptotes parallel to the axis of y can be determined.

EXAMPLES.

Find the real asymptotes in the following curves:—

1. $y^2x - ay^2 = x^3 + ax^2 + b^3$. *Ans.* $x = a$, $y = x + a$, $y + x + a = 0$.
2. $y(x^2 - 3bx + 2b^2) = x^3 - 3ax^2 + a^3$. $x = b$, $x = 2b$, $y + 3a = x + 3b$.
3. $x^2y^2 = a^2(x^2 + y^2)$. $x = \pm a$, $y = \pm a$.
4. $x^2y^2 = a^2(x^2 - y^2)$. $y + a = 0$, $y - a = 0$.
5. $y^2a - y^2x = x^3$. $x = a$.

201. Parabolic Branches.—Suppose the equation $f_0(\mu) = 0$ has equal roots, then $f_0'(\mu_1)$ vanishes along with $f_0(\mu)$, and the corresponding value of ν found from (5) becomes infinite, unless $f_1(\mu)$ vanish at the same time.

Accordingly, the corresponding asymptote is, in general, situated altogether at infinity.

The ordinary parabola, whose equation is of the form

$$(ax + \beta y)^2 = lx + my + n,$$

furnishes the simplest example of this case, having the line at infinity for an asymptote. (Salmon's *Conic Sections*, Art. 254.)

Branches of this latter class belonging to a curve are called *parabolic*, while branches having a finite asymptote are called *hyperbolic*.

202. From the preceding investigation it appears that the asymptotes to a curve of the nth degree depend, in general, only on the terms of the nth and the $(n-1)^{th}$ degrees in its equation. Consequently, *all curves which have the same terms of the two highest degrees have generally the same asymptotes.*

There are, however, exceptions to this rule, one of which will be considered in the next Article.

203. **Parallel Asymptotes.**—We shall now consider the case where $f_0(\mu) = 0$ has a pair of equal roots, each represented by μ_1, and where $f_1(\mu_1) = 0$, at the same time.

In this case the coefficients of x^n and x^{n-1} in (2) both vanish independently of ν, when $\mu = \mu_1$; we accordingly infer that all lines parallel to the line $y = \mu_1 x$ meet the curve in two points at infinity, and consequently are, in a certain sense, asymptotes. There are, however, two lines which are more properly called by that name; for, substituting μ_1 for μ in (2), the two first terms vanish, as already stated, and the coefficient of x^{n-2} becomes

$$\frac{\nu^2}{1 \cdot 2} f_0''(\mu_1) + \nu f_1'(\mu_1) + f_2(\mu_1).$$

Hence, if ν_1 and ν_2 be the roots of the quadratic

$$\frac{\nu^2}{1 \cdot 2} f_0''(\mu_1) + \nu f_1'(\mu_1) + f_2(\mu_1) = 0), \qquad (6)$$

the lines $\quad y = \mu_1 x + \nu_1, \text{ and } y = \mu_1 x + \nu_2,$

are a pair of parallel asymptotes, meeting the curve in three points at infinity.

If the roots of the quadratic be imaginary, the corresponding asymptotes are also imaginary.

Again, if the term u_{n-1} be wanting in the equation, and if $f_0(\mu) = 0$ have equal roots, the corresponding asymptotes are given by the quadratic

$$\frac{\nu^2}{1 \cdot 2} f_0''(\mu_1) + f_2(\mu_1) = 0.$$

In order that these asymptotes should be real, it is necessary that $f_2(\mu_1)$ and $f_0''(\mu_1)$ should have opposite signs.

There is no difficulty in extending the preceding investigation to the case where $f_0(\mu) = 0$ has three or more equal roots.

Examples.

1. $(x+y)^3(x^2+y^2+xy) = a^2y^2 + a^3(x-y).$

Here $f_0(\mu) = (1+\mu)^2(1+\mu+\mu^2), \quad f_1(\mu) = 0, \quad f_2(\mu) = -a^2\mu^2;$

$\therefore \mu_1 = -1, \quad f_0''(\mu_1) = 2, \quad f_2(\mu_1) = -a^2;$

accordingly $\nu_1 = a, \quad \nu_2 = -a,$

and the corresponding asymptotes are

$$y + x - a = 0, \text{ and } y + x + a = 0.$$

The other asymptotes are evidently imaginary.

2. $x^2(x+y)^2 + 2ay^2(x+y) + 8a^2xy + a^3y = 0.$

Here $f_0(\mu) = (1+\mu)^2, \quad f_1(\mu) = 2a\mu^2(1+\mu), \quad f_2(\mu) = 8a^2\mu;$

$\therefore \mu_1 = -1, \quad f_0''(\mu) = 2, \quad f_1'(\mu_1) = 2a, \quad f_2(\mu_1) = -8a^2,$

and the corresponding asymptotes are

$$y + x - 2a = 0, \text{ and } y + x + 4a = 0.$$

204. If the equation to a curve of the nth degree be of the form

$$(y + ax + \beta)\phi_1 + \phi_2 = 0,$$

where the highest terms containing x and y in ϕ_1 are of the degree $n-1$, and those in ϕ_2 are of the degree $n-2$ at most, the line

$$y + ax + \beta = 0$$

is an asymptote to the curve.

For, on substituting $-ax-\beta$ instead of y in the equation, it is evident that the coefficients of x^n and x^{n-1} both vanish; hence, by Art. 198, the line $y + ax + \beta = 0$ is an asymptote.

Conversely, it can be readily seen that if $y + ax + \beta$ be an asymptote to a curve of the nth degree its equation admits of being thrown into the preceding form.

In general, if the equation to a curve of the nth degree be of the form

$$(y + a_1x + \beta_1)(y + a_2x + \beta_2) \ldots (y + a_nx + \beta_n) + \phi_2 = 0, \quad (7)$$

where ϕ_2 contains no term higher than the degree $n-2$, the lines

$$y + a_1x + \beta_1 = 0, \quad y + a_2x + \beta_2 = 0, \ldots y + a_nx + \beta_n = 0$$

are the n asymptotes of the curve.

This follows at once as in the case considered at the commencement of this Article.

For example, the asymptotes to the curve

$$xy(x + y + a_1)(x + y + a_2) + b_1x + b_2y = 0$$

are evidently the four lines

$$x = 0, \quad y = 0, \quad x + y + a_1 = 0, \quad x + y + a_2 = 0.$$

If the curve be of the third degree, ϕ_2 is of the first, and accordingly the equation of such a curve, having three real asymptotes, may be written in the form

$$(y + a_1x + \beta_1)(y + a_2x + \beta_2)(y + a_3x + \beta_3) + lx + my + n = 0. \quad (8)$$

Hence we infer that the *three points in which the asymptotes to a cubic meet the curve lie in the same right line*, viz.,

$$lx + my + n = 0.$$

The student will find a short discussion of a cubic with three real asymptotes in Chapter XVIII.

205. To prove that, in general, *the distance of a point in any branch of a curve from the corresponding asymptote diminishes indefinitely as its distance from the origin increases indefinitely*.

If $y + ax + \beta = 0$ be the equation of an asymptote, then, as in the preceding Article, the equation of the curve may be written in the form

$$(y + ax + \beta)\phi_1 = \phi_2,$$

where ϕ_2 is at least one degree lower than ϕ_1 in x and y.

Hence $$y + ax + \beta = \frac{\phi_2}{\phi_1},$$

and the perpendicular distance of any point (x_0, y_0) on the curve from the line $y + ax + \beta = 0$ is

$$\frac{y_0 + ax_0 + \beta}{\sqrt{1 + a^2}}, \text{ or } \frac{1}{\sqrt{1 + a^2}}\left(\frac{\phi_2}{\phi_1}\right)_0,$$

where the suffix denotes that x_0 and y_0 are substituted for x and y in the functions ϕ_1 and ϕ_2.

Now, when x_0 and y_0 are taken infinitely great, the value of the preceding fraction depends, in general, on the terms of the highest degree (in x and y) in ϕ_1 and ϕ_2; and since the degree of ϕ_2 is *one lower* than that of ϕ_1, it can be easily seen by the method of Ex. 7, Art. 89, that the fraction $\frac{\phi_2}{\phi_1}$ becomes, in general, infinitely small when x and y become infinitely great. Hence, the distance of the line $y + ax + \beta$ from the curve becomes infinitely small at the same time.

It is not considered necessary to go more fully into this discussion here.

The subject of parabolic and other curvilinear asymptotes is omitted as being unsuited to an elementary treatise. Moreover, their discussion, unless in some elementary cases, is both indefinite and unsatisfactory, since it can be easily seen that if a curve has parabolic branches, the number of its parabolic asymptotes is generally infinite. The reader who desires full information on this point, as well as the discussion of the particular parabolas called *osculating*, is referred to a paper by M. Plücker, in Liouville's Journal, vol. i., p. 229.

206. Asymptotes in Polar Co-ordinates.—If a curve be referred to polar co-ordinates, the *directions* of its points at an infinite distance from the origin can be in general determined by making $r = \infty$, or $u = 0$, in its equation, and solving the resulting equation in θ. The *position* of the asymptote corresponding to any such value of θ is obtained by finding the length of the corresponding polar subtangent, i.e., by finding the value of $\frac{d\theta}{du}$ corresponding to $u = 0$.

It should be observed that when $\dfrac{d\theta}{du}$ is positive, the asymptote lies *above* the corresponding radius vector, and when negative, *below* it; as is easily seen from Art. 182.

If we suppose the equation of the curve, when arranged in powers of r, to be

$$r^n f_0(\theta) + r^{n-1} f_1(\theta) + \ldots + r f_{n-1}(\theta) + f_n(\theta) = 0,$$

the transformed equation in u is

$$u^n f_n(\theta) + u^{n-1} f_{n-1}(\theta) + \ldots + u f_1(\theta) + f_0(\theta) = 0 : \qquad (9)$$

consequently, the *directions* of the asymptotes are given by the equation

$$f_0(\theta) = 0. \qquad (10)$$

Again, if we differentiate (9) with respect to θ, it is easily seen that the values of $\dfrac{du}{d\theta}$ corresponding to $u = 0$ are given by the equation

$$f_1(\theta) \frac{du}{d\theta} + f_0'(\theta) = 0, \qquad (11)$$

provided that none of the functions

$$f_1(\theta),\ f_2(\theta),\ \ldots f_n(\theta)$$

become infinite for the values of θ which satisfy equation (10).

Consequently, if a be a root of the equation $f_0(\theta) = 0$, the curve has an asymptote making the angle a with the prime vector, and whose perpendicular distance from the origin is represented by $-\dfrac{f_1(a)}{f_0'(a)}$.

It is readily seen that the equation of the corresponding asymptote is

$$r \sin(a - \theta) + \frac{f_1(a)}{f_0'(a)} = 0.$$

This method will be best explained by applying it to one or two elementary Examples.

Examples.

1. Let the curve be represented by the equation

$$r = a \sec \theta + b \tan \theta.$$

Here
$$u = \frac{\cos \theta}{a + b \sin \theta}.$$

When $\theta = \frac{\pi}{2}$, we have $u = 0$, and $\frac{du}{d\theta} = \frac{-1}{a + b}$.

Accordingly, the corresponding polar subtangent is $a + b$, and hence the line perpendicular to the *prime vector* at the distance $a + b$ from the origin is an asymptote to the curve.

Again, u vanishes also when $\theta = \frac{3\pi}{2}$, and the corresponding value of the polar subtangent is $a - b$; thus giving another asymptote.

2. $$r = a \sec m\theta + b \tan m\theta.$$

Here
$$u = \frac{\cos m\theta}{a + b \sin m\theta}.$$

When $\theta = \frac{\pi}{2m}$, we have $u = 0$, and $\frac{du}{d\theta} = \frac{-m}{a + b}$,

whence we get one asymptote.

Again, when $\theta = \frac{3\pi}{2m}$, $u = 0$, and $\frac{du}{d\theta} = \frac{m}{a - b}$,

which gives a second asymptote.

On making $\theta = \frac{5\pi}{2m}$, we get a third asymptote, and so on.

It may be remarked, that the first, third, . . . asymptotes all touch one fixed circle; and the second, fourth, &c., touch another.

3. Find the equations to the two real asymptotes to the curve

$$r^2 \sin(\theta - a) + ar \sin(\theta - 2a) + a^2 = 0.$$

Ans. $r \sin(\theta - a) = \pm a \sin a.$

207. Asymptotic Circles.—In some curves referred to polar co-ordinates, when θ is infinitely great the value of r tends to a fixed limiting value, and accordingly the curve approaches more and more nearly to the circular form at the same time: in such a case the curve is said to have a circular asymptote.

For example, in the curve

$$r = \frac{a\theta}{\theta + a},$$

so long as θ is positive r is less than a, a being supposed positive; but as θ increases with each revolution, r continually increases, and tends, after a large number of revolutions, to the limit a; hence the circle described with the origin as centre, and radius a, is asymptotic to the curve, which always lies inside the circle for *positive* values of θ. Again, if we assign *negative* values to θ, similar remarks are applicable, and it is easily seen that the same circle is asymptotic to the corresponding branch of the curve; with this difference, that the asymptotic circle lies *within* the curve in the latter case, but outside it in the former. The student will find no difficulty in applying this method to other curves, such as

$$r = \frac{a\theta}{\theta + \sin\theta}, \quad r = \frac{a\theta^2}{\theta^2 + a^2}, \quad r = \frac{a(\theta + \cos\theta)}{\theta + \sin\theta}.$$

Examples.

Find the equations of the real asymptotes to the following curves:—

1. $y(a^2 - x^2) = b^2(2x + c)$.   Ans. $y = 0$, $x + a = 0$, $x - a = 0$.

2. $x^4 - x^2 y^2 + a^2 x^2 + b^4 = 0$. $x + y = 0$, $x - y = 0$, $x = 0$.

3. $x^4 - x^2 y^2 + x^2 + y^2 - a^2 = 0$. $x - 1 = 0$, $x + 1 = 0$, $x - y = 0$, $x + y = 0$.

4. $(a + x)^2(b^2 - x^2) = x^2 y^2$. $x = 0$.

5. $(a + x)^2(b^2 + x^2) = x^2 y^2$. $x = 0$, $y = x + a$, $y + x + a = 0$.

6. $x^3 y - 2x^2 y^2 + xy^3 = a^2 x^2 + b^2 y^2$. $x = 0$, $y = 0$, $x - y = \pm \sqrt{a^2 + b^2}$.

7. $x^3 - 4xy^2 - 3x^2 + 12xy - 12y^2 + 8x + 2y + 4 = 0$.
   Ans. $x + 3 = 0$, $x - 2y = 0$, $x + 2y = 6$.

8. $x^2 y^2 - ax(x + y)^2 - 2a^2 y^2 - a^4 = 0$. $x + 2a = 0$, $x - a = 0$.

9. If the equation to a curve of the third degree be of the form

$$u_3 + u_1 + u_0 = 0,$$

the lines represented by $u_3 = 0$ are its asymptotes.

10. If the asymptotes of a cubic be denoted by $a = 0$, $\beta = 0$, $\gamma = 0$, the equation of the curve may be written in the form

$$\alpha\beta\gamma = A\alpha + B\beta + C\gamma.$$

11. In the logarithmic curve

$$y = a^{\frac{x}{b}},$$

prove that the negative side of the axis of x is an asymptote.

12. Find the asymptotes to the curve

$$r \cos n\theta = a.$$

13. Find the asymptotes to

$$r \cos m\theta = a \cos n\theta.$$

14. Show that the curve represented by

$$x^3 + aby - axy = 0$$

has a parabolic asymptote, $x^2 + bx + b^2 = ay$.

15. Find the circular asymptote to the curve

$$r = \frac{a\theta + b}{\theta + a}.$$

16. Find the condition that the three asymptotes of a cubic should pass through a common point.

Let the equation of the curve be written in the form

$$a_0 + 3b_0 x + 3b_1 y + 3c_0 x^2 + 6c_1 xy + 3c_2 y^2 + d_0 x^3 + 3d_1 x^2 y + 3d_2 xy^2 + d_3 y^3 = 0,$$

then the condition is

$$\begin{vmatrix} d_0, & d_1, & d_2, \\ d_1, & d_2, & d_3, \\ c_0, & c_1, & c_2, \end{vmatrix} = 0.$$

This result can be easily arrived at by substituting $x + a$ and $y + \beta$ instead of x and y in the equation of the cubic, and finding the condition that the part of the second degree in the resulting equation should vanish. See Art. 204.

17. When the preceding condition is satisfied show that the co-ordinates, a and β, of the point of intersection of the three asymptotes, are given by the equations

$$a = \frac{c_1 d_1 - c_0 d_2}{d_0 d_2 - d_1^2}, \qquad \beta = \frac{c_0 d_1 - c_1 d_0}{d_0 d_2 - d_1^2}.$$

18. If from any point, O, a right line be drawn meeting a curve of the nth degree in $R_1, R_2, \ldots R_n$, and its asymptotes in $r_1, r_2, \ldots r_n$, prove that

$$OR_1 + OR_2 + \ldots + OR_n = Or_1 + Or_2 + \ldots Or_n.$$

N.B.—The terms of the nth and $(n-1)^{th}$ degrees are the same for a curve and its asymptotes.

19. If a right line be drawn through the point $(a, 0)$ parallel to the asymptote of the cubic $(x - a)^3 - x^2 y = 0$, prove that the portion of the line intercepted by the axes is bisected by the curve.

20. If from the origin a right line be drawn parallel to any of the asymptotes of the cubic

$$y(ax^2 + 2hxy + by^2 + 2gx + 2fy + c) - x^3 = 0,$$

show that the portion of this line intercepted between the origin and the line $gx + fy + c = 0$ is bisected by the curve.

21. If tangents be drawn to the curve $x^3 + y^3 = a^3$ from any point on the line $y = x$, prove that their points of contact lie on a circle.

22. Show that the asymptotes to the cubic

$$ax^2 y + bxy^2 + a'x^2 + b'y^2 + a''x + b''y = 0$$

are always real, and find their equations.

Ans. $bx + b' = 0, \quad ay + a' = 0,$
$ab(ax + by) - a^2 b' - a' b^2 = 0.$

CHAPTER XIV.

MULTIPLE POINTS ON CURVES.

208. In the following elementary discussion of multiple points of curves the method given by Dr. Salmon in his *Higher Plane Curves* has been followed, as being the simplest, and at the same time the most comprehensive method for their investigation. The discussion here is to be regarded as merely introductory to the more general investigation in that treatise, to which the student is referred for fuller information on this as well as on the entire theory of curves.

We commence with the general equation of a curve of the nth degree, which we shall write in the form

$$a_0$$
$$+ b_0 x + b_1 y$$
$$+ c_0 x^2 + c_1 xy + c_2 y^2$$
$$+ \&c. \qquad + \&c.$$
$$+ l_0 x^n + l_1 x^{n-1} y + \&c. + l_n y^n = 0,$$

where the terms are arranged according to their degrees in ascending order.

When written in the abbreviated form of Art. 175, the preceding equation becomes

$$u_0 + u_1 + u_2 + \ldots + u_{n-1} + u_n = 0.$$

We commence with the equation in its expanded shape, and suppose the axes rectangular. Transforming to polar co-ordinates, by substituting $r\cos\theta$ and $r\sin\theta$ instead of x and y, we get

$$a_0 + (b_0 \cos\theta + b_1 \sin\theta)r$$
$$+ (c_0 \cos^2\theta + c_1 \cos\theta \sin\theta + c_2 \sin^2\theta) r^2 + \ldots$$
$$+ (l_0 \cos^n\theta + l_1 \cos^{n-1}\theta \sin\theta + \ldots + l_n \sin^n\theta) r^n = 0. \quad (1)$$

If θ be considered a constant, the n roots of this equation in r represent the distances from the origin of the n points of intersection of the radius vector with the curve.

If $a_0 = 0$, one of these roots is zero for all values of θ; as is also evident since the origin lies on the curve in this case.

A second root will vanish, if, besides $a_0 = 0$, we have $b_0 \cos\theta + b_1 \sin\theta = 0$. The radius vector in this case meets the curve in two consecutive points* at the origin, and is consequently the tangent at that point.

The direction of this tangent is determined by the equation

$$b_0 \cos\theta + b_1 \sin\theta = 0;$$

accordingly, the equation of the tangent at the origin is

$$b_0 x + b_1 y = 0.$$

Hence we conclude that if the absolute term be wanting in the equation of a curve, it passes through the origin, and *the linear part (u_1) in its equation represents the tangent at that point.*

If $b_0 = 0$, the axis of x is a tangent; if $b_1 = 0$, the axis of y is a tangent.

The preceding, as also the subsequent discussion, equally applies to oblique as to rectangular axes, provided we substitute mr and nr for x and y; where

$$m = \frac{\sin(\omega - \theta)}{\sin\omega}, \text{ and } n = \frac{\sin\theta}{\sin\omega};$$

ω being the angle between the axes of co-ordinates.

From the preceding, we infer at once that the equation of the tangent at the origin to the curve

$$x^2(x^2 + y^2) = a(x - y)$$

---

* Two points which are infinitely close to each other on the *same branch* of a curve are said to be consecutive points on the curve.

is $x - y = 0$, a line bisecting the internal angle between the co-ordinate axes. In like manner, the tangent at the origin can in all cases be immediately determined.

209. **Equation of Tangent at any Point.**—By aid of the preceding method the equation of the tangent at any point on a curve whose equation is algebraic and rational can be at once found. For, transferring the origin to that point, the linear part of the resulting equation represents the tangent in question.

Thus, if $f(x, y) = 0$ be the equation of the curve, we substitute $X + x_1$ for x, and $Y + y_1$ for y, where (x_1, y_1) is a point on the curve, and the equation becomes

$$f(X + x_1, Y + y_1) = 0.$$

Hence the equation of the tangent referred to the new axes is

$$X \left(\frac{df}{dx} \right)_1 + Y \left(\frac{df}{dy} \right)_1 = 0.$$

On substituting $x - x_1$, and $y - y_1$, instead of X and Y, we obtain the equation of the tangent referred to the original axes, viz.

$$(x - x_1) \left(\frac{df}{dx} \right)_1 + (y - y_1) \left(\frac{df}{dy} \right)_1 = 0.$$

This agrees with the result arrived at in Art. 169.

210. **Double Points.**—If in the general equation of a curve we have $a_0 = 0$, $b_0 = 0$, $b_1 = 0$, the coefficient of r is zero for all values of θ, and it follows that all lines drawn through the origin meet the curve in two points, coincident with the origin.

The origin in this case is called a double point.

Moreover, if θ be such as to satisfy the equation

$$c_0 \cos^2\theta + c_1 \cos\theta \sin\theta + c_2 \sin^2\theta = 0, \qquad (2)$$

the coefficient of r^2 will also disappear, and three roots of equation (1) will vanish.

As there are two values of $\tan \theta$ satisfying equation (2), it follows that through a double point two lines can be drawn, each meeting the curve in three coincident points.

Double Points.

The equation (2), when multiplied by r^2, becomes

$$c_0 x^2 + c_1 xy + c_2 y^2 = 0.$$

Hence we infer that the lines represented by this equation connect the double point with consecutive points on the curve, and are, consequently, tangents to the two branches of the curve passing through the double point.

Accordingly, when the lowest terms in the equation of a curve are of the second degree (u_2), the origin is a double point, *and the equation* $u_2 = 0$ *represents the pair of tangents at that point.*

For example, let us consider the Lemniscate, whose equation is

$$(x^2 + y^2)^2 = a^2 (x^2 - y^2).$$

On transforming to polar co-ordinates its equation becomes

$$r^4 = a^2 r^2 (\cos^2 \theta - \sin^2 \theta), \text{ or, } r^2 = a^2 \cos 2\theta.$$

Now, when $\theta = 0$, $r = \pm a$; and, if we confine our attention to the positive values of r, we see that as θ increases from 0 to $\dfrac{\pi}{4}$, r diminishes from a to zero. When $\theta > \dfrac{\pi}{4}$ and $< \dfrac{3\pi}{4}$, r is imaginary, &c.,

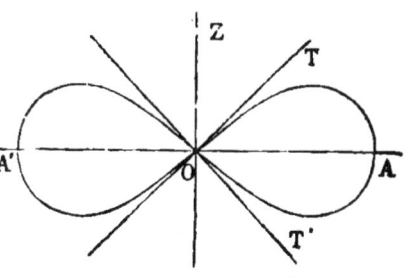

Fig. 18.

and it is evident that the figure of the curve is as annexed, having two branches intersecting at the origin, and that the tangents at that point bisect the angles between the axes. The equations of these tangents are

$$x + y = 0, \text{ and } x - y = 0,$$

results which agree with the preceding theory.

211. **Nodes, Cusps, and Conjugate Points.***—The pair of lines represented by $u_2 = 0$ will be real and distinct, coincident, or imaginary, according as the roots of equation (2) are real and unequal, real and equal, or imaginary.

---

* These have been respectively styled *crunodes*, *spinodes*, and *acnodes*, by Professor Cayley. See Salmon's *Higher Plane Curves*, Art. 38.

S 2

Hence we conclude that there may be one of three kinds of singular point on a curve so far as the vanishing of u_0 and u_1 is concerned.

(1). For real and unequal roots, the tangents at the double point are real and distinct, and the point is called a *node*; arising from the intersection of two real branches of the curve, as in the annexed figure.

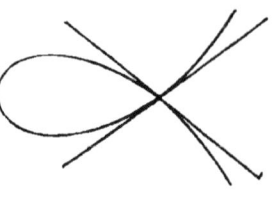

Fig. 19.

(2). If the roots be equal, *i.e.* if u_2 be a perfect square, the tangents coincide, and the point is called a cusp: the two branches of the curve touching each other at the point, as in figure 20.

(3). If the roots of u_2 be imaginary, the tangents are imaginary, and the double point is called a *conjugate* or *isolated point*; the co-ordinates of the point satisfy the equation of the curve, but the curve has no real points consecutive to this point, which lies altogether outside the curve itself.

Fig. 20.

It should be observed also that in some cases of singularities of a higher order, the origin is a conjugate point even when u_2 is a perfect square, as will be more fully explained in a subsequent chapter.

We add a few elementary examples of these different classes for illustration.

EXAMPLES.

1. $$y^2(a^2 + x^2) = x^2(a^2 - x^2).$$

Here the origin is a node, the tangents bisecting the angles between the axes of co-ordinates.

2. $$ay^2 = x^3.$$

In this case the origin is a cusp. Again, solving for y we get

$$y = \pm \frac{x^{\frac{3}{2}}}{a^{\frac{1}{2}}}.$$

Hence, if a be positive, y becomes *imaginary for negative* values of x; and, accordingly, no portion of the curve extends to the negative side of the axis of x. Moreover, for positive values of x, the corresponding values of y have opposite signs. This curve is called the semi-cubical parabola. The form of the curve near the origin is exhibited in Fig. 20.

3. $$y^3 = x^2(x+a).$$
Ans. The origin is a cusp.

4. $$b(x^2 + y^2) = x^3.$$
Ans. The origin is a conjugate point.

5. $$x^3 - 3axy + y^3 = 0.$$
Ans. The two branches at the origin touch the co-ordinate axes.

212. Double Points in General.—In order to seek the double points on any algebraic curve, we transform the origin to a point (x_1, y_1) on the curve; then, if we can determine values of x_1, y_1 for which the linear part disappears from the resulting equation, the new origin (x_1, y_1) is a double point on the curve.

From Art. 209 it is evident that the preceding conditions give

$$\left(\frac{df}{dx}\right)_1 = 0, \text{ and } \left(\frac{df}{dy}\right)_1 = 0;$$

moreover, since the point (x_1, y_1) is situated on the curve, we must have

$$f(x_1, y_1) = 0.$$

As we have but two variables, x_1, y_1, in order that they should satisfy these three equations simultaneously, a condition must evidently exist between the constants in the equation of the curve, viz., the condition arising from the elimination of x_1, y_1 between the three preceding equations.

Again, when the curve has a double point (x_1, y_1), if the origin be transferred to it, the part of the second degree in the resulting equation is evidently

$$x^2 \left(\frac{d^2u}{dx^2}\right)_1 + 2xy \left(\frac{d^2u}{dxdy}\right)_1 + y^2 \left(\frac{d^2u}{dy^2}\right)_1.$$

Accordingly, the lines represented by this quadratic are the tangents at the double point.

The point consequently is a node, a cusp, or a conjugate point, according as

$$\left(\frac{d^2u}{dxdy}\right)_1^2 \text{ is } > = \text{ or } < \left(\frac{d^2u}{dx^2}\right)_1 \left(\frac{d^2u}{dy^2}\right)_1.$$

It may be remarked here that no cubic can have more than one double point; for if it have two, the line joining them must be regarded as cutting the curve in four points, which is impossible.

Again, every line passing through a double point on a cubic must meet the curve in one, and but one, other point; except the line be a tangent to either branch of the cubic at the double point, in which case it cannot meet the curve elsewhere; the points of section being two consecutive on one branch, and one on the other branch.

In many cases the existence of double points can be seen immediately from the equation of the curve. The following are some easy instances:—

Examples.

To find the position and nature of the double points in the following curves:—

1. $$(bx - cy)^2 = (x - a)^5.$$

The point $x = a$, $y = \dfrac{ab}{c}$, is evidently a cusp, at which $bx - cy = 0$ is the tangent, as in the accompanying figure

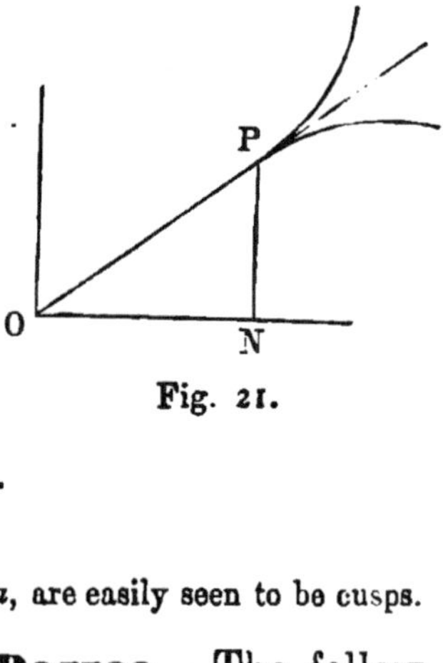

2. $$(y - c)^2 = (x - a)^4 (x - b).$$

The point $x = a$, $y = c$, is a cusp if $a > b$, or if $a = b$; but is a conjugate point if $a < b$.

Fig. 21.

3. $$y^2 = x(x + a)^2.$$

The point $y = 0$, $x = -a$ is a conjugate point.

4. $$x^{\frac{2}{3}} + y^{\frac{2}{3}} = a^{\frac{2}{3}}.$$

The points $x = 0$, $y = \pm a$; and $y = 0$, $x = \pm a$, are easily seen to be cusps.

213. Parabolas of the Third Degree.—The following example* will assist the student towards seeing the distinction, as well as the connexion, between the different kinds of double points.

Let
$$y^2 = (x - a)(x - b)(x - c)$$
be the equation of a curve, where $a < b < c$.

---

* Lacroix, *Cal. Dif.*, pp. 395-7.   Salmon's *Higher Plane Curves*, Art. 39.

Parabolas of the Third Degree.

Here y vanishes when $x = a$, or $x = b$, or $x = c$; accordingly, if distances $OA = a$, $OB = b$, $OC = c$, be taken on the axis of x, the curve passes through the points A, B, and C.

Moreover, when $x < a$, y^2 is negative, and therefore y is imaginary.

,, $x > a$, and $< b$, y^2 is positive, and therefore y is real.

,, $x > b$, and $< c$, y^2 is negative, and therefore y is imaginary.

,, $x > c$, y^2 is positive, and therefore y is real; and increases indefinitely along with x.

Hence, since the curve is symmetrical with respect to the axis of x, it evidently consists of an oval lying between A and B, and an infinite branch passing through C, as in the annexed figure. It is easily shown that the oval is not symmetrical with respect to the perpendicular to AB at its middle point. Again, if $b = c$, the equation becomes

$$y^2 = (x - a)(x - b)^2.$$

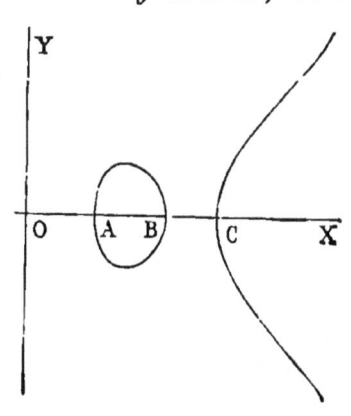

Fig. 22.

In this case the point B coincides with C, the oval has joined the infinite branch, and B has become a double point, as in the annexed figure.

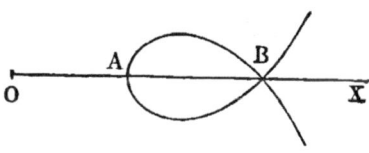

Fig. 23.

On the other hand, let $b = a$, and the equation becomes

$$y^2 = (x - a)^2(x - c);$$

in this case the oval has shrunk into the point A, and the curve is of the annexed form, having A for a conjugate point.

Next, let $a = b = c$, and the equation becomes

$$y^2 = (x - a)^3;$$

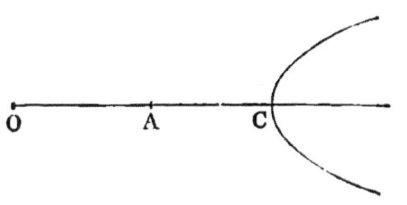

Fig. 24.

here the points A, B, C, have come together, and the curve has a cusp at the point A, as in the annexed figure.

The curves considered in this Article are called parabolas of the third degree.

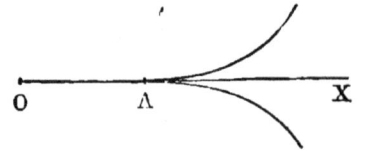

Fig. 25.

As an additional example, we shall investigate the following problem:—

214. *Given the three asymptotes of a cubic, to find its equation, if it have a double point.*

Taking two of its asymptotes as axes of co-ordinates, and supposing the equation of the third to be $ax + by + c = 0$, the equation of the cubic, by Art. 204, is of the form

$$xy(ax + by + c) = lx + my + n.$$

Again, the co-ordinates of the double point must satisfy the equations

$$\frac{du}{dx} = 0, \quad \frac{du}{dy} = 0,$$

or $\quad (2ax + by + c) y = l, \quad (ax + 2by + c) x = m;$

from which l and m can be determined when the co-ordinates of the double point are given.

To find n, we multiply the former equation by x, and the latter by y, and subtract the sum from three times the equation of the curve, and thus we get

$$cxy = 2lx + 2my + 3n;$$

from which n can be found.

In the particular case where the double point is a cusp,* its co-ordinates must satisfy the additional condition

$$\frac{d^2u}{dx^2}\frac{d^2u}{dy^2} = \left(\frac{d^2u}{dxdy}\right)^2,$$

or $\quad (2ax + 2by + c)^2 = 4abxy,$

and consequently the cusp must lie on the conic represented by this equation.

---

* It is essential to notice that the existence of a cusp involves one more relation among the coefficients of the equation of a curve than in the case of an ordinary double point or node.

It can be easily seen that this conic* touches at their middle points the sides of the triangle formed by the asymptotes.

The preceding theorem is due to Plücker,† and is stated by him as follows :—

"The locus of the cusps of a system of curves of the third degree, which have three given lines for asymptotes, is the maximum ellipse inscribed in the triangle formed by the given asymptotes."

It can be easily seen that the double point is a node or a conjugate point, according as it lies outside or inside the above-mentioned ellipse.

215. **Multiple Points of Higher Curves.**—By following out the method of Art. 208, the conditions for the existence of multiple points of higher orders can be readily determined.

Thus, if the lowest terms in the equation of a curve be of the third degree, the origin is a triple point, and the tangents to the three branches of the curve at the origin are given by the equation $u_3 = 0$.

The different kinds of triple points are distinguished, according as the lines represented by $u_3 = 0$ are real and distinct, coincident, or one real and two imaginary.

In general, if the lowest terms in the equation of a curve be of the mth degree, the origin is a multiple point of the mth order, &c.

Again, a point is a triple point on a curve provided that when the origin is transferred to it the terms below the third degree disappear from the equation. The co-ordinates of a triple point consequently must satisfy the equations

$$u = 0, \quad \frac{du}{dx} = 0, \quad \frac{du}{dy} = 0, \quad \frac{d^2u}{dx^2} = 0, \quad \frac{d^2u}{dx\,dy} = 0, \quad \frac{d^2u}{dy^2} = 0.$$

Hence in general, for the existence of a triple point on a curve, its coefficients must satisfy four conditions.

The complete investigation of multiple points is effected

---

* From the form of the equation we see that the lines $x = 0$, $y = 0$ are tangents to the conic, and that $2ax + 2by + c = 0$ represents the line joining the points of contact; but this line is parallel to the third asymptote $ax + by + c = 0$, and evidently passes through the middle points of the intercepts made by this asymptote on the two others.

† *Liouville's Journal*, vol. ii. p. 14.

266   *Multiple Points on Curves.*

more satisfactorily by introducing the method of trilinear co-ordinates. The discussion of curves from this point of view is beyond the limits proposed in this elementary Treatise.

215 (*a*). **Cusps, in General.**—Thus far singular points have been considered with reference to the cases in which they occur most simply. In proceeding to curves of higher degrees they may admit of many complications; for instance ordinary cusps, such as represented in Fig. 20, may be called cusps of the first species, the tangent lying between both branches: the cases in which both branches lie on the same side, as exhibited in the accompanying figure, may be called cusps of the second species. Professor Cayley has shown how this is to be considered as consisting of several singularities happening at a point (Salmon's *Higher Plane Curves*, Art. 58).

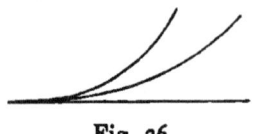

Fig. 26.

Again, both of these classes may be called single cusps, as distinguished from *double cusps* extending on both sides of the point of contact. Double cusps are styled *tacnodes* by Professor Cayley. These points are sometimes called *points of osculation;* however, as the two branches do not in general *osculate* each other, this nomenclature is objectionable. It should be observed that whenever we use the word cusp without limitation, we refer to the ordinary cusp of the first species.

Cusps are called *points de rebroussement* by French writers, and *Rückkehrpunkte* by Germans, both expressing the turning backwards of the point which is supposed to trace out the curve; an idea which has its English equivalent in their name of *stationary points*. A fuller discussion of the different classes of cusps will be given in a subsequent place. We shall conclude this chapter with a few remarks on the multiple points of curves whose equations are given in polar co-ordinates.

EXAMPLES.

1. $(y - x^2)^2 = x^5$.

Here the origin is a cusp; also

$$y = x^2 \pm x^{\frac{5}{2}};$$

hence, when x is less than unity, both values of y are positive, and consequently the cusp is of the *second* species.

2. Show that the origin is a *double cusp* in the curve

$$x^5 + bx^4 - a^3y^2 = 0.$$

216. Multiple Points of Curves in Polar Co-ordinates.

—If a curve referred to polar co-ordinates pass through the origin, it is evident that the direction of the tangent at that point is found by making $r = 0$ in its equation; in this case, if the equation of the curve reduce to $f(\theta) = 0$, the resulting value of θ gives the direction of the tangent in question.

If the equation $f(\theta) = 0$ has two real roots in θ, less than π, the origin is a double point, the tangents being determined by these values of θ.

If these values of θ were equal, the origin would be a cusp; and so on.

In fact, it will be observed that the multiple points on algebraic curves have been discussed by reducing them to polar equations, so that the theory already given must apply to curves referred to polar, as well as to algebraic co-ordinates.

It may be remarked, however, that the *order* of a multiple point cannot, generally, be determined unless with reference to Cartesian co-ordinates, in like manner as the degree of a curve in general is determined only by a similar reference.

For example, in the equation

$$r = a \cos^2\theta - b \sin^2\theta,$$

the tangents at the origin are determined by the equation $\tan \theta = \pm \sqrt{\dfrac{a}{b}}$, and the origin would seem to be only a double point; however, on transforming the equation to rectangular axes, it becomes

$$(x^2 + y^2)^3 = (ax^2 - by^2)^2;$$

from which it appears that the origin is a multiple point of the fourth order, and the curve of the sixth degree. In fact, what is meant by the degree of a curve, or the multiplicity of a point, is the number of intersections of the curve with any right line, or the number of intersections which coincide for every line through such a point, and neither of these are at once evident unless the equation be expressed by line co-ordinates, such as Cartesian, or trilinear co-ordinates; whereas in polar co-ordinates one of the variables is a circular co-ordinate.

EXAMPLES.

1. Determine the tangents at the origin to the curve
$$y^2 = x^2(1-x^2). \qquad Ans.\ x+y=0,\ x-y=0.$$

2. Show that the curve
$$x^4 - 3axy + y^4 = 0$$
touches the axes of co-ordinates at the origin.

3. Find the nature of the origin on the curve
$$x^4 - ax^2y + by^3 = 0.$$

4. Show that the origin is a conjugate point on the curve
$$ay^2 - x^3 + bx^2 = 0$$
when a and b have the same sign; and a node, when they have opposite signs.

5. Show that the origin is a conjugate point on the curve
$$y^2(x^2 - a^2) = x^4.$$

6. Prove that the origin is a cusp on the curve
$$(y - x^2)^2 = x^3.$$

7. In the curve
$$(y - x^2)^2 = x^n,$$
show that the origin is a cusp of the first or second species, according as n is $<$ or $>$ 4.

8. Find the number and the nature of the singular points on the curve
$$x^4 + 4ax^3 - 2ay^3 + 4a^2x^2 - 3a^2y^2 + 4a^4 = 0.$$

9. Show that the points of intersection of the curve
$$\left(\frac{x}{a}\right)^{\frac{2}{3}} + \left(\frac{y}{b}\right)^{\frac{2}{3}} = 1$$
with the axes are cusps.

10. Find the double points on the curve
$$x^4 - 4ax^3 + 4a^2x^2 - b^2y^2 + 2b^3y - a^4 - b^4 = 0.$$

11. Prove that the four tangents from the origin to the curve

$$u_1 + u_2 + u_3 = 0$$

are represented by the equation $4u_1 u_3 = u_2^2$.

12. Show that to a double point on any curve corresponds another double point, of the same kind, on the inverse curve with respect to any origin.

13. Prove that the origin in the curve

$$x^4 - 2ax^2y - axy^2 + a^2y^2 = 0$$

is a cusp of the second species.

14. Show that the cardioid

$$r = a(1 + \cos\theta)$$

has a cusp at the origin.

15. If the origin be situated on a curve, prove that its first pedal passes through the origin, and has a cusp at that point.

16. Find the nature of the origin in the following curves:—

$$r^3 = a^3 \sin 3\theta, \quad r^n = a^n \sin n\theta, \quad r = \frac{a\theta^2}{b\theta + c}.$$

17. Show that the origin is a conjugate point on the curve

$$x^4 - ax^2y + axy^2 + a^2y^2 = 0.$$

18. If the inverse of a conic be taken, show that the origin is a double point on the inverse curve; also that the point is a conjugate point for an ellipse, a cusp for a parabola, and a node for a hyperbola.

19. Show that the condition that the cubic

$$xy^2 + ax^3 + bx^2 + cx + d + 2ey = 0$$

may have a double point is the same as the condition that the equation

$$ax^4 + bx^3 + cx^2 + dx - e^2 = 0$$

may have equal roots.

20. In the inverse of a curve of the nth degree, show that the origin is a multiple point of the nth order, and that the n tangents at that point are parallel to the asymptotes to the original curve.

CHAPTER XV.

ENVELOPES.

217. Method of Envelopes.—If we suppose a series of different values given to a in the equation

$$f(x, y, a) = 0, \qquad (1)$$

then for each value we get a distinct curve, and the above equation may be regarded as representing an indefinite number of curves, each of which is determined when the corresponding value of a is known, and varies as a varies.

The quantity a is called a variable *parameter*, and the equation $f(x, y, a) = 0$ is said to represent a *family of curves*; a single determinate curve corresponding to each distinct value of a; provided a enters into the equation in a rational form only.

If now we regard a as varying continuously, and suppose the two curves

$$f(x, y, a) = 0, \quad f(x, y, a + \Delta a) = 0$$

taken, then the co-ordinates of their points of intersection satisfy each of these equations, and therefore also satisfy the equation

$$\frac{f(x, y, a + \Delta a) - f(x, y, a)}{\Delta a} = 0.$$

Now, in the limit, when Δa is infinitely small, the latter equation becomes

$$\frac{df(x, y, a)}{da} = 0; \qquad (2)$$

and accordingly the points of intersection of two infinitely near curves of the system satisfy each of the equations (1) and (2).

The locus of the points of *ultimate* intersection for the entire system of curves represented by $f(x, y, a) = 0$, is obtained by eliminating a between the equations (1) and (2). This locus is called the *envelope* of the system, and it can be easily seen that it is touched by every curve of the system.

For, if we consider three consecutive curves, and suppose P_1 to be the point of intersection of the first and second, and P_2 that of the second and third, the line $P_1 P_2$ joins two infinitely near points on the envelope as well as on the intermediate of the three curves; and hence is a tangent to each of these curves.

This result appears also from analytical considerations, thus:—the direction of the tangent at the point x, y, to the curve $f(x, y, a) = 0$, is given by the equation

$$\frac{df}{dx} + \frac{df}{dy}\frac{dy}{dx} = 0;$$

in which a is considered a constant.

Again, if the point x, y be on the envelope, since then a is given in terms of x and y by equation (2), the direction of the tangent to the envelope is given by the equation

$$\frac{df}{dx} + \frac{df}{dy}\frac{dy}{dx} + \frac{df}{da}\left(\frac{da}{dx} + \frac{da}{dy}\frac{dy}{dx}\right) = 0,$$

or
$$\frac{df}{dx} + \frac{df}{dy}\frac{dy}{dx} = 0,$$

since $\dfrac{df}{da} = 0$ for the point on the envelope.

Consequently, the values of $\dfrac{dy}{dx}$ are the same for the two curves at their common point, and hence they have a common tangent at that point.

One or two elementary examples will help to illustrate this theory.

The equation $x \cos a + y \sin a = p$, in which a is a variable parameter, represents a system of lines situated at the same perpendicular distance p from the origin, and consequently all touching a circle.

This result also follows from the preceding theory; for we have

$$f(x, y, a) = x \cos a + y \sin a - p = 0,$$

$$\frac{df(x, y, a)}{da} = -x \sin a + y \cos a = 0,$$

and, on eliminating a between these equations, we get

$$x^2 + y^2 = p^2,$$

which agrees with the result stated above.

Again, to find the envelope of the line

$$y = ax + \frac{m}{a},$$

where a is a variable parameter.

Here

$$f(x, y, a) = y - ax - \frac{m}{a} = 0,$$

$$\frac{df(x, y, a)}{da} = -x + \frac{m}{a^2} = 0; \therefore a = \sqrt{\frac{m}{x}}.$$

Substituting this value for a, we get for the envelope

$$y^2 = 4mx,$$

which represents a parabola.

218. Envelope of $La^2 + 2Ma + N = 0$. Suppose L, M, N, to be known functions of x and y, and a a parameter, then

$$f(x, y, a) = La^2 + 2Ma + N = 0,$$

$$\frac{df}{da} = 2La + 2M = 0;$$

accordingly, the envelope of the curve represented by the preceding expression is the curve

$$LN = M^2.$$

Hence, when L, M, N are linear functions in x and y, this envelope is a conic touching the lines L, N, and having M for the chord of contact.

Conversely, the equation to any tangent to the conic $LN = M^2$ can be written in the form

$$La^2 + 2Ma + N = 0,{}^*$$

where a is an arbitrary parameter.

219. Undetermined Multipliers applied to Envelopes.—In many cases of envelopes the equation of the moving curve is given in the form

$$f(x, y, a, \beta) = c_1, \qquad (3)$$

where the parameters a, β are connected by an equation of the form

$$\phi(a, \beta) = c_2. \qquad (4)$$

In this case we may regard β in (3) as a function of a by reason of equation (4); hence, differentiating both equations, the points of intersection of two consecutive curves must satisfy the two following equations:

$$\frac{df}{da} + \frac{df}{d\beta}\frac{d\beta}{da} = 0, \text{ and } \frac{d\phi}{da} + \frac{d\phi}{d\beta}\frac{d\beta}{da} = 0.$$

Consequently
$$\frac{\dfrac{df}{da}}{\dfrac{d\phi}{da}} = \dfrac{\dfrac{df}{d\beta}}{\dfrac{d\phi}{d\beta}}.$$

If each of these fractions be equated to the undetermined quantity λ, we get

$$\left.\begin{array}{l} \dfrac{df}{da} = \lambda \dfrac{d\phi}{da} \\[1em] \dfrac{df}{d\beta} = \lambda \dfrac{d\phi}{d\beta} \end{array}\right\}; \qquad (5)$$

---

* Salmon's *Conics*, Art. 248.

and the required envelope is obtained by eliminating α, β, and λ between these and the two given equations.

The advantage of this method is especially found when the given equations are homogeneous functions in α and β; for suppose them to be of the forms

$$f(x, y, \alpha, \beta) = c_1, \quad \phi(\alpha, \beta) = c_2,$$

where the former is homogeneous of the nth degree, and the latter of the mth, in α and β. Multiply the former equation in (5) by α, and the latter by β, and add; then, by Euler's theorem of Art. 102, we shall have

$$nc_1 = mc_2\lambda, \quad \text{or } \lambda = \frac{nc_1}{mc_2}, \qquad (6)$$

by means of which value we can generally eliminate α and β from our equations.

EXAMPLES.

1. To find the envelope of a line of given length (a) whose extremities move along two fixed rectangular axes.

Taking the given lines for axes of co-ordinates, we have the equations

$$\frac{x}{\alpha} + \frac{y}{\beta} = 1, \quad \alpha^2 + \beta^2 = a^2.$$

Hence
$$\frac{x}{\alpha^2} = \lambda\alpha, \quad \frac{y}{\beta^2} = \lambda\beta,$$

from which we get
$$\lambda = \frac{1}{a^2};$$

$$\therefore \alpha = (a^2 x)^{\frac{1}{3}}, \quad \beta = (a^2 y)^{\frac{1}{3}},$$

and the required locus is represented by

$$x^{\frac{2}{3}} + y^{\frac{2}{3}} = a^{\frac{2}{3}}.$$

2. To find the envelope of a system of concentric and coaxal ellipses of constant area.

Here
$$\frac{x^2}{\alpha^2} + \frac{y^2}{\beta^2} = 1, \quad \alpha\beta = c;$$

hence
$$\frac{x^2}{\alpha^3} = \lambda\beta, \quad \frac{y^2}{\beta^3} = \lambda\alpha; \quad \therefore 2\lambda c = 1,$$

and the required envelope is the equilateral hyperbola

$$2xy = c.$$

Examples.

3. To find the envelope of all the normals to an ellipse.

Here we have the equations

$$a^2 \frac{x}{\alpha} - b^2 \frac{y}{\beta} = a^2 - b^2, \text{ and } \frac{\alpha^2}{a^2} + \frac{\beta^2}{b^2} = 1,$$

where α and β are the co-ordinates of any point on the ellipse.

Hence, $\qquad \dfrac{a^2 x}{\alpha^2} = \lambda \dfrac{\alpha}{a^2}, \quad \dfrac{b^2 y}{\beta^2} = -\lambda \dfrac{\beta}{b^2};$

consequently $\qquad \lambda = a^2 - b^2,$

and we get $\qquad a^4 x = (a^2 - b^2)\alpha^3, \quad b^4 y = -(a^2 - b^2)\beta^3;$

$$\therefore \frac{\alpha}{a} = \left(\frac{ax}{a^2 - b^2}\right)^{\frac{1}{3}}, \quad \frac{\beta}{b} = -\left(\frac{by}{a^2 - b^2}\right)^{\frac{1}{3}}.$$

Substituting in the equation of the ellipse, we get for the required envelope,

$$(ax)^{\frac{2}{3}} + (by)^{\frac{2}{3}} = (a^2 - b^2)^{\frac{2}{3}}.$$

This equation represents the *evolute* of the ellipse.

4. Find the envelope of the line $\dfrac{x}{\alpha} + \dfrac{y}{\beta} = 1$, where α and β are connected by the equation

$$\alpha^m + \beta^m = c^m. \qquad Ans. \ x^{\frac{m}{m+1}} + y^{\frac{m}{m+1}} = c^{\frac{m}{m+1}}.$$

220. The preceding method can be readily extended to the general case in which the equation of the enveloping curve contains any number, n, of variable parameters, which are connected by $n-1$ independent equations. The method of procedure is the same as that already considered in Chapter XI. on maxima and minima, and does not require a separate investigation here.

T 2

Examples.

1. Prove that the envelope of the system of lines $\dfrac{x}{l} + \dfrac{y}{m} = 1$, where l and m are connected by the equation $\dfrac{l}{a} + \dfrac{m}{b} = 1$, is the parabola

$$\left(\dfrac{x}{a}\right)^{\frac{1}{2}} + \left(\dfrac{y}{b}\right)^{\frac{1}{2}} = 1.$$

2. One angle of a triangle is fixed in position, find the envelope of the opposite side when the area is given. *Ans.* A hyperbola.

3. Find the envelope of a right line when the sum of the squares of the perpendiculars on it from two given points is constant.

4. Find the envelope of a right line, when the rectangle under the perpendiculars from two given points is constant.
Ans. A conic having the two points as foci.

5. From a point P on the hypothenuse of a right-angled triangle, perpendiculars PM, PN are drawn to the sides; find the envelope of the line MN.

6. Find the envelope of the system of circles whose diameters are the chords drawn parallel to the axis-minor of a given ellipse.

7. Find the envelope of the circle

$$x^2 + y^2 - 2aex + a^2 - b^2 = 0,$$

where a is an arbitrary parameter; and find when the contact between the circle and the envelope is real, and when imaginary.

(*a*). Show from this example that the focus of an ellipse may be regarded as an infinitely small circle having double contact with the ellipse, the directrix being the chord joining the points of contact.

8. Show that the envelope of the system of conics

$$\dfrac{x^2}{a} + \dfrac{y^2}{a-h} = 1,$$

where a is a variable parameter, is represented by the equation

$$(x \pm \sqrt{h})^2 + y^2 = 0.$$

Hence show that a system of conics having the same foci may be regarded as inscribed in the same imaginary quadrilateral.

9. Find the envelope of the line

$$x a^m + y \beta^m = a^{m+1},$$

where the parameters a and β are connected by the equation

$$a^n + \beta^n = b^n.$$

Ans. $x^{\frac{n}{n-m}} + y^{\frac{n}{n-m}} = b\left(\dfrac{a^{m+1}}{b^m}\right)^{\frac{n}{n-m}}.$

10. On any radius vector of a curve as diameter a circle is described: prove geometrically that the envelope of all such circles is the first pedal of the curve with respect to the origin.

11. If circles be described on the focal radii vectores of a conic as diameters, prove that their envelope is the circle described on the axis major of the conic as diameter.

12. Prove that the envelope of the circles described on the central radii of an ellipse as diameters is a Lemniscate.

13. Find the envelope of semicircles described on the radii of the curve

$$r^n = a^n \cos n\theta$$

as diameters.

14. If perpendiculars be drawn at each point on a curve to the radii vectores drawn from a given point, prove that their envelope is the reciprocal polar of the inverse of the given curve, with respect to the given point.

15. Find the envelope of a circle whose centre moves along the circumference of a fixed circle, and which touches a given right line.

16. Ellipses are described with coincident centre and axes, and having the sum of their semiaxes constant; find their envelope.

17. Find the equation of the envelope of the line $\lambda x + \mu y + \nu = 0$, where the parameters are connected by the equation

$$a\lambda^2 + b\mu^2 + c\nu^2 + 2f\mu\nu + 2g\nu\lambda + 2h\lambda\mu = 0.$$

Ans. $\begin{vmatrix} a, & h, & g, & x \\ h, & b, & f, & y \\ g, & f, & c, & 1 \\ x, & y, & 1, & 0 \end{vmatrix} = 0.$

18. At any point of a parabola a line is drawn making with the tangent an angle equal to the angle between the tangent and the ordinate at the point; prove that the envelope of the line is the first negative pedal, with regard to the focus, of the parabola; and hence that its equation is $r^{\frac{1}{3}} \cos \frac{1}{3}\theta = a^{\frac{1}{3}}$, the focus being pole.

N.B.—This curve is the *caustic by reflexion* for rays perpendicular to the axis of the parabola.

19. Join the centre, O, of an equilateral hyperbola to any point, P, on the curve, and at P draw a line, PQ, making with the tangent an angle equal to the angle between OP and the tangent. Show that the envelope of PQ is the first negative pedal of the curve

$$r^2 = 2a^2 \sin \frac{2}{3}\theta \sin \frac{4}{3}\theta,$$

the centre being pole, and axis minor *prime vector*.

N.B.—This gives the *caustic by reflexion* of the equilateral hyperbola, the centre being the radiant point.

20. A right line revolves with a uniform angular velocity, while one of its points moves uniformly along a fixed right line; find its envelope.

Ans. A cycloid.

CHAPTER XVI.

CONVEXITY AND CONCAVITY. POINTS OF INFLEXION.

221. Convexity and Concavity.—If the tangent be drawn at any point on a curve, the neighbouring portion of the curve generally lies altogether on one side of the tangent, and is convex with respect to all points lying at the opposite side of that line, and concave for points at the same side.

Thus, in the accompanying figure, the portion QPQ' is convex towards all points lying below the tangent, and concave for points above.

If the curve be referred to the co-ordinate axes OX and OY, then whenever the ordinates of points near to P on the curve are greater than those of the points on the tangent corresponding to the same abscissæ, the curve is said to be concave towards the positive direction of Y.

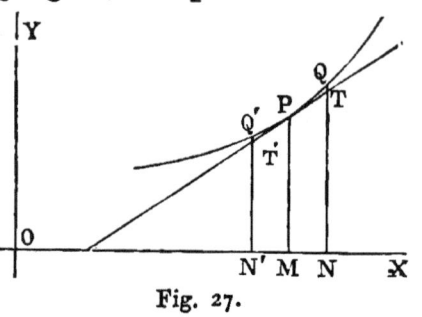

Fig. 27.

Now, suppose $y = \phi(x)$ to be the equation of the curve, then that of the tangent at a point x, y, by Art. 168, is

$$Y - y = (X - x)\frac{dy}{dx}.$$

Let P be the point x, y, and $MN = h = MN'$, $QN = y_1$, $TN = Y_1$, and we have

$$y_1 = \phi(x + h) = \phi(x) + h\phi'(x) + \frac{h^2}{1 \cdot 2}\phi''(x) + \frac{h^3}{1 \cdot 2 \cdot 3}\phi'''(x) + \&c.$$

$$Y_1 = y + h\phi'(x) = \phi(x) + h\phi'(x);$$

$$\therefore y_1 - Y_1 = \frac{h^2}{1 \cdot 2}\phi''(x) + \frac{h^3}{1 \cdot 2 \cdot 3}\phi'''(x) + \&c. \quad (1)$$

When h is very small, the sign of the right-hand side of this equation is the same in general as that of its first term, and accordingly the sign of $y_1 - Y_1$, or of QT, is the same as that of $\phi''(x)$.

Hence, for a point above the axis of x, the curve is convex towards that axis when $\phi''(x)$ is positive, and concave when negative.

We accordingly see that the convexity or concavity at any point depends on the sign of $\phi''(x)$ or $\dfrac{d^2y}{dx^2}$, at the point.

222. Points of Inflexion.—If, however, $\phi''(x) = 0$ at the point P, we shall have

$$y_1 - Y_1 = \frac{h^3}{1.2.3}\phi'''(x) + \frac{h^4}{1.2.3.4}\phi^{iv}(x) + \&c. \qquad (2)$$

Now, provided $\phi'''(x)$ be not zero, $y_1 - Y_1$ changes its sign with h, i.e. if $MN' = MN = h$, and if Q lies above T, the corresponding point Q' lies below T', and the portions of the curve near to P lie at opposite sides of the tangent, as in the figure.

Consequently, the tangent at such a point cuts the curve, as well as touches it, at its point of contact. Such points on a curve are called points of inflexion.

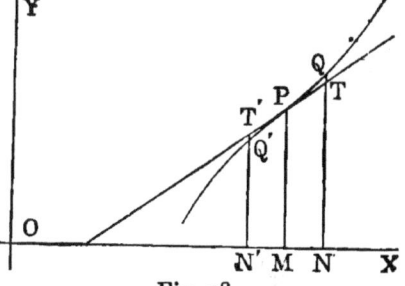

Fig. 28.

Again, if $\phi'''(x)$ as well as $\phi''(x)$ vanish at the point P, we shall have

$$y_1 - Y_1 = \frac{h^4}{1.2.3.4}\phi^{iv}(x) + \&c.;$$

and, provided $\phi^{iv}(x)$ be not zero at the point, $y_1 - Y_1$ does not change sign with h, and accordingly the tangent does not intersect the curve at its point of contact.

Generally, the tangent does or does not cut the curve at its point of contact, according as the first derived function which does not vanish is of an odd, or of an even order; as can be easily seen by the preceding method.

From the foregoing discussion it follows that at a point of inflexion the curve changes from convex to concave with respect to the axis of x, or conversely.

On this account such points are called points of *contrary flexure* or of *inflexion*.

223 The subject of inflexion admits also of being treated by the method of Art. 196, as follows:—The points of intersection of the line $y = \mu x + \nu$ with the curve $y = \phi(x)$ are evidently determined by the equation

$$\phi(x) = \mu x + \nu. \qquad (3)$$

Suppose A, B, C, D, &c., to represent the points of section in question, and let $x_1, x_2, \ldots x_n$ be the roots of equation (3); then the line becomes a tangent, if two of these roots are equal, i.e., if

Fig. 29.

$\phi'(x_1) = \mu$, where x_1 denotes the value of x belonging to the point of contact.

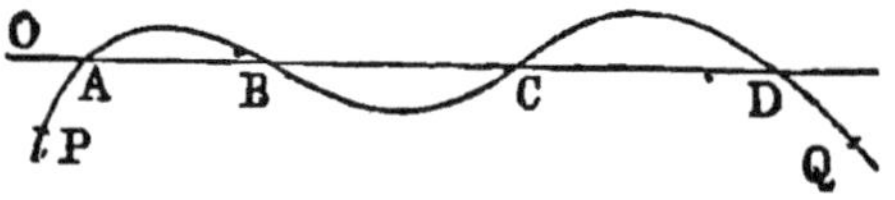

Again, three of the roots become equal if we have in addition $\phi''(x_1) = 0$; in this case the tangent meets the curve in three consecutive points, and evidently cuts the curve at its point of contact; for in our figure the portions PA and CD of the curve lie at opposite sides of the cutting line, but when the points A, B, C become coincident, the portions AB and BC become evanescent, and the curve is evidently cut as well as touched by the line.

In like manner, if $\phi'''(x_1)$ also vanish, the tangent must be regarded as cutting the curve in four consecutive points: such a point is called a *point of undulation*.

It may be observed, that if a right line cut a continuous branch of a curve in three points, A, B, C, as in our figure, the curve must change from convex to concave, or conversely, between the extreme points A and C, and consequently it must have a point of inflexion between these points; and so on for additional points of section.

Again, the tangent to a curve of the nth degree at a point of inflexion cannot intersect the curve in more than $n - 3$ other points: for the point of inflexion counts for three among the points of section. For example, the tangent to a curve of the third degree at a point of inflexion cannot meet the curve in any other point. Consequently, if a point of inflexion on a cubic be taken as origin, and the tangent at it as axis of x, the equation of the curve must be of the form

$$x^3 + y\phi = 0,$$

where ϕ represents an expression of the second and lower degrees in x and y. For, when $y = 0$, the three roots of the resulting equation in x must be each zero, as the axis of x meets the curve in three points coincident with the origin.

The preceding equation is of the form

$$u_3 + u_2 + u_1 = 0,$$

or, when written in full,

$$x^3 + y(ax^2 + 2hxy + by^2) + y(2gx + 2fy + c) = 0. \quad (4)$$

Now, supposing tangents drawn from the origin to the curve, their points of contact, by Art. 176, lie on the curve

$$u_2 + 2u_1 = 0,$$

i.e. on the curve

$$(gx + fy + c)y = 0.$$

The factor $y = 0$ corresponds to the tangent at the point of inflexion, and the other factor $gx + fy + c = 0$ passes through the points of contact of the three other tangents to the curve.

Hence, we infer that *from a point of inflexion on a cubic but three tangents can be drawn to the curve, and their three points of contact lie in a right line.*

It can be shown that this right line cuts harmonically every radius vector of the curve which passes through the point of inflexion.

For, transforming equation (4) to polar co-ordinates, and dividing by r, it becomes of the form

$$Ar^2 + Br + C = 0.$$

If r', r'' be the roots of this quadratic, we have

$$\frac{1}{r'} + \frac{1}{r''} = -\frac{B}{C}.$$

Now, if ρ be the harmonic mean between r' and r'', this gives

$$\frac{2}{\rho} = \frac{1}{r'} + \frac{1}{r''} = -\frac{B}{C} = -\frac{2g\cos\theta + 2f\sin\theta}{c}.$$

Hence the equation of the locus of the extremities of the harmonic means is

$$gx + fy + c = 0. \qquad Q.E.D.$$

This theorem is due to Maclaurin (*De Lin. Geom. Prop. Gen.*, Sec. III. Prop. 9).

From this property the line is called the *harmonic polar* of the point of inflexion. This line holds a fundamental place in the general theory of cubics.*

224. Stationary Tangents.—Since the tangent at a point of inflexion may be regarded as meeting the curve in three consecutive points, it follows that at such a point the tangent does not alter its position as its point of contact passes to the consecutive point, and hence the tangent in this case is called a *stationary* tangent.

The equation $\dfrac{d^2y}{dx^2} = 0$ follows immediately from the last consideration; for when the tangent is stationary we must have $\dfrac{d\phi}{dx} = 0$, where ϕ, as in Art. 171, denotes the angle which the tangent makes with the axis of x; but $\tan\phi = \dfrac{dy}{dx}$, hence $\dfrac{d^2y}{dx^2} = 0$, which is the same condition for a point of inflexion as that before arrived at.

---

* Chasles, *Aperçu Historique*, note xx.; Salmon's *Higher Plane Curves*, Art. 179.

Examples.

1. Show that the origin is a point of inflexion on the curve
$$a^3y = bxy + cx^3 + dx^4.$$

2. The origin is a point of inflexion on the cubic $u_3 + u_1 = 0$?

3. In the curve
$$a^{m-1}y = x^m,$$
prove that the origin is a point of inflexion if m be greater than 2.

4. In the system of curves
$$y^n = kx^m,$$
find under what circumstances the origin is (a) a point of inflexion, (b) a cusp.

5. Find the co-ordinates of the point of inflexion on the curve
$$x^3 - 3bx^2 + a^2y = 0. \qquad Ans.\ x = b,\ y = \frac{2b^3}{a^2}.$$

6. If a curve of an odd degree has a centre, prove that it is a point of inflexion on the curve.

7. Prove that the origin is a point of undulation on the curve
$$u_1 + u_4 + u_5 + \&c., + u_n = 0.$$

8. Show that the points of inflexion on curves referred to polar co-ordinates are determined by aid of the equation
$$u + \frac{d^2u}{d\theta^2} = 0, \quad \text{where } u = \frac{1}{r}.$$

9. In the curve $r\theta^m = a$, prove that there is a point of inflexion when
$$\theta = \sqrt{m(1-m)}.$$

10. In the curve $y = c \sin \dfrac{x}{a}$, prove that the points in which the curve meets the axis of x are all points of inflexion.

11. Show geometrically that to a node on any curve corresponds a line touching its reciprocal polar in two distinct points; and to a cusp corresponds a point of inflexion.

12. If the origin be a point of inflexion on the curve

$$u_1 + u_2 + u_3 + \ldots + u_n = 0,$$

prove that u_2 must contain u_1 as a factor.

13. Show that the points of inflexion of the cubical parabola

$$y^2 = (x - a)^2 (x - b)$$

lie on the line

$$3x + a = 4b:$$

and hence prove that if the cubic has a node, it has no real point of inflexion; but if it has a conjugate point, it has two real points of inflexion, besides that at infinity.

14. Prove that the points of inflexion on the curve $y^2 = x^2(x^2 + 2px + q)$ are determined by the equation $2x^3 + 6px^2 + 3(p^2 + q)x + 2pq = 0$.

15. If $y^2 = f(x)$ be the equation of a curve, prove that the abscissæ of its points of inflexion satisfy the equation

$$\{f'(x)\}^2 = 2f(x) \cdot f''(x).$$

16. Show that the maximum and minimum ordinates of the curve

$$y = 2f(x) f''(x) - \{f'(x)\}^2$$

correspond to the points of intersection of the curve $y^2 = f(x)$ with the axis of x.

17. When $y^2 = f(x)$ represents a cubic, prove that the biquadratic in x which determines its points of inflexion has one, and but one, pair of real roots. Prove also that the lesser of these roots corresponds to no real point of inflexion, while the greater corresponds, in general, to two.

18. Prove that the point of inflexion of the cubic

$$ay^3 + 3bxy^2 + 3cx^2y + dx^3 + 3ex^2 = 0$$

lies in the right line $ay + bx = 0$, and has for its co-ordinates

$$x = -\frac{3a^2e}{G}, \text{ and } y = \frac{3abe}{G},$$

where G is the same as in Example 32, p. 190.

19. Find the nature of the double point of the curve

$$y^2 = (x - 2)^2 (x - 5),$$

and show that the curve has two real points of inflexion, and that they subtend a right angle at the double point.

20. The co-ordinates of a point on a curve are given in terms of an angle θ by the equations

$$x = \sec^3 \theta, \quad y = \tan \theta \sec^2 \theta;$$

prove that there are two finite points of inflexion on the curve, and find the values of θ at these points.

CHAPTER XVII.

RADIUS OF CURVATURE. EVOLUTES. CONTACT. RADII OF CURVATURE AT A DOUBLE POINT.

225. Curvature. Angle of Contingence.—Every continuous curve is regarded as having a determinate curvature at each point, this curvature being greater or less according as the curve deviates more or less rapidly from the tangent at the point.

The total curvature of an arc of a plane curve is measured by the angle through which it is bent between its extremities—that is, by the external angle between the tangents at these points, assuming that the arc in question has no point of inflexion on it. This angle is called the *angle of contingence* of the arc.

The curvature of a circle is evidently the same at each of its points.

To compare the curvatures of different circles, let the arcs AB and ab of two circles be of equal length, then the total curvatures of these arcs are measured by the angles between their tangents, or by the angles ACB and acb at their centres: but

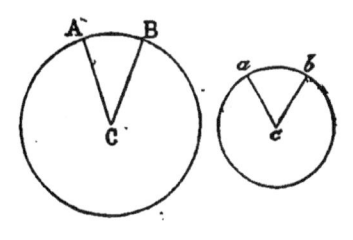

Fig. 30.

$$\angle ACB : \angle acb = \frac{\text{arc } AB}{AC} : \frac{\text{arc } ab}{ac} = \frac{1}{AC} : \frac{1}{ac}.$$

Consequently, the curvatures of the two circles are to each other inversely as their radii; or the curvature of a circle varies inversely as its radius.

Also if Δs represent any arc of a circle of radius r, and $\Delta \phi$ the angle between the tangents at its extremities, we have

$$r = \frac{\Delta s}{\Delta \phi}.$$

The curvature of a curve at any point is found by determining the circle which has the same curvature as that of an indefinitely small elementary arc of the curve taken at the point.

226. Radius of Curvature.—Let ds denote an infinitely small element of a curve at a point, $d\phi$ the corresponding angle of contingence expressed in circular measure, then $\dfrac{ds}{d\phi}$ evidently represents the radius of the circle which has the same curvature as that of the given curve at the point.

This radius is called the *radius of curvature* for the point, and is usually denoted by the letter ρ.

To find an expression for ρ, let the curve be referred to rectangular axes, and suppose x and y to be the co-ordinates of the point in question; then if ϕ denote the angle which the tangent makes with the axis of x, we have

$$\tan\phi = \frac{dy}{dx}; \quad \therefore \frac{d.\tan\phi}{dx} = \frac{d^2y}{dx^2},$$

or

$$\sec^2\phi \frac{d\phi}{dx} = \frac{d^2y}{dx^2}.$$

Again,

$$\frac{d\phi}{ds} = \frac{d\phi}{dx}\frac{dx}{ds} = \cos\phi \frac{d\phi}{dx} = \cos^3\phi \frac{d^2y}{dx^2}.$$

Hence

$$\rho = \frac{1}{\dfrac{d\phi}{ds}} = \frac{\sec^3\phi}{\dfrac{d^2y}{dx^2}} = \frac{\left(1 + \left(\dfrac{dy}{dx}\right)^2\right)^{\frac{3}{2}}}{\dfrac{d^2y}{dx^2}}. \tag{1}$$

At a point of inflexion $\dfrac{d^2y}{dx^2} = 0$: accordingly the radius of curvature at such a point is infinite: this is otherwise evident since the tangent in this case meets the curve in three consecutive points. (Art. 222.)

Again, as the expression $\left(1 + \left(\dfrac{dy}{dx}\right)^2\right)^{\frac{3}{2}}$ has always two values, the one positive and the other negative, while the curve can have in general but one definite circle of curvature at any point, it is necessary to agree upon which sign is to be taken. We shall adopt the positive sign, and regard ρ as being positive when $\dfrac{d^2y}{dx^2}$ is positive; i.e. when the curve is convex at the point with respect to the axis of x.

227. Other Expressions for ρ.—It is easy to obtain other forms of expression for the radius of curvature; thus by Art. 178 we have

$$\cos\phi = \frac{dx}{ds}, \quad \sin\phi = \frac{dy}{ds}.$$

Hence, if the arc be regarded as the independent variable, we get

$$-\sin\phi\,\frac{d\phi}{ds} = \frac{d^2x}{ds^2}, \quad \cos\phi\,\frac{d\phi}{ds} = \frac{d^2y}{ds^2},$$

from which, if we square and add, we obtain

$$\frac{1}{\rho^2} = \left(\frac{d\phi}{ds}\right)^2 = \left(\frac{d^2x}{ds^2}\right)^2 + \left(\frac{d^2y}{ds^2}\right)^2. \qquad (2)$$

Again, the equations $dx = \cos\phi\,ds$, $dy = \sin\phi\,ds$, give by differentiation (substituting $\dfrac{ds}{\rho}$ for $d\phi$),

$$d^2x = \cos\phi\,d^2s - \sin\phi\,\frac{(ds)^2}{\rho}, \quad d^2y = \sin\phi\,d^2s + \cos\phi\,\frac{(ds)^2}{\rho}. \qquad (3)$$

Whence, squaring and adding, we obtain

$$(d^2x)^2 + (d^2y)^2 = (d^2s)^2 + \frac{(ds)^4}{\rho^2},$$

or

$$\rho = \frac{ds^2}{\sqrt{(d^2x)^2 + (d^2y)^2 - (d^2s)^2}}. \qquad (4)$$

Again, if the former equation in (3) be multiplied by $\sin \phi$, and the latter by $\cos \phi$, we obtain on subtraction,

$$\cos \phi\, d^2y - \sin \phi\, d^2x = \frac{ds^2}{\rho}, \quad \text{or} \quad dx\, d^2y - dy\, d^2x = \frac{ds^3}{\rho}.$$

Hence
$$\rho = \frac{(dx^2 + dy^2)^{\frac{3}{2}}}{dx\, d^2y - dy\, d^2x}. \tag{5}$$

The *independent* variable is undetermined in formulæ (4) and (5), and may be any quantity of which both x and y are functions.

For example, in the motion of a particle along a curve, when the *time* is taken as the independent variable, we get from (4) an important result in Mechanics.

Examples.

1. To find the radius of curvature at any point on the parabola $x^2 = 4my$.

Here
$$2m\frac{dy}{dx} = x, \quad 2m\frac{d^2y}{dx^2} = 1, \quad 1 + \left(\frac{dy}{dx}\right)^2 = 1 + \frac{x^2}{4m^2} = 1 + \frac{y}{m};$$

$$\therefore \rho = \frac{2(m+y)^{\frac{3}{2}}}{m^{\frac{1}{2}}}.$$

2. Find the radius of curvature in the catenary

$$y = \frac{a}{2}\left(e^{\frac{x}{a}} + e^{-\frac{x}{a}}\right).$$

Here
$$\frac{dy}{dx} = \frac{1}{2}\left(e^{\frac{x}{a}} - e^{-\frac{x}{a}}\right), \quad \frac{d^2y}{dx^2} = \frac{y}{a^2}; \quad \therefore \rho = -\frac{y^2}{a}.$$

Hence the radius of curvature is equal to the part of the normal intercepted by the axis of x, but measured in the opposite direction (Ex. 4, Art. 171).

3. In the cubical parabola $3a^2y = x^3$, we have

$$a^2\frac{dy}{dx} = x^2, \quad a^2\frac{d^2y}{dx^2} = 2x; \quad \left\{1 + \left(\frac{dy}{dx}\right)^2\right\}^{\frac{3}{2}} = \frac{(a^4 + x^4)^{\frac{3}{2}}}{a^6}; \quad \therefore \rho = \frac{(a^4 + x^4)^{\frac{3}{2}}}{2a^4x}.$$

General Expression for Radius of Curvature. 289

4. To find the radius of curvature in the ellipse $\dfrac{x^2}{a^2} + \dfrac{y^2}{b^2} = 1$.

Let $x = a \cos \phi$, then $y = b \sin \phi$, and we have

$$dx = - a \sin \phi \, d\phi, \quad d^2x = - a \cos \phi \, d\phi^2 - a \sin \phi \, d^2\phi,$$

$$dy = b \cos \phi \, d\phi, \quad d^2y = - b \sin \phi \, d\phi^2 + b \cos \phi \, d^2\phi.$$

Hence by formula (5) we obtain

$$\rho = \frac{(a^2 \sin^2 \phi + b^2 \cos^2 \phi)^{\frac{3}{2}}}{ab}.$$

5. In the hypocycloid $x^{\frac{2}{3}} + y^{\frac{2}{3}} = a^{\frac{2}{3}}$, let $x = a \cos^3 \phi$, then $y = a \sin^3 \phi$, and regarding ϕ as the *independent* variable, we have

$$dx = - 3a \cos^2\phi \sin \phi \, d\phi, \qquad d^2x = 3a \cos \phi \, d\phi^2 (2 \sin^2 \phi - \cos^2 \phi),$$

$$dy = 3a \sin^2\phi \cos \phi \, d\phi, \qquad d^2y = 3a \sin \phi \, d\phi^2 (2 \cos^2 \phi - \sin^2 \phi),$$

whence

$$(dx^2 + dy^2)^{\frac{1}{2}} = 3a \sin \phi \cos \phi \, d\phi, \text{ and } dx\,d^2y - dy\,d^2x = - 9a^2 \sin^2 \phi \cos^2 \phi \, d\phi^3,$$

from which we obtain

$$\rho = - 3 (axy)^{\frac{1}{3}}.$$

6. Find the radius of curvature at any point of the curve

$$e^{\frac{y}{a}} = \sec \left(\frac{x}{a}\right). \qquad\qquad Ans. \; \rho = a \sec \left(\frac{x}{a}\right).$$

228. General Expression for Radius of Curvature.—The value of ρ becomes usually difficult of determination from formula (1) whenever y is not given explicitly in terms of x, that is, when the equation of the curve is of the form

$$u = f(x, y) = 0.$$

We proceed to show how the equation for ρ is to be transformed in this case. Suppose

$$\frac{du}{dx} = L, \quad \frac{du}{dy} = M, \quad \frac{d^2u}{dx^2} = A, \quad \frac{d^2u}{dx\,dy} = B, \quad \frac{d^2u}{dy^2} = C;$$

then, by Art. 100, we have

$$L + M\frac{dy}{dx} = 0.$$

U

Again, differentiating this equation with respect to x, regarding y as a function of x in consequence of the given equation, and observing that

$$\frac{d}{dx}(L) = \frac{dL}{dx} + \frac{dL}{dy}\frac{dy}{dx}, \quad \frac{d}{dx}(M) = \frac{dM}{dx} + \frac{dM}{dy}\frac{dy}{dx},$$

we obtain

$$\frac{dL}{dx} + \frac{dL}{dy}\frac{dy}{dx} + \left(\frac{dM}{dx} + \frac{dM}{dy}\frac{dy}{dx}\right)\frac{dy}{dx} + M\frac{d^2y}{dx^2} = 0,$$

or
$$A + 2B\frac{dy}{dx} + C\frac{dy^2}{dx^2} + M\frac{d^2y}{dx^2} = 0; \qquad (6)$$

whence, on substituting $-\dfrac{L}{M}$ for $\dfrac{dy}{dx}$, we obtain

$$\frac{d^2y}{dx^2} = -\frac{AM^2 - 2BLM + CL^2}{M^3}.$$

Consequently

$$\rho = \pm \frac{(L^2 + M^2)^{\frac{3}{2}}}{AM^2 - 2BLM + CL^2}. \qquad (7)$$

Or, on replacing L, M, A, B, C, by their values,

$$\rho = \pm \frac{\left\{\left(\dfrac{du}{dx}\right)^2 + \left(\dfrac{du}{dy}\right)^2\right\}^{\frac{3}{2}}}{\dfrac{d^2u}{dx^2}\left(\dfrac{du}{dy}\right)^2 - 2\dfrac{d^2u}{dxdy}\dfrac{du}{dx}\dfrac{du}{dy} + \dfrac{d^2u}{dy^2}\left(\dfrac{du}{dx}\right)^2}.$$

The result in (6) enables us to determine the second differential coefficient of an implicit function in general; a process which is sometimes required in analysis.

229. **The Centre of Curvature is the point of intersection of two Consecutive Normals.**—We shall next proceed to consider the subject from a geometrical point of view.

As a circle which passes through two infinitely near points on a curve is said to have *contact of the first order* with the curve, so the circle which passes through three infinitely near points on a curve is said to have *contact of the second order* with it, and is called the circle of curvature, or the *osculating circle* at the point.

Again, the centre of the circle which passes through three points, P, Q, R, is the intersection of the perpendiculars drawn at the middle points of PQ and QR; but when P, Q, R become infinitely near points on a curve, the perpendiculars become normals, and the centre of the circle becomes the *limiting position of the intersection of two infinitely near normals to the curve*. (Compare Art. 37, note.)

From this it is easily seen that we obtain $\dfrac{ds}{d\phi}$ for the length of the radius of the circle in the limit, as before.

230. **Newton's Method of investigating Radii of Curvature.**—When the equation of the curve is algebraic and rational it is easy to obtain an expression for its radius of curvature* at any point.

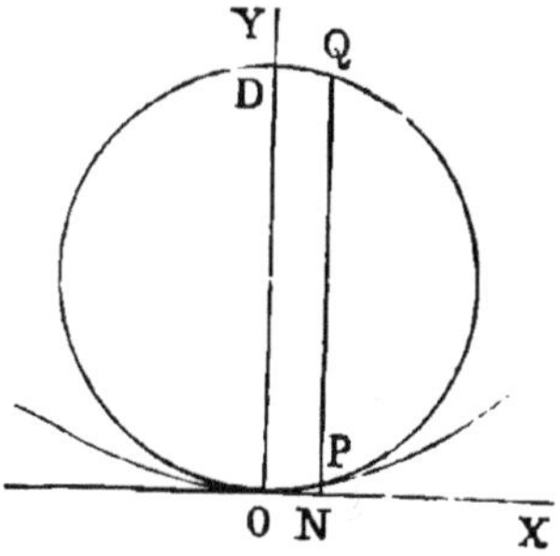

For, take the origin O at the point, and the tangent and normal for co-ordinate axes; let P be a point on the curve near to O, and describe a circle through P and O touching the axis of x; draw PN perpendicular to OX and produce it to meet the circle in Q; then we have

Fig. 31.

$$ON^2 = PN \cdot NQ.$$

Hence, if x and y be the co-ordinates of P, we get

$$NQ = \frac{ON^2}{PN} = \frac{x^2}{y}.$$

But when P is infinitely near to O, NQ becomes OD, the

---

* This method of finding the radius of curvature is indicated by Newton (*Principia*, Lib. I., Sect. 1., *Lemma* xi.), and has been adopted in a more or less modified form by many subsequent writers.

diameter of the circle of curvature, and if ρ be its radius, we have

$$2\rho = \text{limit of } \frac{x^2}{y} \text{ when } x \text{ is infinitely small.}$$

Again, since the axis of x is the tangent at the origin, the equation of the curve, by Art. 208, is of the form

$$b_1 y = c_0 x^2 + 2c_1 xy + c_2 y^2 + \text{ terms of the third and higher degrees}$$

$$= c_0 x^2 + 2c_1 xy + c_2 y^2 + u_3 + u_4 + \&c. \tag{9}$$

On dividing by y we obtain

$$b_1 = c_0 \frac{x^2}{y} + 2c_1 x + c_2 y + \frac{u_3}{y} + \&c.$$

Again, when x is infinitely small, $\frac{x^2}{y}$ becomes 2ρ, and each* of the other terms at the right-hand side becomes infinitely small; hence

$$\rho = \frac{b_1}{2c_0}.$$

Thus, for example, the radius of curvature at the origin in the curve

$$6y = 2x^2 + 3xy - 4y^2 + x^3$$

is $\frac{3}{2}$, the axes being rectangular.

---

* We have assumed above that the terms $\frac{u_3}{y}$, $\frac{u_4}{y}$, &c., become evanescent along with x; this can be readily established as follows:—

Let $$u_3 = \alpha x^3 + \beta x^2 y + \gamma x y^2 + \delta y^3,$$

then $$\frac{u_3}{y} = \alpha \frac{x^3}{y} + \beta x^2 + \gamma x y + \delta y^2;$$

each of the terms after the first vanishes with x, while the first becomes $\alpha \frac{x^2}{y} x$, or $2\alpha\rho x$, which also vanishes with x, when ρ is finite.

Similar reasoning is applicable to the terms, $\frac{u_4}{y}$, &c.

From the preceding it follows that when the axis of x is a tangent at the origin, the length of the radius of curvature at that point is independent of all the coefficients except those of y and x^2.

231. **Case of Oblique Axes.**—If the co-ordinate axes be oblique, and intersect at an angle ω, then PQ no longer passes through the centre of the circle in the limit, but becomes the chord of the circle of curvature which makes the angle ω with the tangent; accordingly, we have in this case

$$2\rho \sin \omega = \frac{ON^2}{PN} = \frac{x^2}{y}, \text{ in the limit.}$$

Hence, in the case of oblique axes, we have

$$\rho \sin \omega = \frac{b_1}{2c_0}. \tag{10}$$

If b_1 and c_0 have opposite signs, ρ is negative; this indicates that the centre of curvature lies below the axis of x, towards the negative side of the axis of y.

The preceding results show that the radius of curvature at the origin is the same as that of the parabola, $b_1 y = c_0 x^2$, at the same point; and also that the system of curves obtained by varying all the coefficients in (9), except those of y and x^2, have the same osculating circle, in oblique as well as in rectangular co-ordinates.

Again, as in Art. 223, the osculating circle, since it meets the curve in three consecutive points, cuts the curve at the point, in general, as well as touches it.

If $c_0 = 0$ in the equation of the curve, and b_1 be not zero, the radius of curvature becomes infinite, and the origin is a point of inflexion. This is also evident from the form of the equation, since the axis of x meets the curve in this case in *three* consecutive points.

232. In general, the equation of a curve referred to any rectangular axes, when the origin is on the curve, may be written in the form

$$2b_0 x + 2b_1 y = c_0 x^2 + 2c_1 xy + c_2 y^2 + u_3 + \&c.$$

Here $b_0 x + b_1 y = 0$ is the equation of the tangent at the origin; and the length of the perpendicular PN from the point (x, y) on this tangent is

$$\frac{b_0 x + b_1 y}{\sqrt{b_0^2 + b_1^2}}.$$

Also, $OP^2 = x^2 + y^2$, and $OP^2 = 2\rho \cdot PN$ in the limit.

Accordingly, we have, when x and y are infinitely small,

$$\frac{1}{\rho} = \frac{2PN}{OP^2} = \frac{2b_0 x + 2b_1 y}{(x^2 + y^2)\sqrt{b_0^2 + b_1^2}}$$

$$= \frac{c_0 x^2 + 2c_1 xy + c_2 y^2}{(x^2 + y^2)\sqrt{b_0^2 + b_1^2}} + \frac{u_3}{(x^2 + y^2)\sqrt{b_0^2 + b_1^2}} + \&c.$$

(since the point x, y is on the curve).

Again, the terms contained in $\dfrac{u_3}{x^2 + y^2}$, &c., become evanescent in the limit, as before (see note, Art. 230).

Hence we have

$$\frac{1}{\rho} = \frac{c_0 x^2 + 2c_1 xy + c_2 y^2}{(x^2 + y^2)\sqrt{b_0^2 + b_1^2}} = \frac{c_0 + 2c_1 \dfrac{y}{x} + c_2 \left(\dfrac{y}{x}\right)^2}{\left(1 + \left(\dfrac{y}{x}\right)^2\right)\sqrt{b_0^2 + b_1^2}}.$$

But for points infinitely near the origin we have

$$b_0 x + b_1 y = 0, \text{ or } \frac{y}{x} = -\frac{b_0}{b_1}.$$

Substituting this value instead of $\dfrac{y}{x}$ in the preceding equation, it becomes

$$\frac{1}{\rho} = \frac{c_0 b_1^2 - 2b_0 b_1 c_1 + c_2 b_0^2}{(b_0^2 + b_1^2)^{\frac{3}{2}}}. \tag{11}$$

The student will find no difficulty in showing the identity of this result with that given in (7).

233. Radii of Curvature of Inverse Curves.

—It may be convenient to state here that if two curves be inverse to each other with respect to any origin, their osculating circles at two inverse points are also inverse to each other with respect to the same origin.

This property is evident geometrically from the consideration that a circle is determined when three points on it are given.

Again, since the centres of the two inverse circles are *in directum* with the origin, we can construct the centre of curvature at any point on a curve, when that for the corresponding point on the inverse curve is known.

Also, if the osculating circle at any point on a curve pass through the origin, the corresponding point is a point of inflexion on the inverse curve.

We shall next proceed to establish another expression for the radius of curvature, which is of extensive application in curves referred to polar co-ordinates.

234. Radius of Curvature in terms of r and p.—

Let PN and PC be the tangent and normal at any point P on a curve, $P'N'$ and $P'C$ those at the infinitely near point P', then C is the centre of curvature corresponding to the point P. Let O be the origin.

Join OC, and let $OC = \delta$, $OP = r$, $OP' = r'$, $ON = p$, $ON' = p'$, $CP = CP' = \rho$; then we have

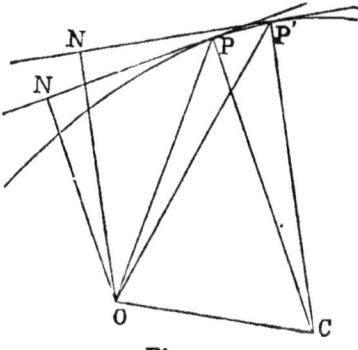

Fig. 32.

$$OC^2 = OP^2 + CP^2 - 2OP \cdot CP \cdot \cos OPC,$$

or
$$\delta^2 = r^2 + \rho^2 - 2\rho p.$$

In like manner we have

$$\delta^2 = r'^2 + \rho^2 - 2\rho p'.$$

Subtracting, we get

$$r'^2 - r^2 = 2\rho(p' - p), \text{ or } \frac{r' - r}{p' - p} = \frac{2\rho}{r + r'}.$$

Hence we have
$$\frac{dr}{dp} = \frac{\rho}{r}, \text{ or } \rho = r\frac{dr}{dp}. \tag{12}$$

This formula can also be deduced immediately from Art. 193: thus

$$r \cos \psi = PN = \frac{dp}{d\omega} = \frac{dp}{ds}\frac{ds}{d\omega} = \rho\frac{dp}{ds} = \rho\frac{dp}{dr}\frac{dr}{ds} = \rho \cos \psi \frac{dp}{dr};$$

$$\therefore r = \rho\frac{dp}{dr}, \text{ or } \rho = r\frac{dr}{dp}.$$

235. Chord of Curvature through the Origin.—Let γ denote half the intercept made on the line OP by the circle of curvature, and we evidently have

$$\gamma = \rho \sin OPN = \rho\frac{p}{r} = p\frac{dr}{dp}. \tag{13}$$

This and the preceding formula are of importance whenever we can express the equation of the curve in terms of the lines represented by r and p.

Their use will be illustrated by the following elementary examples:—

EXAMPLES.

1. To find the radius of curvature at any point on a parabola.
Taking the focus as pole, the equation of the curve in terms of r and p evidently is $p^2 = 2mr$.

Hence $\quad \rho = r\dfrac{dr}{dp} = \dfrac{pr}{m} = \left(\dfrac{2r^3}{m}\right)^{\frac{1}{2}}$; also, $\gamma = p\dfrac{dr}{dp} = \dfrac{p^2}{m} = 2r.$

2. To find the radius of curvature in an ellipse.
Taking the centre as origin, the equation of the curve is

$$a^2 + b^2 - r^2 = \frac{a^2b^2}{p^2};$$

$$\therefore \rho = r\frac{dr}{dp} = \frac{a^2b^2}{p^3}.$$

3. To find the radius of curvature in the Lemniscate.
Here, by Ex. 3, Art. 190, we have $r^3 = a^2p$;

$$\therefore 3r^2\frac{dr}{dp} = a^2; \text{ hence } \rho = \frac{a^2}{3r}; \text{ also, } \gamma = \frac{r}{3}.$$

4. To find the chord of curvature which passes through the origin in the Cardioid
$$r = a(1 + \cos\theta).$$
In this case, we have $r^3 = 2ap^2$.

Hence
$$\gamma = p\frac{dr}{dp} = \frac{2}{3}r.$$

5. To find the radius of curvature at any point on the curve $r^m = a^m \cos m\theta$.

Here $r^{m+1} = a^m p$, by Art. 190.

Hence
$$\rho = \frac{a^m}{(m+1)r^{m-1}} = \frac{r^2}{(m+1)p}; \text{ also, } \gamma = \frac{r}{m+1}.$$

This result furnishes a simple geometrical method of finding the centre of curvature in all curves included under this equation.

236. To prove that $\rho = p + \dfrac{d^2p}{d\omega^2}$. If p and ω have the same signification as in Art. 192, the formula of that Art. becomes
$$\rho = \frac{ds}{d\omega} = p + \frac{d^2p}{d\omega^2}. \qquad (14)$$

EXAMPLES.

1. In a central ellipse prove that
$$p = \sqrt{a^2 \cos^2\omega + b^2 \sin^2\omega},$$
and hence deduce an expression for the radius of curvature at any point on the curve.

2. In a parabola referred to its focus as pole, prove that $p = m \sec\omega$, and hence show that $\rho = 2m \sec^3\omega$.

237. Evolutes and Involutes.—If the centre of curvature for each point on a curve be taken, we get a new curve called the *evolute* of the original one. Also, the original curve, when considered with respect to its evolute, is called an *involute*.

To investigate the connexion between these curves, let P_1, P_2, P_3, &c., represent a series of infinitely near points on a curve; C_1, C_2, C_3, &c., the corresponding centres of curvature, then the lines P_1C_1, P_2C_2, P_3C_3, &c., are normals to the curve, and the lines C_1C_2, C_2C_3, C_3C_4, &c., may be regarded in the limit as consecutive *elements* of the evolute; also, since

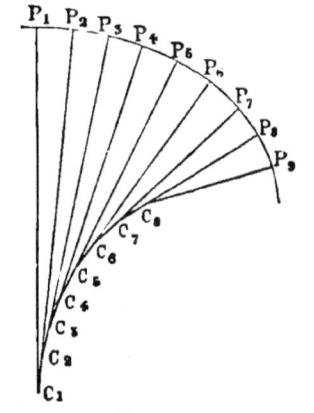

Fig. 33.

each of the normals P_1C_1, P_2C_2, P_3C_3, &c., passes through two consecutive points on the evolute, they are tangents to that curve in the limit.

Again, if ρ_1, ρ_2, ρ_3, ρ_4, &c., denote the lengths of the radii of curvature at the points P_1, P_2, P_3, P_4, &c., we have

$$\rho_1 = P_1C_1, \quad \rho_2 = P_2C_2, \quad \rho_3 = P_3C_3, \quad \rho_4 = P_4C_4, \quad \&c.\ ;$$

$$\therefore \quad \rho_1 - \rho_2 = P_1C_1 - P_2C_2 = P_2C_1 - P_2C_2 = C_1C_2;$$

also $\quad \rho_2 - \rho_3 = C_2C_3, \quad \rho_3 - \rho_4 = C_3C_4, \ldots \rho_{n-1} - \rho_n = C_{n-1}C_n\ ;$

hence by addition we have

$$\rho_1 - \rho_n = C_1C_2 + C_2C_3 + C_3C_4 + \ldots + C_{n-1}C_n.$$

This result still holds when the number n is increased indefinitely, and we infer that the *length of any arc of the evolute is equal, in general, to the difference between the radii of curvature at its extremities*.

It is evident that the curve may be generated from its evolute by the motion of the extremity of a stretched thread, supposed to be wound round the evolute and afterwards unrolled; in this case each point on the string will describe a different *involute* of the curve.

The names evolute and involute are given in consequence of the preceding property.

It follows, also, that while a curve has but one evolute, it can have an infinite number of involutes; for we may regard each point on the stretched string as generating a separate involute.

The curves described by two different points on the moving line are said to be *parallel;* each being got from the other by cutting off a constant length on its normal measured from the curve.

238. Evolutes regarded as Envelopes.—From the preceding it also follows that the determination of the evolute of a curve is the same as the finding the envelope of all its normals. We have already, in Ex. 3, Art. 219, investigated the equation of the evolute of an ellipse from this point of view.

239. Evolute of a Parabola.—We proceed to determine the evolute of the parabola in the same manner.

Evolute of Ellipse.

Let the equation of the curve be $y^2 = 2mx$, then that of its normal at a point (x, y) is

$$(Y - y)\frac{m}{y} + X - x = 0,$$

or $\qquad y^3 + 2my(m - X) - 2m^2 Y = 0.$

The envelope of this line, where y is regarded as an arbitrary parameter, is got by eliminating y between this equation and its derived equation

$$3y^2 + 2m(m - X) = 0.$$

Accordingly, the equation of the required envelope is obtained by substituting $\dfrac{3m}{2}\dfrac{Y}{m - X}$ instead of y in the latter equation.

Hence, we get for the required evolute, the semi-cubical parabola

$$27m Y^2 = 8(X - m)^3.$$

The form of this evolute is exhibited in the annexed figure, where $VN = m = 2VF$. If P, P', represent the points of intersection of the evolute with the curve, it is easily seen that

$$VM = 4VN = 4m.$$

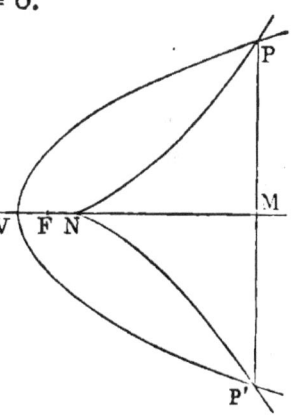

Fig. 34.

240. Evolute of an Ellipse.—The form of the evolute of an ellipse, when e is greater than $\frac{1}{2}\sqrt{2}$, is exhibited in the accompanying figure; the points M, N, M', N', are evidently cusps on the curve, and are the centres of curvature corresponding to the four vertices of the ellipse. In general, if a curve be symmetrical at both sides of a point on it, the osculating circle cannot intersect

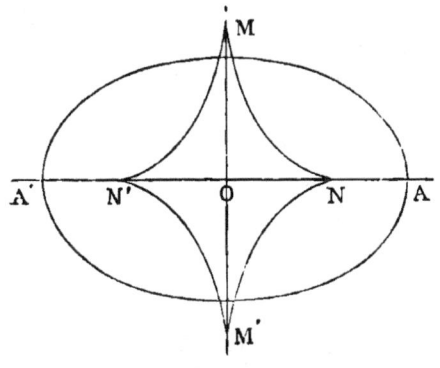

Fig. 35.

the curve at the point; accordingly, the radius of curvature is a maximum or a minimum at such a point, and the corresponding point on the evolute is a cusp.

It can be easily seen geometrically that through any point four real normals, or only two, can be drawn to an ellipse, according as the point is inside or outside the evolute.

It may be here observed that to a point of inflexion on any curve corresponds plainly an asymptote to its evolute.

241. Evolute of an Equiangular Spiral.—We shall next consider the equiangular or logarithmic spiral, $r = a^\theta$.

Let P and Q be two points on the curve, O its pole, PC, QC the normals at P and Q; join OC. Then by the fundamental property of the curve (Art. 181), the angles OPC and OQC are equal, and consequently the four points, O, P, Q, C, lie on a circle: hence $\angle QOC = \angle QPC$; but in the limit when P and Q are coincident, the angle QPC becomes a right angle, and C becomes the centre of curvature belonging to the point P; hence POC also becomes a right angle, and the point C is immediately determined.

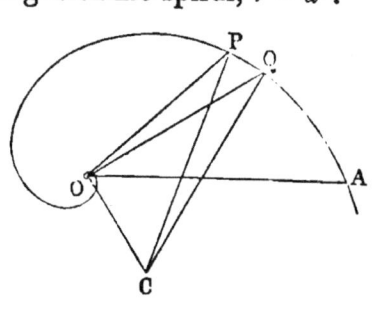

Fig. 36.

Again, $\angle OCP = \angle OQP$; but, in the limit, the angle OQP is constant; $\therefore \angle OCP$ is also constant; and since the line CP is a tangent to the evolute at C, it follows that the tangent makes a constant angle with the radius vector OC. From this property it follows that the evolute in question is another logarithmic spiral. Again, as the constant angle is the same for the curve and for its evolute, it follows that the latter curve is the same spiral turned round through a known angle (whose circular measure is $\frac{\pi}{2} - \log_a M$).

241 (a). Involute of a Circle.—As an example of involutes, suppose APQ to represent a portion of an involute of the circle BAC, whose centre is O. Let
$$OC = a, \quad \angle COA = \phi,$$
and CA the length of the string unrolled; then
$$CP = CA = a\phi.$$

Draw ON perpendicular to the tangent at P, and let $ON = p$, then we have

$$p = a\phi.$$

Hence, since

$$\angle BON = \angle COA = \phi,$$

the pedal of the curve APQ is a spiral of Archimedes.

Also, since

$$OP^2 = OC^2 + CP^2,$$

we have

$$r^2 = p^2 + a^2,$$

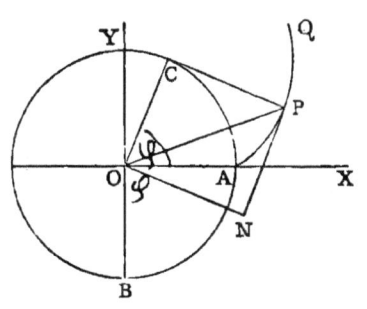

Fig. 37.

which gives the equation to the involute of a circle in terms of the co-ordinates r and p.

Again, if $AP = s$, we have

$$\frac{ds}{d\phi} = CP = a\phi;$$

from which it is easily seen that

$$s = \frac{a\phi^2}{2}.$$

242. Radius of Curvature, and Points of Inflexion, in Polar Co-ordinates.—We shall first find an expression for ρ in terms of u (the reciprocal of the radius vector) and θ.

By Article 185 we have

$$\frac{1}{p^2} = u^2 + \left(\frac{du}{d\theta}\right)^2;$$

hence

$$-\frac{1}{p^3}\frac{dp}{du} = u + \frac{d^2u}{d\theta^2}.$$

Also

$$\rho = r\frac{dr}{dp} = -\frac{1}{u^3}\frac{du}{dp};$$

consequently $\rho\left(u + \dfrac{d^2u}{d\theta^2}\right) = \dfrac{1}{p^3 u^3} = \left\{1 + \left(\dfrac{du}{u\,d\theta}\right)^2\right\}^{\frac{3}{2}};$

$$\therefore \rho = \dfrac{\left\{1 + \left(\dfrac{du}{u\,d\theta}\right)^2\right\}^{\frac{3}{2}}}{u + \dfrac{d^2u}{d\theta^2}}. \qquad (15)$$

Again, since $u = \dfrac{1}{r}$, we have $\dfrac{du}{d\theta} = -\dfrac{1}{r^2}\dfrac{dr}{d\theta}$,

and $\dfrac{d^2u}{d\theta^2} = \dfrac{2}{r^3}\left(\dfrac{dr}{d\theta}\right)^2 - \dfrac{1}{r^2}\dfrac{d^2r}{d\theta^2};$

$$\therefore \rho = \dfrac{\left\{r^2 + \left(\dfrac{dr}{d\theta}\right)^2\right\}^{\frac{3}{2}}}{r^2 - r\dfrac{d^2r}{d\theta^2} + 2\left(\dfrac{dr}{d\theta}\right)^2}. \qquad (16)$$

This result can also be established in another manner, as follows:—

On reference to the figure of Art. 180, it is obvious that $\phi = \theta + \psi$; where ϕ is the angle the tangent at P makes with the prime vector OX.

Hence $\dfrac{d\phi}{d\theta} = 1 + \dfrac{d\psi}{d\theta}$, or $\dfrac{d\phi}{ds}\dfrac{ds}{d\theta} = 1 + \dfrac{d\psi}{d\theta};$

$$\therefore \dfrac{1}{\rho} = \dfrac{d\phi}{ds} = \dfrac{1 + \dfrac{d\psi}{d\theta}}{\dfrac{ds}{d\theta}}.$$

Again, denoting $\dfrac{dr}{d\theta}$ and $\dfrac{d^2r}{d\theta^2}$ by r' and r'', we have $\tan\psi = \dfrac{r}{r'}$; and hence

$\dfrac{d\psi}{d\theta} = \cos^2\psi\,\dfrac{r'^2 - rr''}{r'^2} = \dfrac{r'^2 - rr''}{r^2 + r'^2};$

$\therefore 1 + \dfrac{d\psi}{d\theta} = \dfrac{r^2 - rr'' + 2r'^2}{r^2 + r'^2};$ also $\dfrac{ds}{d\theta} = (r^2 + r'^2)^{\frac{1}{2}}.$

Hence, we get $\rho = \dfrac{(r^2 + r'^2)^{\frac{3}{2}}}{r^2 - rr'' + 2r'^2}$.

Or, replacing r' and r'' by their values,

$$\rho = \dfrac{\left(r^2 + \left(\dfrac{dr}{d\theta}\right)^2\right)^{\frac{3}{2}}}{r^2 - r\dfrac{d^2r}{d\theta^2} + 2\left(\dfrac{dr}{d\theta}\right)^2}.$$

Again, since $\rho = \infty$ at a point of inflexion, we infer that the points of intersection of the curve represented by the equation

$$r^2 - r\dfrac{d^2r}{d\theta^2} + 2\left(\dfrac{dr}{d\theta}\right)^2 = 0,$$

with the original curve, determine in general its points of inflexion.

In some cases the points of inflexion can be easier found by aid of (15), which gives, when $\rho = \infty$,

$$\dfrac{d^2u}{d\theta^2} + u = 0.$$

Examples.

1. Find the radius of curvature at any point in the spiral of Archimedes, $r = a\theta$.
   Ans. $a\dfrac{(1 + \theta^2)^{\frac{3}{2}}}{2 + \theta^2}$.

2. Find the radius of curvature of the logarithmic spiral $r = a^\theta$.
   Ans. $r(1 + (\log a)^2)^{\frac{1}{2}}$.

3. Find the points of inflexion on the curve
   $$r = 29 - 11 \cos 2\theta.$$
   Ans. $\cos 2\theta = \dfrac{9}{11}$.

4. Prove that the circle $r = 10$ intersects the curve
   $$r = 11 - 2 \cos 5\theta$$
   in its points of inflexion.

5. Prove that the curve
   $$r = a + b \cos n\theta$$
   has no real points of inflexion unless a is $> b$ and $< (1 + n^2)b$. When a lies between these limits, prove that all the points of inflexion lie on a circle; and show how to determine the radius of the circle.

242 (a). Intrinsic Equation of a Curve.—In many cases the equation of a curve is most simply expressed in terms of the length, s, of the curve, measured from a fixed point on it, and the angle, ϕ, through which it is bent, *i.e.* the angle of deviation of the tangent at any point from the tangent at the fixed point, taken as origin. These are styled the *intrinsic* elements of the curve by Dr. Whewell,* to whom this method of discussing curves is due.

The relation between the length s and the deviation ϕ for any curve is called its *intrinsic equation*.

If this relation be represented by the equation

$$s = f(\phi),$$

then if ρ be the radius of curvature at any point, we have

$$\rho = \frac{ds}{d\phi} = f'(\phi).$$

Also, if s_1 denote the length of the evolute, from Art. 237 it is easily seen that the equation of the evolute is of the form

$$s_1 = f'(\phi) + const.$$

From this it follows that the series of successive evolutes are in this case easily determined by successive differentiation.

The simplest case of an intrinsic equation is that of the circle, in which case we have

$$s = a\phi.$$

Again, from Art. 241 (a), the intrinsic equation of the involute of a circle is reducible to the form

$$s = \frac{a\phi^2}{2}.$$

We shall meet with further examples of intrinsic equations subsequently.

243. Contact of Different Orders.—As already stated, the tangent to a curve has a contact of the first order with the curve at its point of contact, and the osculating circle a contact of the second order. We now proceed to distinguish more fully the different orders of contact between two curves.

---

* *Cambridge Philosophical Transactions*, Vols. VIII. and IX.

Suppose the curves to be represented by the equations

$$y = f(x), \text{ and } y = \phi(x),$$

and that x_1 is the abscissa of a point common to both curves, then we have

$$f(x_1) = \phi(x_1).$$

Again, substituting $x_1 + h$, instead of x in both equations, and supposing y_1 and y_2 the corresponding ordinates of the two curves, we have

$$y_1 = f(x_1 + h) = f(x_1) + hf'(x_1) + \frac{h^2}{1 \cdot 2} f''(x_1) + \&c.,$$

$$y_2 = \phi(x_1 + h) = \phi(x_1) + h\phi'(x_1) + \frac{h^2}{1 \cdot 2} \phi''(x_1) + \&c.$$

Subtracting, we get

$$y_1 - y_2 = h\{f'(x_1) - \phi'(x_1)\} + \frac{h^2}{1 \cdot 2} \{f''(x_1) - \phi''(x_1)\} + \&c. \quad (17)$$

Now, suppose $f'(x_1) = \phi'(x_1)$, or that the curves have a common tangent at the point, then

$$y_1 - y_2 = \frac{h^2}{1 \cdot 2} \{f''(x_1) - \phi''(x_1)\} + \frac{h^3}{1 \cdot 2 \cdot 3} \{f'''(x_1) - \phi'''(x_1)\} + \&c.$$

In this case the curves have a contact of the first order; and when h is small, the difference between the ordinates is a small quantity of the second order, and as $y_1 - y_2$ does not change sign with h, the curves do not cross each other at the point.

If, in addition

$$f''(x_1) = \phi''(x_1),$$

then $\quad y_1 - y_2 = \dfrac{h^3}{1 \cdot 2 \cdot 3} \{f'''(x_1) - \phi'''(x_1)\} + \&c.$

In this case the difference between the ordinates is an infinitely small magnitude of the *third* order when h is taken an infinitely small magnitude of the first; the curves are then said to have a *contact of the second order*, and approach infinitely nearer to each other at the point of contact than in the former case. Moreover, since $y_1 - y_2$ changes its sign with h, the curves cut each other at the point as well as touch.

If we have in addition $f'''(x_1) = \phi'''(x_1)$, the curves are said to have a contact of the *third order:* and, in general, if all the derived functions, up to the nth inclusive, be the same for both curves when $x = x_1$, the curves have a contact of the nth order, and we have

$$y_1 - y_2 = \frac{h^{n+1}}{\underline{|n+1}} \{f^{(n+1)}(x_1) - \phi^{(n+1)}(x_1)\} + \&c. \qquad (18)$$

Also, if the contact be of an even order, $n + 1$ is odd, and consequently h^{n+1} changes its sign with h, and hence the curves cut each other at their point of contact; for whichever is the lower at one side of the point becomes the upper at the other side.

If the curves have a contact of an odd order, they do not cut each other at their point of contact.

From the preceding discussion the following results are immediately deduced:—

(1). If two curves have a contact of the nth order, no curve having with either of them a contact of a lower order can fall between the curves near their point of contact.

(2). Two curves which have a contact of the nth order at a point are infinitely closer to one another near that point than two curves having a contact of an order lower than the nth.

(3). If any number of curves have a contact of the second order at a point, they have the same osculating circle at the point.

244. **Application to Circle.**—It can be easily verified that the circle which has a contact of the second order with a curve at a point is the same as the osculating circle determined by the former method.

For, let
$$(X - a)^2 + (Y - \beta)^2 = R^2$$

be the equation of a circle having contact of the second order at the point (x, y) with a given curve; then, by the preceding, the values of $\dfrac{dy}{dx}$ and $\dfrac{d^2y}{dx^2}$ must be the same for the circle and for the curve at the point in question.

Differentiating the equation of the circle twice, and substituting x and y for X and Y, we get

$$x - a + (y - \beta)\frac{dy}{dx} = 0, \qquad (19)$$

and

$$1 + (y - \beta)\frac{d^2y}{dx^2} + \left(\frac{dy}{dx}\right)^2 = 0. \qquad (20)$$

Hence $\quad y - \beta = -\dfrac{1 + \left(\dfrac{dy}{dx}\right)^2}{\dfrac{d^2y}{dx^2}},\ x - a = \dfrac{\dfrac{dy}{dx}\left[1 + \left(\dfrac{dy}{dx}\right)^2\right]}{\dfrac{d^2y}{dx^2}};\quad (21)$

$$\therefore R^2 = (x - a)^2 + (y - \beta^2) = \frac{\left[1 + \left(\dfrac{dy}{dx}\right)^2\right]^3}{\left(\dfrac{d^2y}{dx^2}\right)^2}.$$

This agrees with the expression for the radius of curvature found in Art. 226.

The co-ordinates a, β of the centre of curvature can be found by aid of equations (21); and the equation of the evolute by the elimination of x and y between these equations and that of the curve.

In practice, the following equations are often more useful: thus, by differentiation with respect to x, we get from (19),

$$\beta \frac{d^2y}{dx^2} = 1 + \frac{d}{dx}\left(y\frac{dy}{dx}\right). \qquad (22)$$

In like manner, from the equation

$$(y - \beta) + (x - a)\frac{dx}{dy} = 0,$$

we obtain

$$a \frac{d^2x}{dy^2} = 1 + \frac{d}{dy}\left(x\frac{dx}{dy}\right). \qquad (23)$$

245. Centre of Curvature, and Evolute of Ellipse.—As an illustration, we shall apply these equations to determine the co-ordinates of the centre of curvature, and the equation of the evolute of the ellipse

$$\frac{x^2}{a^2} + \frac{y^2}{b^2} = 1.$$

Here $\quad y\dfrac{dy}{dx} = -\dfrac{b^2}{a^2}x, \quad x\dfrac{dx}{dy} = -\dfrac{a^2}{b^2}y;$

$$\therefore \frac{d}{dx}\left(y\frac{dy}{dx}\right) = -\frac{b^2}{a^2}, \quad \frac{d}{dy}\left(x\frac{dx}{dy}\right) = -\frac{a^2}{b^2}.$$

Hence $\quad y\dfrac{d^2y}{dx^2} = -\dfrac{b^2}{a^2} - \left(\dfrac{dy}{dx}\right)^2 = -\dfrac{b^2}{a^2} - \dfrac{b^4}{a^4}\dfrac{x^2}{y^2}$

$$= -\frac{b^2}{a^2}\left\{1 + \frac{b^2 x^2}{a^2 y^2}\right\} = -\frac{b^4}{a^2 y^2}.$$

In like manner, we have

$$x\frac{d^2x}{dy^2} = -\frac{a^4}{b^2 x^2}.$$

Substituting in (22) and (23), we obtain for the co-ordinates of the centre of curvature

$$\beta = -\frac{(a^2 - b^2)y^3}{b^4}, \quad \alpha = \frac{(a^2 - b^2)x^3}{a^4}. \tag{24}$$

Again, substituting the values of x and y given by these equations, in the equation $\dfrac{x^2}{a^2} + \dfrac{y^2}{b^2} = 1$, we get for the equation of the evolute

$$(a\alpha)^{\frac{2}{3}} + (\beta b)^{\frac{2}{3}} = (a^2 - b^2)^{\frac{2}{3}}.$$

246. It may be noticed that *the osculating circle cuts the curve in general, as well as touches it.* This follows from Article 243, since the circle has a contact of the second order at the point.

At the points of maximum and minimum curvature the

osculating circle has a contact of the *third order* with the curve; for example, at any of the four vertices of an ellipse the osculating circle has a contact of the third order, and does not cut the curve at its point of contact (Art. 240).

247. Osculating Curves.—When the equation of a curve contains a number, n, of arbitrary coefficients, we can in general determine their values so that the curve shall have a contact of the $(n-1)^{th}$ order with a given curve at a given point; for the n arbitrary constants can be determined so that the n quantities

$$y, \frac{dy}{dx}, \frac{d^2y}{dx^2}, \ldots \frac{d^{n-1}y}{dx^{n-1}},$$

shall be the same at the point in the proposed as in the given curve, and thus the curves will have a contact of the $(n-1)^{th}$ order.

The curve thus determined, which has with a given curve a contact of the highest possible order, is called an *osculating curve*, as having a closer contact than any other curve of the same species at the point.

For instance, as the equation of a circle contains but three arbitrary constants, the osculating circle has a contact of the *second* order, and cannot, in general, have contact of a higher order; similarly, the osculating parabola has a contact of the *third* order; and, since the general equation of a conic contains five arbitrary constants, the general osculating conic has a contact of the *fourth* order. In general, if the greatest number of constants which determine a curve of a given species be n, the osculating curve of that species has a contact of the $(n-1)^{th}$ order.

248. Geometrical Method.—The subject of contact admits also of being considered in a geometrical point of view; thus two curves have a contact of the first order, when they intersect in *two consecutive points*; of the second, if they intersect in *three*; of the nth, if in $n+1$. For a simple investigation of the subject in this point of view the student is referred to Salmon's *Conic Sections*, Art. 239.

249. Curvature at a Double Point.—We now proceed to consider the method of finding the radii of curvature of the two branches of a curve at a double point.

In this case the ordinary formula (8) becomes indeterminate, since

$$\frac{du}{dx} = 0, \text{ and } \frac{du}{dy} = 0$$

at a double point. The question admits, however, of being treated in a manner analogous to that already employed in Art. 230: we commence with the case of a node.

250. Radii of Curvature at a Node.—Suppose the origin transferred to the node, and the tangents to the two branches of the curve taken as co-ordinate axes, ω representing the angle between them.

By Art. 210, the equation of the curve is in this case of the form

$$2hxy = ax^3 + \beta x^2 y + \gamma xy^2 + \delta y^3 + u_4 + \&c.:$$

dividing by xy we obtain

$$2h = a\frac{x^2}{y} + \beta x + \gamma y + \delta\frac{y^2}{x} + \frac{u_4}{xy} + \&c.$$

Now, let ρ_1 and ρ_2 be the radii of curvature at the origin for the branches of the curve which touch the axes of x and y, respectively; then, by Art. 231, we have

$$2\rho_1 \sin \omega = \frac{x^2}{y}, \text{ and } 2\rho_2 \sin \omega = \frac{y^2}{x}, \text{ in the limit.}$$

Again, it can be readily seen, as in the note to Art. 230, that the terms in $\dfrac{u_4}{xy}$, &c., become evanescent along with x and y, and accordingly the limiting values of $\dfrac{x^2}{y}$ and $\dfrac{y^2}{x}$ can be separately found, as in the Article referred to.

Hence we obtain

$$\rho_1 = \frac{h}{a \sin \omega}, \quad \rho_2 = \frac{h}{\delta \sin \omega}. \qquad (25)$$

Also, if $a = 0$, we get $\rho_1 = \infty$, and the corresponding branch of the curve has a point of inflexion at the origin. Similarly, if $\delta = 0$, $\rho_2 = \infty$.

If $a = 0$, and $\delta = 0$, the origin is a point of inflexion on both branches. This appears also immediately from the consideration that in this case u_3 contains u_2 as a factor.

If the equation of a curve when the origin is at a node contain no terms of the third degree, the origin is a point of inflexion on both branches. An example of this is seen in the Lemniscate, Art. 210.

EXAMPLES.

1. Find the radii of curvature at the origin of the two branches of the curve

$$ax^3 - 2bxy + cy^3 = x^4 + y^4,$$

the axes being rectangular. $\qquad Ans.\ \dfrac{b}{a}$ and $\dfrac{b}{c}$.

2. Find the radii of curvature at the origin in the curve

$$a(y^2 - x^2) = x^3.$$

Transforming the equation to the internal and external bisectors of the angle between the axes, it becomes

$$4axy\sqrt{2} = (x-y)^3;$$

hence the radii of curvature are $2a\sqrt{2}$ and $-2a\sqrt{2}$, respectively.

251. Radii of Curvature at a Cusp.—The preceding method fails when applied to a cusp, because the angle ω vanishes in that case. It is easy, however, to supply an independent investigation: for, if we take the tangent and normal at the cusp for the axes of x and y, respectively, the equation of the curve, by the method of Art. 210, may be written in the form

$$y^2 = ax^3 + \beta x^2 y + \gamma xy^2 + \delta y^3 + u_4 + \&c. \qquad (26)$$

Now in this, as in every case, the curvature at the origin depends on the form of the portion of the curve indefinitely near to that point; consequently, in investigating this form we may neglect $y^2 x$, y^3, &c., in comparison with y^2; and x^4, $x^3 y$, &c., in comparison with x^3.

Accordingly, the curvature at the origin is the same, in general, as that of the cubic

$$y^2 = ax^3 + \beta x^2 y. \qquad (27)$$

Dividing by x^2, we get

$$\frac{y^2}{x^2} = ax + \beta y.$$

Hence, in immediate proximity to the origin, $\frac{y}{x}$ becomes very small, i.e. y is very small in comparison with x. Accordingly, the form of the curve near the origin is represented by the equation

$$y^2 = ax^3.$$

From this we infer that the form of any algebraic curve near a cusp is, in general, a semi-cubical parabola (*see* Ex. 2, Art. 211).

Again, since

$$\frac{x^4}{y^2} = \frac{x}{a},$$

we have, by Art. 230,

$$\rho = \pm \tfrac{1}{2} \sqrt{\frac{x}{a}};$$

from which we see that ρ vanishes along with x, and accordingly the *radii of curvature are zero for both branches at the origin.*

This result can also be arrived at by differentiation, by aid of formula (1).

252. Case where the Coefficient of x^3 is wanting.—Next, suppose that the term containing x^3 disappears, or $a = 0$, then the equation of the curve is of the form

$$y^2 = \beta x^2 y + \gamma x y^2 + \delta y^3 + a' x^4 + \&c.;$$

and proceeding as before, the curvature at the origin is the same as in the curve

$$y^2 = \beta x^2 y + a' x^4. \qquad (28)$$

The two branches of this curve are determined by the equation

$$y = \frac{\beta}{2} x^2 \pm \frac{x^2}{2} \sqrt{\beta^2 + 4a'}. \qquad (29)$$

The nature of the origin depends on the sign of $\beta^2 + 4a'$, and the discussion involves three cases.

(1). If $\beta^2 + 4a'$ be *positive*, it is evident that the curve extends at both sides of the origin, and that point is a *double cusp* (Art. 215(*a*)).

On dividing equation (28) by y^2, and substituting 2ρ for $\dfrac{x^2}{y}$, we get

$$1 = 2\beta\rho + 4a'\rho^2. \qquad (30)$$

The roots of this quadratic determine the radii of curvature of the two branches at the cusp.

These branches evidently lie at the same, or at opposite sides of the axis of x, according as the radii of curvature have the same or opposite signs: *i.e.* according as a' has a negative or positive sign.

These results also appear immediately from the circumstance, that in this case the form of the curve very near the origin becomes that of the two parabolas represented by equation (29).

(2). If $\beta^2 + 4a'$ be *negative*, y becomes imaginary, and the origin is a *conjugate point*.

(3). If $\beta^2 + 4a' = 0$, the equation (30) becomes a perfect square: we proceed to prove that in this case the origin is a *cusp of the second species*.

To investigate the form of the curve near the origin, it is necessary in this case to take into account the terms of the fifth degree in x (y being regarded as of the second): this gives

$$(y - \frac{\beta}{2} x^2)^2 = \gamma x y^2 + \beta' x^3 y + a'' x^5 = x(\gamma y^2 + \beta' x^2 y + a'' x^4). \qquad (31)$$

It will be observed that the right-hand side changes its sign with x; accordingly the origin is a cusp. Also, the cusp is of the second species, for the two roots of the equation in y plainly have the same sign, viz., that of β; and consequently both branches of the curve at the origin lie at the same side of the axis of x.

Moreover, as equation (30) has equal roots in this case, *the radii of curvature of the two branches are equal*, and the branches have a contact of the second order.

We conclude that when the term involving x^3 in equation (28) disappears, the origin is a *double cusp, a cusp of the second species*, or *a conjugate point*, according as $\beta^2 + 4a' >$ = or < 0.

Moreover, if $a' = 0$, one root of the quadratic (30) is infinite, and the other is $\dfrac{1}{2\beta}$. The origin in this case is a double cusp, and is also a point of inflexion on one branch. Such a point is called a point of *oscul-inflexion* by Cramer.

If $\beta = 0$ in addition to $a' = 0$, the origin is a cusp of the first species, but having the radii of curvature infinite for both branches.

It is easy to see from other considerations that the radii of curvature at a cusp of the first species are always either zero or infinite.

For, since the two branches of the curve in this case turn their convexities in opposite directions, $\dfrac{d^2y}{dx^2}$ must have opposite signs at both sides of the cusp, and consequently it must *change its sign* at that point; but this can happen only in its passage through zero, or through infinity.

It should be observed that the preceding discussion applies to the case of a curve referred to oblique axes of co-ordinates, provided that we substitute γ instead of ρ; where γ is half the chord intercepted on the axis of y by the osculating circle at the origin.

253. **Recapitulation.**—The conclusions arrived at in the two preceding Articles may be briefly stated as follows:—

(1). Whenever the equation of a curve can be transformed into the shape $y^2 = ax^3$ + terms of the third and higher degrees, the origin is a cusp of the first species; both radii of curvature being zero at the point.

(2). When the coefficient of x^3 vanishes,* the origin is

---

* In this case, if v_1 be the equation of the tangent at the cusp, the equation of the curve is of the form

$$v_1^2 + v_1 v_2 + v_4 + \&c. = 0.$$

This is also evident from geometrical considerations.

generally either a double cusp, a conjugate point, or a cusp of the second species. In the latter case the two branches of the curve have the same centre of curvature, and consequently have a contact of the second order with each other.

(3). If the lowest term in x (independent of y) be of the 5th degree, the origin is a point of oscul-inflexion.

If, however, the coefficient of $x^2 y$ also vanish, the origin is not only a cusp of the first species, but also a point of inflexion on both branches of the curve.

254. **General Investigation of Cusps.**—The preceding results admit of being established in a somewhat more general manner as follows:—

By the method already given, the equation which determines the form of an algebraic curve near to a cusp may be written in the following general shape:

$$y^2 = 2Ax^a y + Bx^b + Cx^c, \qquad (32)$$

where $2Ax^a$ is the lowest term in the coefficient of y, and Bx^b, $Cx^{c\,*}$ are the lowest terms independent of y.

By hypothesis, a, b, c are positive integers, and $a > 1$, $b > 2$, $c > 3$; now, solving for y, we obtain

$$y = Ax^a \pm \sqrt{A^2 x^{2a} + Bx^b + Cx^c},$$

which represents two parabolas† osculating the two branches at the origin.

The discussion of the preceding form for y resolves itself into three cases, according as $2a$ is $>$ $=$ or $<$ b.

(1). Let $2a = b + h$, then

$$y = Ax^{\frac{b+h}{2}} \pm x^{\frac{b}{2}} \sqrt{B + A^2 x^h + Cx^{c-b}}.$$

(α). If b be *odd*, $x^{\frac{b}{2}}$ becomes imaginary for negative values of x, and accordingly the origin is a cusp of the *first species* in this case.

---

* This term is retained, as it is necessary in the case of a cusp of the second species.

† The word parabola is here employed in its more extensive signification.

(β). If b be *even*, and B positive, y is real for all values of x near the origin; accordingly that point is a *double cusp*.

(γ). If b be *even*, and B negative, the origin is a *conjugate point*.

(2). If $2a = b$, we have

$$y = Ax^a \pm x^a \sqrt{(A^2 + B) + Cx^{c-b}}.$$

In this case, the origin is either a double cusp, or a conjugate point, according as $A^2 + B$ is positive or negative.

Again, if $A^2 + B = 0$, we have

$$y = x^a \left(A + x^{\frac{c-b}{2}} \sqrt{C}\right).$$

(a). If $c - b$ be an odd number, the origin is a cusp of the second species.

(β). If $c - b$ be even, the origin is a double cusp or a conjugate point according as C is positive or negative.

(3). $\qquad 2a < b$, or $b = 2a + h$.

Here $\qquad y = Ax^a \pm x^a \sqrt{A^2 + Bx^h + Cx^{c-2a}}$,

and the curve evidently extends at both sides of the origin, which accordingly is a *double cusp*.

This method of investigating curvature is capable of being modified so as to apply to the case of multiple points of a higher order; the discussion, however, is neither sufficiently elementary, nor sufficiently important, to be introduced here.

255. **Points on Evolute corresponding to Cusps on Curve.**—In connexion with evolutes and involutes, the preceding results lead to a few interesting conclusions.

(1). If a curve has a cusp of the first species, its evolute *in general* passes through the cusp. However, if in addition the cusp be a point of inflexion, the tangent at it is an asymptote to the evolute.

(2). To a cusp of the second species corresponds in general a point of inflexion on the evolute: in some cases the point of inflexion lies altogether at infinity.

(3). To a double cusp corresponds a double tangent to the evolute.

256. Equation of the Osculating Conic.

As an additional illustration of the principles involved in the preceding investigation, it is proposed to discuss the question of the conic which *osculates* an algebraic curve at a given point. Transferring the origin to the point, and taking the tangent as axis of x, the equation of the curve may be written in the form

$$ay = x^2 + a_1 xy + a_2 y^2 + b_0 x^3 + b_1 x^2 y + b_2 xy^2 + b_3 y^3$$
$$+ c_0 x^4 + c_1 x^3 y + \&c. + d_0 x^5 + \&c. \qquad (33)$$

In considering the form of the curve near the origin, as a first approximation we may, as in Art. 251, neglect xy, y^2, &c., in comparison with y; and x^3, x^4, &c., in comparison with x^2; thus the equation reduces to the form

$$ay = x^2. \qquad (34)$$

Hence the form to which every curve of finite curvature approximates in the limit is that of the common parabola, as already seen in Art. 231.

To proceed to the next approximation, we retain terms of the *third* order (remembering that when x is a very small quantity of the first order, y is one of the *second*), and the equation becomes

$$ay = x^2 + a_1 xy + b_0 x^3.$$

On substituting ay instead of x^2 in the term $b_0 x^3$, the preceding equation becomes

$$ay = x^2 + (a_1 + b_0 a) xy. \qquad (35)$$

This represents a conic having contact of the *third* order with the proposed curve at the origin. When $a_1 + b_0 a = 0$, the parabola $ay = x^2$ has a contact of the *third* order at the origin, and accordingly so also has the osculating circle.

In proceeding to the next and final approximation, we retain terms of the fourth order, and we get

$$ay = x^2 + a_1 xy + a_2 y^2 + b_0 x^3 + b_1 x^2 y + c_0 x^4. \qquad (36)$$

Moreover, from the preceding approximation we have

$$b_0 a x y = b_0 x^3 + b_0 x^2 y (a_1 + a b_0).$$

Hence, we get for the equation of the conic having a contact of the closest kind with the given curve

$$ay = x^2 + (a_1 + b_0 a)\, xy + [a_2 + a(b_1 - a_1 b_0) + a^2(c_0 - b_0^2)] y^2. \quad (37)$$

This conic, since it has the closest contact possible with the given curve at the origin, is *the osculating conic* (Art. 246) for that point.

In like manner the parabola

$$ay = x^2 + (a_1 + b_0 a)\, xy + \frac{(a_1 + b_0 a)^2}{4} y^2, \quad (38)$$

since it has the closest contact possible for a parabola, is *the osculating parabola* at the point.

Examples.

1. Prove that the radius of curvature at the vertex of a parabola is equal to its semi-latus rectum.

2. Find the length of the radius of curvature at the origin in the curve

$$y^4 + x^3 + a(x^2 + y^2) = a^2 y. \qquad Ans. \ \frac{a}{2}.$$

3. Find the radius of curvature at the origin in the curve

$$a^2 y = bx^3 + cx^2 y. \qquad Ans. \ \infty.$$

4. Prove that the locus of the centre of a conic having contact of the third order with a given curve at a common point is a right line.

5. Prove that the locus of the centres of equilateral hyperbolas, which have contact of the second order with a given curve at a fixed point, is a circle, whose radius is half that of the circle of curvature at the point.

6. Prove geometrically that the centre of curvature at any point on an ellipse is the pole of the tangent at the point, with respect to the confocal hyperbola which passes through that point.

7. The locus of the centres of ellipses whose axes have a given direction, and which have a contact of the second order with a given curve at a common point, is an equilateral hyperbola passing through the point?

8. Prove that the locus of the focus of a parabola, which has a contact of the second order with a given curve at a given point, is a circle.

9. Prove that the radius of curvature of the curve $a^{m-1} y = x^m$ at the origin is zero, $\frac{a}{2}$, or infinity, according as m is $< = $ or > 2 : m being assumed to be greater than unity.

10. Two plane closed curves have the same evolute : what is the difference between their perimeters ?

$Ans.$ $2\pi d$, where d is the distance between the curves.

11. Find the radius of curvature at the origin in the curve

$$3y = 4x - 15x^2 - 3x^3:$$

find also at what points the radius of curvature is infinite.

12. Apply the principles of investigating maxima and minima to find the greatest and least distances of a point from a given curve; and show that the problem is solved by drawing the normals to the curve from the given point.

(a). Prove that the distance is a minimum, if the given point be nearer to the curve than the corresponding centre of curvature, and a maximum if it be further.

(b). If the given point be on the evolute, show that the solution arrived at is neither a maximum nor a minimum; and hence show that the circle of curvature cuts as well as touches the curve at its point of contact.

13. Find an expression for the whole length of the evolute of an ellipse.

$$\text{Ans. } 4\frac{a^3 - b^3}{ab}.$$

14. Find the radii of curvature at the origin of the two branches of the curve

$$x^4 - \frac{5}{2}ax^2y - axy^2 + a^2y^2 = 0. \qquad \text{Ans. } a \text{ and } \frac{a}{4}.$$

15. Prove that the evolute of the hypocycloid

$$x^{\frac{2}{3}} + y^{\frac{2}{3}} = a^{\frac{2}{3}}$$

is the hypocycloid

$$(a + \beta)^{\frac{2}{3}} + (a - \beta)^{\frac{2}{3}} = 2a^{\frac{2}{3}}.$$

16. Find the radius of curvature at any point on the curve

$$y + \sqrt{x(1-x)} = \sin^{-1}\sqrt{x}.$$

17. If the angle between the radius vector and the normal to a curve has a maximum or a minimum value, prove that $\gamma = r$; where γ is the semi-chord of curvature which passes through the origin.

18. If the co-ordinates of a point on a curve be given by the equations

$$x = c \sin 2\theta (1 + \cos 2\theta), \qquad y = c \cos 2\theta (1 - \cos 2\theta),$$

find the radius of curvature at the point. $\qquad$ Ans. $4c \cos 3\theta$

19. Show that the evolute of the curve

$$r^2 - a^2 = mp^2$$

has for its equation

$$r^2 - (1 - m) a^2 = mp^2.$$

20. If α and β be the co-ordinates of the point on the evolute corresponding to the point (x, y) on a curve, prove that

$$\frac{dy}{dx}\frac{d\alpha}{d\beta} + 1 = 0.$$

21. If ρ be the radius of curvature at any point on a curve, prove that the radius of curvature at the corresponding point in the evolute is $\dfrac{d\rho}{d\omega}$; where ω is the angle the radius of curvature makes with a fixed line.

22. In a curve, prove that

$$\frac{1}{\rho} = \frac{d}{dx}\left(\frac{dy}{ds}\right).$$

23. Find the equation of the evolute of an ellipse by means of the eccentric angle.

24. Prove that the determination of the equation of the evolute of the curve $y = kx^n$ reduces to the elimination of x between the equations

$$a = \frac{n-2}{n-1}x - \frac{k^2 n^2}{n-1}x^{2n-1}, \text{ and } \beta = \frac{2n-1}{n-1}kx^n + \frac{1}{kn(n-1)x^{n-2}}.$$

25. In figure, Art. 239, if the tangent to the evolute at P meet the parabola in a point H, prove that HN is perpendicular to the axis of the parabola.

26. If on the tangent at each point on a curve a constant length measured from the point of contact be taken, prove that the normal to the locus of the points so found passes through the centre of curvature of the proposed curve.

27. In general, if through each point of a curve a line of given length be drawn making a constant angle with the normal, the normal to the curve locus of the extremities of this line passes through the centre of curvature of the proposed. (Bertrand, *Cal. Dif.*, p. 573.)

This and the preceding theorem can be immediately established from geometrical considerations.

28. If from the points of a curve perpendiculars be drawn to one of its tangents, and through the foot of each a line be drawn in a fixed direction, proportional to the length of the corresponding perpendicular; the locus of the extremity of this line is a curve touching the proposed at their common point. Find the ratio of the radii of curvature of the curves at this point.

29. Find an expression for the radius of curvature in the curve $p = \dfrac{mr}{\sqrt{m^2 - r^2}}$, p being the perpendicular on the tangent.

30. Being given any curve and its osculating circle at a point, prove that the portion of a parallel to their common tangent intercepted between the two curves is a small quantity of the second order, when the distances of the point of contact from the two points of intersection are of the first order.

Prove that, under the same circumstances, the intercept on a line drawn parallel to the common normal is a small quantity of the third order.

31. In a curve referred to polar co-ordinates, if the origin be taken on the curve, with the tangent at the origin as prime vector, prove that the radius of curvature at the origin is equal to one-half the value of $\dfrac{r}{\theta}$ in the limit.

32. Hence find the length of the radius of curvature at the origin in the curve $r = a \sin n\theta$. *Ans.* $\rho = \dfrac{na}{2}$.

33. Find the co-ordinates of the centre of curvature of the catenary; and show that the radius of curvature is equal, but opposite, to the normal.

34. If ρ, ρ' be the radii of curvature of a curve and of its pedal at corresponding points, show that

$$\rho'(2r^2 - p\rho) = r^3.$$

Ind. Civ. Ser. Exam., 1878.

CHAPTER XVIII.

ON TRACING OF CURVES.

257. Tracing Algebraic Curves.—Before concluding the discussion of curves, it seems desirable to give a brief statement of the mode of tracing curves from their equations.

The usual method in the case of algebraic curves consists in assigning a series of different values to one of the co-ordinates, and calculating the corresponding series of values of the other; thus determining a definite number of points on the curve. By drawing a curve or curves of continuous curvature through these points, we are enabled to form a tolerably accurate idea of the shape of the curve under discussion.

In curves of degrees beyond the second, the preceding process generally involves the solution of equations beyond the second degree: in such cases we can determine the series of points only approximately.

258. The following are the principal circumstances to be attended to:—

(1). Observe whether from its equation the curve is symmetrical with respect to either axis; or whether it can be made so by a transformation of axes. (2). Find the points in which the curve is met by the co-ordinate axes. (3). Determine the positions of the asymptotes, if any, and at which side of an asymptote the corresponding branches lie. (4). Determine the double points, or multiple points of higher orders, if any belong to the curve, and find the tangents at such points by the method of Art. 212. (5). The existence of ovals can be often found by determining for what values of either co-ordinate the other becomes imaginary. (6). If the curve has a multiple point, its tracing is usually simplified by taking that point as origin, and transforming to polar co-ordinates: by assigning a series of values to θ we can usually determine the corresponding values of r, &c. (7). The points where the y ordinate is a maximum or a minimum are found from the equation $\frac{dy}{dx} = 0$: by this means the limits of the curve can be often assigned. (8). Determine when possible the points of inflexion on the curve.

259. **To trace the Curve** $y^2 = x^2(x - a)$; a being supposed positive.

In this case the origin is a conjugate point, and the curve cuts the axis of x at a distance $OA = a$. Again, when x is less than a, y is imaginary, consequently no portion of the curve lies to the left-hand side of A.

The points of inflexion, I and I', are easily determined from the equation $\frac{d^2y}{dx^2} = 0$; the

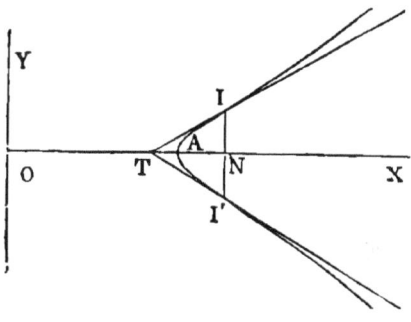

Fig. 38.

corresponding value of x is $\frac{4a}{3}$; accordingly $AN = \frac{OA}{3}$.

Again, if TI be the tangent at the point of inflexion I, it can readily be seen that $TA = \frac{a}{9} = \frac{AN}{3}$.

This curve has been already considered in Art. 213, and is a cubical parabola having a conjugate point.

260. **Cubic with three Asymptotes.**—We shall next consider the curve*

$$y^2x + ey = ax^3 + bx^2 + cx + d, \qquad (1)$$

where a is supposed positive.

The axis of y is an asymptote to the curve (Art. 200), and the directions of the two other asymptotes are given by the equation

$$y^2 - ax^2 = 0, \quad \text{or } y = \pm x\sqrt{a}.$$

---

* This investigation is principally taken from Newton's *Enumeratio Linearum Tertii Ordinis*.

If the term bx^2 be wanting, these lines are asymptotes; if b be not zero, we get for the equation of the asymptotes

$$y = x\sqrt{a} + \frac{b}{2\sqrt{a}}, \quad y + x\sqrt{a} + \frac{b}{2\sqrt{a}} = 0.$$

On multiplying the equations of the three asymptotes together, and subtracting the product from the equation of the curve, we get

$$ey = \left(c - \frac{b^2}{4a}\right)x + d:$$

this is the equation of the right line which passes through the three points in which the cubic meets its asymptotes. (Art. 204.)

Again, if we multiply the proposed equation by x, and solve for xy, we get

$$xy = -\frac{e}{2} \pm \sqrt{ax^4 + bx^3 + cx^2 + dx + \frac{e^2}{4}}: \qquad (2)$$

from which a series of points can be determined on the curve corresponding to any assigned series of values for x.

It also follows that all chords drawn parallel to the axis of y are bisected by the hyperbola $xy + \frac{e}{2} = 0$: hence we infer that the middle points of all chords drawn parallel to an asymptote of the cubic lie on a hyperbola.

The form of the curve depends on the roots of the biquadratic under the radical sign. (1). Suppose these roots to be all real, and denoted by a, β, γ, δ, arranged in order of increasing magnitude, and we have

$$xy = -\frac{e}{2} \pm \sqrt{a(x-a)(x-\beta)(x-\gamma)(x-\delta)}.$$

Now when x is $< a$, y is real; when $x > a$ and $< \beta$, y is imaginary; when $x > \beta$ and $< \gamma$, y is real; when $x > \gamma$ and $< \delta$, y is imaginary; when $x > \delta$, y is real.

Asymptotes.

We infer that the curve consists of three branches, extending to infinity, together with an oval lying between the values β and γ for x.

The accompanying figure* represents such a curve.

Again, if either the two greatest roots or the two least roots become equal, the corresponding point becomes a *node*.

If the intermediate roots become equal, the *oval* shrinks into a *conjugate point* on the curve.

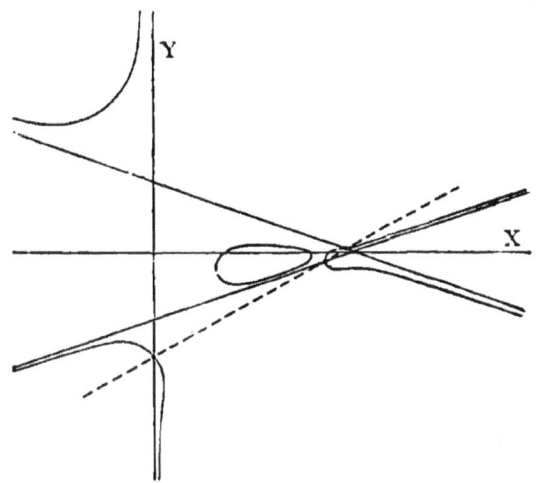

Fig. 39.

If three roots be equal, the corresponding point is a *cusp*.

If two of the roots be impossible and the other two unequal, the curve can have neither an *oval* nor a *double point*.

If the sign of a be negative, the curve has but one real asymptote.

261. **Asymptotes.**—In the preceding figure the student will observe that to each asymptote correspond two infinite branches; this is a general property of algebraic curves, of which we have a familiar instance in the common hyperbola.

By the student who is acquainted with the elementary principles of conical projection the preceding will be readily apprehended; for if we suppose any line drawn cutting a closed oval curve in two points at which tangents are drawn, and if the figure be so projected that the intersecting line is sent to infinity, then the tangents will be projected into asymptotes, and the oval becomes a curve in two portions, each having two infinite branches, a pair for each asymptote, as in the hyperbola.

---

* The figure is a tracing of the curve

$$9xy^2 + 108y = (x-5)(x-11)(x-12).$$

It should also be observed that the points of contact at infinity on the asymptote in the opposite directions along it must be regarded as being *one* and the *same* point, since they are the projection of the same point. That the points at infinity in the two opposite directions on any line must be regarded as a single point is also evident from the consideration that a right line is the limiting state of a circle of infinite radius.

The property admits also of an analytical proof; for if the asymptote be taken as the axis of x, the equation of the curve (Art. 204) is of the form

$$y\phi_1 + \phi_2 = 0, \quad \text{or } y = -\frac{\phi_2}{\phi_1},$$

where ϕ_2 is at least one degree lower than ϕ_1 in x and y. Now, when x is infinitely great, the fraction $\frac{\phi_2}{\phi_1}$ becomes in general infinitely small, whether x be *positive* or *negative*; and consequently the axis is asymptotic to the curve in both directions.

262. To trace the Curve

$$a^3 y^2 = bx^4 + x^5,$$

where a and b are both positive.

Here $\quad ya^{\frac{3}{2}} = \pm x^2 (x + b)^{\frac{1}{2}}.$

The curve is symmetrical with respect to the axis of x, and has two infinite branches; the origin is a double cusp. The shape of the curve is exhibited in the figure annexed.

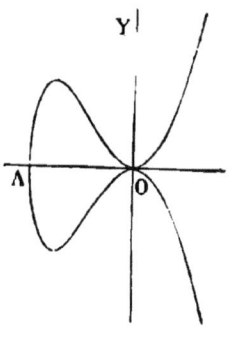

Fig. 40.

If b were negative, we should have

$$ya^{\frac{3}{2}} = \pm x^2 (x - b)^{\frac{1}{2}}.$$

Here y becomes imaginary for values of x less than b; accordingly, the origin is a conjugate point in this case: the curve has two infinite branches as in the former case

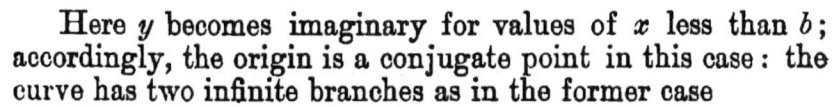

263. To trace the Curve

$$a^3 y^2 = 2abx^2 y + x^5.$$

From the form of its equation we see that the origin is a point of *oscul*-inflexion (Art. 251).

Solving for y, we can easily determine any number of points on the curve we please. It has two infinite branches at opposite sides of the axis of x, and a loop at the negative side of that axis, as exhibited in the figure.

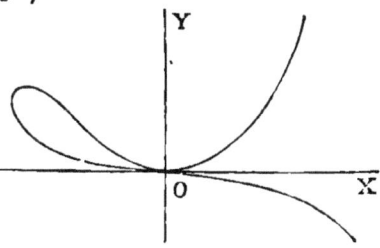

264. **To trace the Curve**

$$x^4 + x^2 y^2 + y^4 = x(ax^2 - by^2).$$

(1). Let a and b have the same sign, then the origin is a triple point, having for its tangents the lines

$$x = 0, \; x\sqrt{a} + y\sqrt{b} = 0,$$

and $\quad x\sqrt{a} - y\sqrt{b} = 0.$

Moreover, since the curve has no real asymptote, it is a finite or closed curve with three loops passing through the origin; and it is easily seen that its shape is that represented in the accompanying figure.

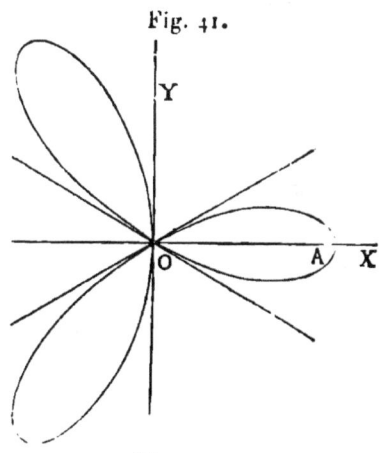

(2). If a and b have opposite signs, the lines represented by $ax^2 - by^2 = 0$ become imaginary. The curve in this case consists of a single oval as in the figure.

This and the preceding figure were traced for the case where $b = 3a$: if the value of $\dfrac{b}{a}$ be altered, the shape of the curve will alter at the same time. If a be greater than b, the curve (2) will lie inside the tangent at the point X.

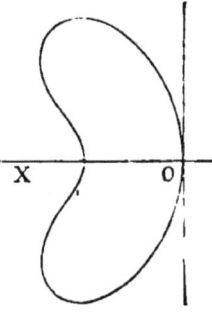

265. **Form of Curve near a Double Point.**—Whenever the curve has a node or a cusp, by transforming the origin to that point, the shape of the curve for the branches

passing through the point admits of being investigated by the method explained in Arts. 250, 251. It is unnecessary to enter into detail on this subject here, as it has been already discussed in the articles referred to.

266. In connexion with the tracing and the discussion of curves there is an elementary general principle which may be introduced here.

If the equation of a curve be of the form

$$LL' - MM' = 0,$$

where L, M, L', M' are each functions of the co-ordinates x and y, the curve evidently passes through all the points of intersection of the curves represented by the equations $L = 0$ and $M = 0$; similarly it passes through the intersections of $L = 0$ and $M' = 0$; and also those of $M = 0$ and $L' = 0$; and of $L' = 0$ and $M' = 0$. Moreover, if L and L' become identical, the points of intersection coincide in pairs, and the equation of the curve becomes of the form $L^2 - MM' = 0$; which represents a curve touching the curves $M = 0$, $M' = 0$, at their points of intersection with the curve $L = 0$.

This principle admits of easy extension; but as the subject belongs properly to the method of trilinear co-ordinates, it is not considered necessary to enter more fully into it here.

267. **On Tracing Curves given in Polar Co-ordinates.**—The mode of procedure in this case does not differ essentially from that for Cartesian co-ordinates. We have already, in Arts. 206 and 207, considered the method of finding the asymptotes and asymptotic circles in such cases. It need scarcely be observed that the number and variety of curves whose discussion more properly comes under the method of polar co-ordinates are indefinite. We propose to confine our attention to a few varieties of the class of curves represented by the equation

$$r^m = a^m \cos m\theta.$$

268. **On the Curves** $r^m = a^m \cos m\theta$.—In this case, since the equation is unaltered when θ is changed into $-\theta$, the curve is symmetrical with respect to the prime vector: again, when $\theta = 0$, we have $r = a$; and as θ increases from zero

to $\frac{\pi}{2m}$, r diminishes from a to zero. When m is a positive integer, it is easily seen that the curve consists of m similar loops.

There are many familiar curves included under this equation. Thus, when $m = 1$, we have $r = a \cos \theta$, which represents a circle: again, if $m = -1$, the equation gives $r \cos \theta = a$, which represents a right line. Also, if $m = 2$, we have $r^2 = a^2 \cos 2\theta$, a Lemniscate (Art. 210). If $m = -2$, we get $r^2 \cos 2\theta = a^2$, an equilateral hyperbola.

If $m = \frac{1}{2}$ we get $r^{\frac{1}{2}} = a^{\frac{1}{2}} \cos \frac{\theta}{2}$, whence $r = \frac{a}{2}(1 + \cos \theta)$, a cardioid (Ex. 4, p. 232); with $m = -\frac{1}{2}$, it is $r^{\frac{1}{2}} \cos \frac{\theta}{2} = a^{\frac{1}{2}}$, a parabola (Ex. 1, p. 231); and so on. As already observed, if we change m into $-m$ we get a new curve, inverse of the original. Also, the reciprocal polar is obtained by substituting $-\frac{m}{m+1}$ instead of m.

The tangent and normal can be immediately drawn at any point on a curve of this class by aid of the results arrived at in Art. 190. The radius of curvature at any point has been determined in Ex. 5, Art. 235. The method of finding the equations of the successive *pedals*, both positive and negative, has been also already explained.

A few examples in the case of fractional indices are here added.

Example 1.
$$r^{\frac{1}{2}} = a^{\frac{1}{2}} \cos \frac{\theta}{3}.$$

Here when $\theta = 0$, we have $r = a$, and the curve cuts the prime vector at a distance OA equal to a: again, when $\theta = \frac{\pi}{2}$, $r = \frac{3a\sqrt{3}}{8}$: also when $\theta = \pi$, $r = \frac{a}{8}$, or $OB = \frac{a}{8}$.

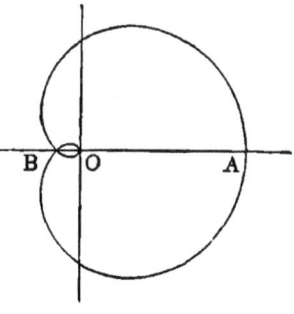

Fig. 44.

The shape of the curve is given in the accompanying figure. This curve is the inverse of the *caustic* considered in Example 18, p. 277.

Ex. 2. Ex. 3. Ex. 4.

$r^{\frac{3}{4}} = a^{\frac{3}{4}} \cos \frac{3}{4}\theta.$ $r^{\frac{4}{5}} = a^{\frac{4}{5}} \cos \frac{4}{5}\theta.$ $r^{\frac{5}{3}} = a^{\frac{5}{3}} \cos \frac{5}{3}\theta.$

In Ex. 2, as θ increases from zero to 120°, r diminishes from a to zero: when θ increases from 120° to 240°, r increases from zero to a: when θ increases from 240° to 360°, r diminishes from a to zero. By assigning negative values to θ, the remaining part of the curve is seen to be symmetrical with that traced as above. The same result plainly follows by continuing the values for θ from 360° up to 720°. The form of the curve is exhibited in the annexed figure.

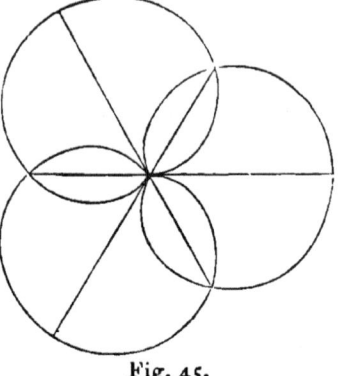

Fig. 45.

In Ex. 3, according as $\cos \frac{4}{5}\theta$ is positive or negative, we get equal and opposite *real* values, or *imaginary* values, for r. Hence it is easily seen that for values of θ between $\pm \frac{5}{8}\pi$ the radius vector traces out two symmetrical portions of the curve: again, between $\frac{15}{8}\pi$ and $\frac{25}{8}\pi$ we get two other

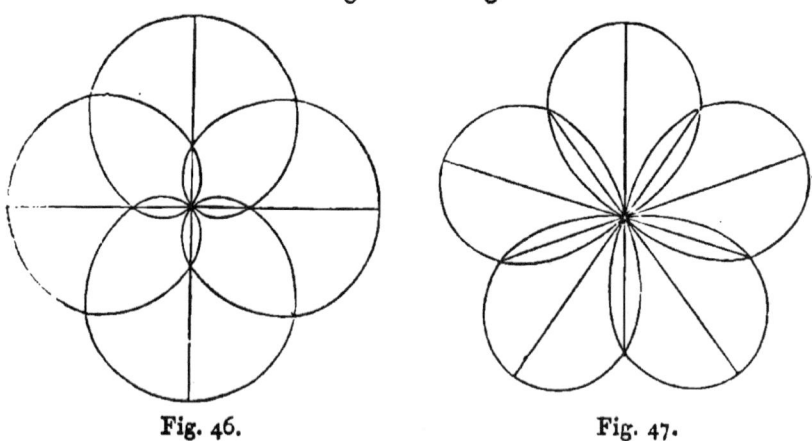

Fig. 46. Fig. 47.

symmetrical portions. The shape is that given in the former of the two accompanying figures.

The latter figure represents the curve in Ex. 4; it consists of *five* symmetrical portions ranged round the origin.

The results above stated admit of generalization, and it can be shown, without difficulty, that in general the curve $r^{\frac{p}{q}} = a^{\frac{p}{q}} \cos \frac{p\theta}{q}$ consists of p similar portions arranged about the origin; and that the entire curve is included within a circle of radius a when p is positive, but lies altogether outside it when p is negative.

Many curves can be best traced by aid of some simple geometrical property. We shall terminate the Chapter with one or two examples of such curves.

269. **The Limaçon.**—The inverse of a conic section with respect to a focus is called a Limaçon. From the polar equation of a conic, its focus being origin, it is evident that the equation of its inverse may be written in the form

$$r = a \cos \theta + b,$$

where a and b are constants.

It is easily seen that $\frac{a}{b}$ is the eccentricity of the conic.

The curve can be readily traced by drawing from a fixed point on a circle any number of chords, and taking off a constant length on each of these lines, measured from the circumference of the circle.

If a be less than b, the curve is the inverse of an ellipse, and lies altogether outside the circle.

If a be greater than b, the curve is the inverse of a hyperbola, and its form can be easily seen to be that exhibited in the annexed figure, where $OD = a - b$, and the point O is a node on the curve.

If $b = a$, the curve becomes the inverse of the parabola, and is called a cardioid. The inner loop disappears in this case, and the origin is a cusp on the curve.

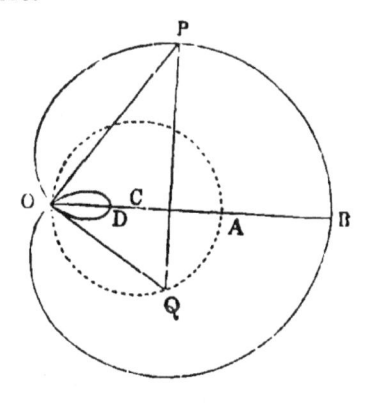

Fig. 48.

When $a = 2b$, the Limaçon is called the **Trisectrix**; a curve by aid of which any given angle can be readily trisected.

270. The Conchoid of Nicomedes.—If through any fixed point A a secant P_1AP be drawn meeting a fixed right line LM in R, and RP and RP_1 be taken each of the same constant length; then the locus of P and P_1 is called the conchoid.

This curve is easily traced from the foregoing geometrical property, and it consists of two branches, having the right line LM for a common asymptote. Moreover, if the perpendicular distance AB of A from the fixed line be less than RP, the curve has a loop with a node at A, as in the annexed figure.

It is easily seen that when $AB = RP$, the point A is a cusp on the curve; and when AB is greater than RP, A is a conjugate point.

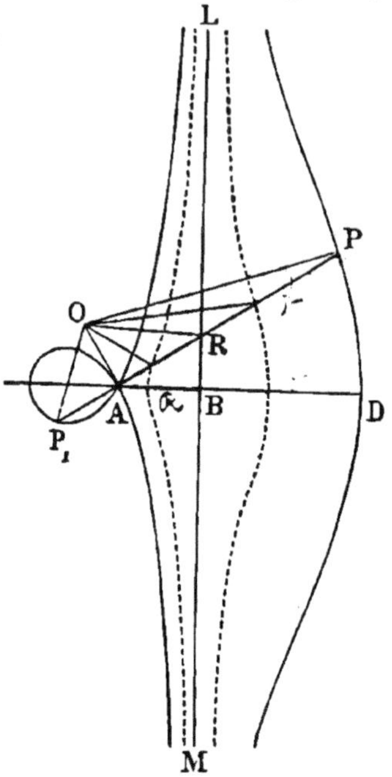

The form of the curve in the latter case is represented by the dotted lines in the figure.

Fig. 49.

If $AB = a$, $RP = b$, the polar equation of the curve is
$(r \pm b) \cos \theta = a$.

When transformed to rectangular co-ordinates, this equation becomes

$$(x^2 + y^2)(a - x)^2 = b^2 x^2.$$

The method of drawing the normal, and finding the centre of curvature, at any point, will be exhibited in the next Chapter.

EXAMPLES.

1. Trace the curve $y = (x-1)(x-2)(x-3)$, and find the position of its point of inflexion.

2. Trace the curve $y^3 - 3axy + x^3 = 0$, drawing its asymptote.
This curve is called the Folium of Descartes.

3. Trace the curve $a^2x = y(b^2 + x^2)$, and find its points of inflexion, and points of greatest and least distance from the axis of x.

4. If an asymptote to a curve meets it in a real finite point, show that the corresponding branch of the curve must have a point of inflexion on it.

5. Find the position of the asymptotes and the form of the curve

$$x^4 - y^4 + 2axy^2 = 0.$$

6. Show that the curve $r = a \cos 2\theta$ consists of four loops, while the curve $r = a \cos 3\theta$ consists of but three. Prove generally that the curve $r = a \cos n\theta$ has n or $2n$ loops according as n is an odd or even integer.

7. Trace the curve

$$y^2(x-a)(x-b) = c^2(x+a)(x+b).$$

8. Show that the curve $x^2y^2 + x^4 = a^2(x^2 - y^2)$ consists of two loops passing through the origin, and find the form of the curve.

9. Trace the curve $y(x+a)^4 = b^2x(x+c)^2$, showing the positions of its asymptotes and infinite branches.

10. Trace the curve whose polar equation is

$$r = a \cos\theta + b \cos 2\theta,$$

and show that it consists of four loops passing through the origin.

11. Given the base and the rectangle under the sides of a triangle, find the equation of the locus of the vertex (an oval of Cassini). Exhibit the different forms of the curve obtained by varying the constants, and find in what case the curve becomes a Lemniscate.

12. Trace the curve $y^2 = ax^3 + 3bx^2 + 3cx + d$, and find its points of greatest and least distance from the axis of x.

Show that two of these points become imaginary when the roots of the cubic in x are all real.

13. Given the base and area of a triangle, prove that the equation of the locus of the centre of a circle touching its three sides is of the form

$$x^2y - a(x^2 + y^2) - b^2(y-a) = 0.$$

14. Prove that all curves of the third degree are reducible to one or other of the forms

(1). $xy^2 + ey = ax^3 + bx^2 + cx + d$.

(2). $xy = ax^3 + bx^2 + cx + d$.

(3). $y^2 = ax^3 + bx^2 + cx + d$.

(4). $y = ax^3 + bx^2 + cx + d$.

Newton, *Enum. Linear. Ter. Ordinis.*

15. Prove that all curves of the third degree can be obtained by projection from the parabolas contained in class (3) in the preceding division. [Newton.]

For every cubic has at least one real point of inflexion: accordingly, if the curve be projected so that the tangent at the point of inflexion is projected to infinity, the harmonic polar (Art. 223) will bisect the system of parallel chords passing through this point at infinity. Hence the projected curve is of the class (3). [This proof is taken from Chasles, *Histoire de la Géométrie, note* xx.]

16. Trace the curve $r = \dfrac{a\theta^2}{\theta^2 - 1}$, and show that it has a point of inflexion when $\theta^2 = 3$; find also its asymptotes and asymptotic circle.

17. Trace the curve $y = a \sin \dfrac{x}{a}$, and show how to draw its tangent at any point. (This is called the curve of sines.)

18. The base of a triangle is fixed in position; find the equation of the locus of its vertex, when the vertical angle is double one of the base angles.

Trace the locus in question, finding the position of its asymptote.

19. Show geometrically that the first pedal of a circle with respect to a point on its circumference is a cardioid.

20. Show in like manner that the Limaçon is the first pedal of a circle with respect to any point.

21. Trace the curve

$$y^4 + 2axy^2 = ax^3 + x^4,$$

and find the equations of its asymptotes, and of the tangents at the origin.

Ind. Civ. Ser. Ex., 1876.

CHAPTER XIX.

ROULETTES.

271. **Roulettes.**—When one curve rolls without sliding upon another, any point invariably connected with the rolling curve describes another curve, called a *roulette*.

The curve which rolls is called the *generating curve*, the fixed curve on which it rolls is called the directing curve, or the *base*, and the point which describes the roulette, the *tracing point*. We shall commence with the simplest example of a roulette: viz., the cycloid.

272. **The Cycloid.**—This curve is the path described by a point on the circumference of a circle, which is supposed to roll upon a fixed right line.

The cycloid is the most important of transcendental curves, as well from the elegance of its properties as from its numerous applications in Mechanics.

We shall proceed to investigate some of the most elementary properties of the curve.

Let LPO be any position of the rolling circle, P the generating point, O the point of contact of the circle with the fixed line. Take the length AO equal to the arc PO, then, from the mode of generation of the curve, A is the position of the generating point when in contact with the fixed line; also, if AA' be equal to the circumference of the circle, A' will be the position of the point at the end of one complete revolution of the circle. Bisect AA' in D, and draw DB perpendicular to it and equal to the diameter of the circle, then B is evidently the highest point in the cycloid. Draw PN perpendicular to AA', and let $PN = y$, $AN = x$, $\angle PCO = \theta$, $OC = a$, and we get

$$x = AO - NO = a(\theta - \sin\theta), \quad y = PN = a(1 - \cos\theta). \quad (1)$$

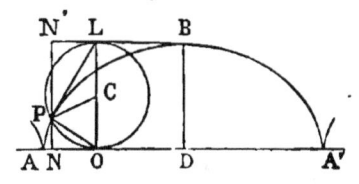

Fig. 50.

The position of any point on the cycloid is determined by these equations when the angle θ is known, *i.e.* the angle through which the circle has rolled, starting from the position for which the generating point is upon the directing line.

273. Cycloid referred to its Vertex.—It is often convenient to refer the cycloid to its vertex as origin, and to the tangent and normal at that point as axes of co-ordinates. In the preceding figure let

$$x = BN', \quad y = PN', \quad \angle PCL = \theta' = \pi - \theta;$$

then we have

$$x = BN' = a(\theta' + \sin \theta'), \quad y = PN' = a(1 - \cos \theta'). \quad (2)$$

274. Tangent and Normal to Cycloid.—It can be easily seen that the line PO is normal at P to the cycloid; for the motion of each point on the circle at the instant is one of rotation about the point O, i.e. each point may be regarded as describing at the instant an infinitely small circular* arc whose centre is at O: and hence PO is normal to the curve.

This result can also be established from the values of x and y in (1): for

$$\frac{dx}{d\theta} = a(1 - \cos \theta), \quad \frac{dy}{d\theta} = a \sin \theta: \quad (3)$$

$$\therefore \frac{dy}{dx} = \frac{\sin \theta}{1 - \cos \theta} = \cot \frac{\theta}{2} = \cot PLO;$$

and, accordingly, PL is the tangent, and PO the normal to the curve at P.

Again, if we square and add the values of $\dfrac{dx}{d\theta}$ and $\dfrac{dy}{d\theta}$, we obtain

$$\left(\frac{ds}{d\theta}\right)^2 = a^2\{(1 - \cos \theta)^2 + \sin^2 \theta\} = 4a^2 \sin^2 \tfrac{1}{2}\theta;$$

---

* This method of finding the normal to a cycloid is due to Descartes, and evidently applies equally to all roulettes.

The Cycloid.

hence
$$\frac{ds}{d\theta} = 2a \sin\frac{\theta}{2} = PO. \qquad (4)$$

275. Radius of Curvature and Evolute of Cycloid. —Let ρ denote the radius of curvature at the point P, and $\angle POA = \phi = \dfrac{\theta}{2}$;

then
$$\rho = \frac{ds}{d\phi} = 2\frac{ds}{d\theta} = 4a \sin\frac{\theta}{2} = 2PO; \qquad (5)$$

or the radius of curvature is double the normal. From this value of ρ the evolute of the curve can be easily determined. For, produce PO until $OP' = OP$, then P' is the centre of curvature belonging to the point P. Again, produce LO until $OO' = OL$, and describe a circle through O, P' and O'; this circle evidently touches AA', and is equal to the generating circle LPO.

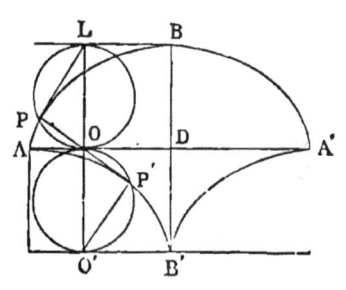

Fig. 51.

Also, the arc $OP' = $ arc $OP = AO$;

∴ arc $O'P' = O'P'O - P'O = AD - AO = OD = B'O'$.

Hence the locus of P' is the cycloid got by the rolling of this new circle along the line $B'O'$; and accordingly the evolute of a cycloid is another cycloid. It is evident that the evolute of the cycloid ABA' is made up of the two semi-cycloids, AB' and $B'A'$, as in figure 51. Conversely, the cycloid ABA' is an involute of the cycloid $AB'A'$.

The position of the centre of curvature for a point P on a cycloid can also be readily determined geometrically, as follows:—

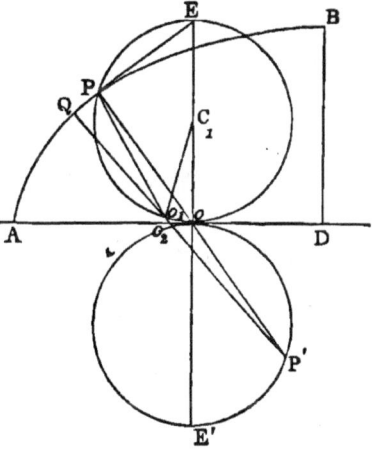

Fig. 52.

Suppose O_1 a point on the circle infinitely near to O, and take $OO_2 = OO_1$. Let P'

z

be the centre of curvature required, and draw PO_1 and $P'O_2$. Now suppose the circle to roll until O_1 and O_2 coincide, then CO_2 becomes perpendicular to AD, and PO_1 and $P'O_2$ will lie in directum (since P' is the point of intersection of two consecutive normals to the cycloid). Hence

$$\angle OCO_1 = \angle PO_1Q = \angle OPO_1 + \angle OP'O_1,$$

since each side of the equation represents the angle through which the circle has turned.

But $\quad\quad \angle OCO_1 = 2 \angle OPO_1.\quad$ (Euclid, III. 20.)

Hence $\quad\quad \angle OPO_1 = \angle OP'O_1;$

$$\therefore PO_1 = P'O_1;$$

and consequently in the limit we have

$$PO = P'O,$$

as before.

We shall subsequently see that a similar method enables us to determine the centre of curvature for a point in any roulette.

276. **Length of Arc of Cycloid.**—Since $AP'B'$ (Fig. 51) is the evolute of the cycloid APB, it follows, from Art. 237, that the arc AP' of the cycloid is equal in length to the line PP', or to twice $P'O$; hence, as A is the highest point in the cycloid $AP'B'$, it follows that the arc AP' measured from the highest point of a cycloid is double the intercept $P'O$, made on the tangent at the point by the tangent at the highest point of the curve.

Hence, denoting the length of the arc AP' by s, we have

$$s = 4a \sin P'OD = 4a \sin \phi. \quad\quad (6)$$

This gives the *intrinsic equation* of the cycloid (see Art. 242 (a)). Hence, also, the whole arc AB' is four times the radius of the generating circle: and accordingly the entire length ABA' of a cycloid is eight times the radius of its generating circle.

Again, if the distance of P' from AA' be represented by y, we shall have

$$P'O^2 = OO' \times y = 2ay.$$

Hence $\quad\quad s^2 = 4P'O^2 = 8ay. \quad\quad (7)$

This relation is of importance in the applications of the cycloid in Mechanics.

Again, since $AO = $ arc OP', if we represent AO by ν, we have*

$$\nu = 2a\phi. \tag{8}$$

277. Trochoids.—In general, if a circle roll on a right line, any point in the plane of the circle carried round with it describes a curve. Such curves are usually styled trochoids. When the tracing point is inside the circle, the locus is called a prolate trochoid; when outside, an oblate. Their forms are exhibited in the accompanying figure.

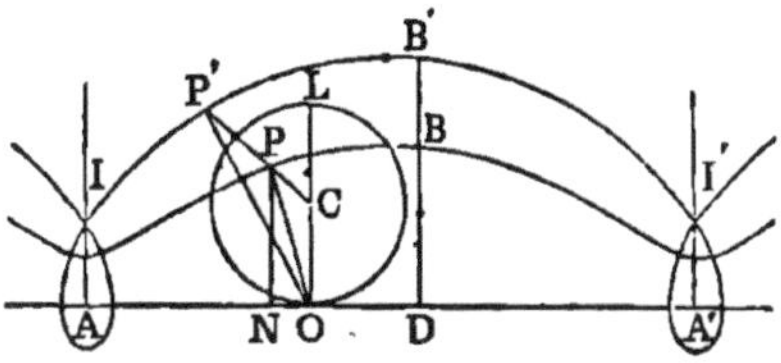

Fig. 53.

Their equations are easily determined; for, let x, y be the co-ordinates of a tracing point P, referred to the axes AD, and AI (A being the position for which the moving radius CP is perpendicular to the fixed line).

Then, if $CO = a$, $CP = d$, $\angle OCP = \theta$, we have

$$\left.\begin{array}{l} x = AN = AO - ON = a\theta - d \sin \theta, \\ y = PN = a - d \cos \theta. \end{array}\right\} \tag{9}$$

278. Epicycloids† and Hypocycloids.—The investi-

---

* This is called, by Professor Casey, the *tangential equation* of the cycloid, and by aid of it he has arrived at some remarkable properties of the curve ("On a New Form of Tangential Equation," *Philosophical Transactions*, 1877). "In general, if a variable line, in any of its positions, make an intercept ν on the axis of x, and an angle ϕ with it; then the equation of the line is

$$x + y \cot \phi - \nu = 0;$$

and ν, ϕ, the quantities which determine the position of the line may be called its co-ordinates. From this it follows that any relation between ν and ϕ, such as

$$\nu = f(\phi),$$

will be the tangential equation of a curve, which is the envelope of the line." For applications, the reader is referred to Professor Casey's Memoir. See also *Dub. Exam. Papers*, Graves, Lloyd Exhibition, 1847.

† I have in this edition adopted the correct definition of these curves as given by Mr. Proctor in his *Geometry of Cycloids*. I have thus avoided the anomaly existing in the ordinary definition, according to which every epicycloid

gation of the properties of the cycloid naturally gave rise to the discussion of the more general case of a circle rolling on a fixed circle. In this case the curve generated by any point on the circumference of the rolling circle is called an *epicycloid*, or a *hypocycloid, according as the rolling circle touches the outside, or the inside of the circumference of the fixed circle*. We shall commence with the former case.

Let P be the position of the generating point at any instant, A its position when on the fixed circle; then the arc OA = arc OP.

Again, let C and C' be the centres of the circles, a and b their radii, $\angle ACO = \theta$, $\angle OC'P = \theta'$; then, since arc OA = arc OP, we have $a\theta = b\theta'$.

Now, suppose C taken as the origin of rectangular co-ordinates, and CA as the axis of x; draw PN and $C'L$ perpendicular, and PM parallel, to CA, and we have

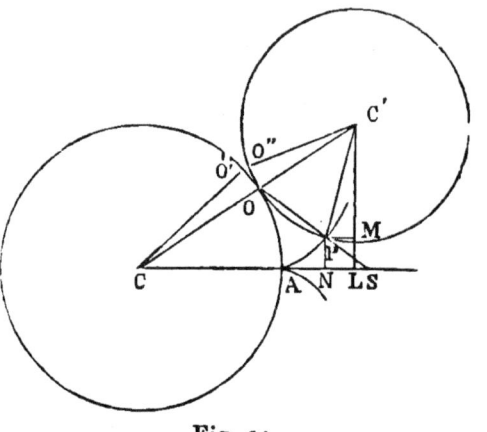

Fig. 54.

$$x = CN = CL - NL = (a + b) \cos \theta - b \cos (\theta + \theta'),$$

$$y = PN = C'L - C'M = (a + b) \sin \theta - b \sin (\theta + \theta');$$

or, substituting $\dfrac{a}{b} \theta$ for θ',

$$\left. \begin{array}{l} x = (a + b) \cos \theta - b \cos \dfrac{a+b}{b} \theta, \\ y = (a + b) \sin \theta - b \sin \dfrac{a+b}{b} \theta. \end{array} \right\} \quad (10)$$

---

is a hypocycloid, but only some hypocycloids are epicycloids. While according to the correct definition no epicycloid is a hypocycloid, though each can be generated in two ways, as will be proved in Art. 280.

When the radius of the rolling circle is a submultiple of that of the fixed circle, the tracing point, after the circle has rolled once round the circumference of the fixed circle, evidently returns to the same position, and will trace the same curve in the next revolution. More generally, if the radii of the circles have a commensurable ratio, the tracing point, after a certain number of revolutions, will return to its original position: but if the ratio be incommensurable, the point will never return to the same position, but will describe an infinite series of distinct arcs. As, however, the successive portions of the curve are in every respect equal to each other, the path described by the tracing point, from the position in which it leaves the fixed circle until it returns to it again, is often taken instead of the complete epicycloid, and the middle point of this path is called the *vertex* of the curve.

In the case of the hypocycloid, the generating circle rolls on the interior of the fixed circle, and it can be easily seen that the expressions for x and y are derived from those in (10) by changing the sign of b; hence we have

$$\left.\begin{aligned} x &= (a-b)\cos\theta + b\cos\frac{a-b}{b}\theta, \\ y &= (a-b)\sin\theta - b\sin\frac{a-b}{b}\theta. \end{aligned}\right\} \quad (11)$$

The properties of these curves are best investigated by aid of the simultaneous equations contained in formulas (10) and (11).

It should be observed that the point A, in Fig. 54, is a cusp on the epicycloid; and, generally, every point in which the tracing point P meets the fixed circle is a cusp on the roulette. From this it follows that if the radius of the rolling circle be the nth part of that of the fixed, the corresponding epi- or hypo-cycloid has n cusps: such curves are, accordingly, designated by the number of their cusps: such as the three-cusped, four-cusped, &c. epi- or hypo-cycloids.

Again, as in the case of the cycloid, it is evident from Descartes' principle that the instantaneous path of the point P is an elementary portion of a circle having O as centre; accordingly, the tangent to the path at P is perpendicular to the line PO, and that line is *the normal to the curve at P*. These results can also be deduced, as in the case of the cycloid, by differentiation from the expressions for x and y. We leave this as an exercise for the student.

To find an expression for an element ds of the curve at the point P; take O', O'', two points infinitely near to O on the circles, and such that $OO' = OO''$; and suppose the generating circle to roll until these points coincide :* then the lines CO' and $C'O''$ will lie in directum, and the circle will have turned through an angle equal to the sum of the angles OCO' and $OC'O''$; hence, denoting these angles by $d\theta$ and $d\theta'$, respectively, we have

$$ds = OP\,(d\theta + d\theta') = OP\left(1 + \frac{a}{b}\right)d\theta; \qquad (12)$$

since
$$d\theta' = \frac{a}{b}\,d\theta.$$

279. Radius of Curvature of an Epicycloid.— Suppose ω to be the angle OSN between the normal at P and the fixed line CA, then

$$\omega = C'OS - C'CS = \frac{\pi}{2} - \frac{\theta'}{2} - \theta;\ \therefore\ d\omega = -d\theta\left\{1 + \frac{a}{2b}\right\}.$$

Hence, if ρ be the radius of curvature corresponding to the point P, we get

$$\rho = -\frac{ds}{d\omega} = OP\,\frac{2(a+b)}{a+2b}. \qquad (13)$$

Accordingly, the radius of curvature in an epicycloid is in a constant ratio to the chord OP, joining the generating point to the point of contact of the circles.

---

* It may be observed that $O'O''$ is infinitely small in comparison with OO'; hence the space through which the point O moves during a small displacement is infinitely small in comparison with the space through which P moves. It is in consequence of this property that O may be regarded as being at rest for the instant, and every point connected with the rolling circle as having a circular motion around it.

280. Double Generation of Epicycloids and Hypocycloids.—In an Epicycloid, it can be easily shown that the curve can be generated in a second manner. For, suppose the rolling circle incloses the fixed circle, and join P, any position of the tracing point, to O, the corresponding point of contact of the two circles; draw the diameter OED, and join $O'E$ and PD; connect C, the centre of the fixed circle, to O', and produce CO' to meet DP produced in D', and describe a circle round the triangle $O'PD'$; this circle plainly touches the fixed circle; also the segments standing on OP, $O'P$, and OO' are obviously similar; hence, since $OP = OO' + O'P$, we have

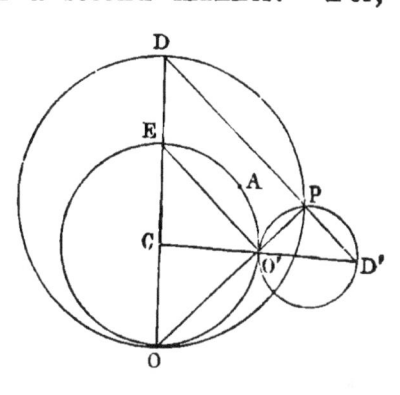

$$\text{arc } OP = \text{arc } OO' + \text{arc } O'P.$$

Fig. 55.

If the arc $OO'A$ be taken equal to the arc OP, we have arc $O'A = $ arc $O'P$; accordingly, the point P describes the same curve, whether we regard it as on the circumference of the circle OPD rolling on the circle $OO'E$, or on the circumference of $O'PD'$ rolling on the same circle; provided the circles each start from the position in which the generating point coincides with the point A. Moreover, it is evident that the *radius of the latter circle is the difference between the radii of the other two.*

Next, for the Hypocycloid, suppose the circle OPD to roll inside the circumference of $OO'E$, and let C be the centre of the fixed circle; join OP, and produce it to meet the circumference of the fixed circle in O'; draw $O'E$ and PD, join CO', intersecting PD in D', and describe a circle round the triangle $PD'O'$. It is evident, as before, that this circle touches the

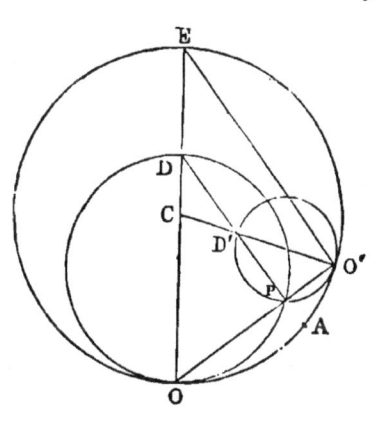

Fig. 56.

larger circle, and that its radius is equal to the difference between the radii of the two given circles. Also, for the same reason as in the former case, we have

$$\text{arc } OO' = \text{arc } OP + \text{arc } O'P.$$

If the arc OA be taken equal to OP, we get arc $O'P$ = arc $O'A$; consequently, the point P will describe the same hypocycloid on whichever circle we suppose it to be situated, provided the circles each set out from the position for which P coincides with A.

The particular case, when the radius of the rolling circle is half that of the fixed circle, may be noticed. In this case the point D coincides with C, and P becomes the middle point of OO', and A that of the arc OO'. From this it follows immediately that the hypocycloid described by P becomes the diameter CA of the fixed circle. This result will be proved otherwise in Art. 285.

The important results of this Article were given by Euler (*Acta. Petrop.*, 1781). By aid of them all epicycloids can be generated by the rolling of a circle *outside* another circle; and all hypocycloids by the rolling of a circle whose radius is less than half that of the fixed circle.

281. **Evolute of an Epicycloid.**—The evolute of an epicycloid can be easily seen to be a similar epicycloid.

For, let P be the tracing point in any position, A its position when on the fixed circle; join P to O, the point of contact of the circles, and produce PO until $PP' = OP \dfrac{2a + 2b}{a + 2b}$, then P is the centre of curvature by (13); hence

$$OP' = OP \frac{a}{a + 2b}.$$

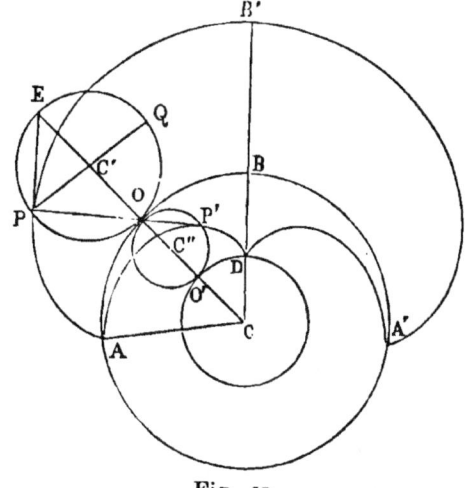

Fig. 57.

Next, draw $P'O'$ perpendicular to $P'O$; circumscribe the

triangle $OP'O'$ by a circle; and describe a circle with C as centre, and CO' as radius: it evidently touches the circle $OP'O'$.

Then $\quad OO' : OE = OP' : OP = a : a + 2b = CO : CE;$

$$\therefore\ CO - OO' : CE - OE = CO : CE,$$

or $\quad\quad CO' : CO = CO : CE;$

that is, the lines CE, CO, and CO' are in geometrical proportion.

Again, join C to B', the vertex of the epicycloid; let CB' meet the inner circle in D, and we have

$$\text{arc } O'D : \text{arc } OB = CO' : CO = CO : CE = O'O : EO$$

$$= \text{arc } P'O' : \text{arc } OQ.$$

But arc OB = arc OQ; $\quad\therefore$ arc $O'D$ = arc $P'O'$.

Accordingly, the path described by P' is that generated by a point on the circumference of the circle $OP'O'$ rolling on the inner circle, and starting when P' is in contact at D. Hence the evolute of the original epicycloid is another epicycloid. The form of the evolute is exhibited in the figure.

Again, since $CO : OE = CO' : O'O$, the ratio of the radii of the fixed and generating circles is the same for both epicycloids, and consequently the *evolute is a similar epicycloid*.

Also, from the theory of evolutes (Art. 237), the line PP' is equal in length to the arc $P'A$ of the interior epicycloid; or the length of $P'A$, the arc measured from the vertex A of the curve, is equal to

$$\frac{2(a+b)}{a}OP' = 2OP'\frac{CC'}{CO} = 2OP'\frac{CC''}{CO'}.$$

Hence, *the length* of any portion of the curve measured from its vertex is to the corresponding chord of the generating circle as twice the sum of the radii of the circles to the radius of the fixed circle.*

---

* The length of the arc of an epicycloid, as also the investigation of its evolute, were given by Newton (*Principia*, Lib. I., Props. 49, 50):

With reference to the outer epicycloid in Fig. 57, this gives

$$\text{arc } PB' = 2PE \cdot \frac{CC'}{CO}. \tag{14}$$

The corresponding results for the hypocycloid can be found by changing the sign of the radius b of the rolling circle in the preceding formulæ.

The investigation of the properties of these curves is of importance in connexion with the proper form of toothed wheels in machinery.

282. Pedal of Epicycloid.—The equation of the pedal, with respect to the centre of the fixed circle, admits of a very simple expression. For let P be the generating point, and, as before, take arc OA = arc OP, and make $AB = 90°$. Join CA, CB, CP, and draw CN perpendicular to DP. Let $\angle PDO = \phi$, $\angle BCN = \omega$, $\angle ACO = \theta$, $CN = p$.

Then since $AO = PO$, we have

$$a\theta = 2b\phi; \quad \therefore \theta = \frac{2b}{a}\phi.$$

Again, $\omega = 90° - ACN = \theta + \phi$

$$= \phi\left(1 + \frac{2b}{a}\right);$$

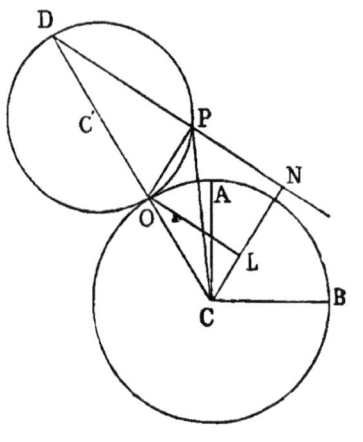

Fig. 58.

hence
$$\phi = \frac{a\omega}{a + 2b}. \tag{15}$$

Also $\qquad CN = CD \sin \phi;$

$$\therefore p = (a + 2b) \sin \frac{a\omega}{a + 2b}, \tag{16}$$

which is the equation of the required pedal.

283. Equation of Epicycloid in terms of r and p.—Again, draw OL parallel to DN, and let $CP = r$, and we have

$$r^2 - p^2 = PN^2 = OL^2 = OC^2 - CL^2 = a^2 - \left(\frac{a}{a + 2b}\right)^2 p^2;$$

hence
$$r^2 = a^2 + \frac{4b(a+b)}{(a+2b)^2} p^2. \tag{17}$$

Also, from (16) it is plain that the equation of DN, the tangent to the epicycloid (referred to CB and CA as axes of x and y respectively), is

$$x \cos \omega + y \sin \omega = (a + 2b) \sin \frac{a\omega}{a + 2b}. \tag{18}$$

The corresponding formulæ for the hypocycloid are obtained by changing the sign of b in the preceding equations.

Again, it is plain that the envelope of the right line represented by equation (18) is an epicycloid. And, in general, *the envelope of the right line*

$$x \cos \omega + y \sin \omega = k \sin m\omega,$$

regarding ω as an arbitrary parameter, is an epicycloid, or a hypocycloid, according as m is less or greater than unity. For examples of this method of determining the equations of epi- and hypo-cycloids the student is referred to Salmon's *Higher Plane Curves*, Art. 310.

284. **Epitrochoids and Hypotrochoids.**—In general, when one circle rolls on another, every point connected with the rolling circle describes a distinct curve. These curves are called *epitrochoids* or *hypotrochoids*, according as the rolling circle touches the exterior or the interior of the fixed circle.

If d be the constant distance of the generating point from the centre of the rolling circle, there is no difficulty in proving, as in Art. 278, that we have in the epitrochoid the equations

$$\left. \begin{array}{l} x = (a+b) \cos \theta - d \cos \dfrac{a+b}{b} \theta, \\[2mm] y = (a+b) \sin \theta - d \sin \dfrac{a+b}{b} \theta. \end{array} \right\} \tag{19}$$

In the case of the hypotrochoid, changing the signs of b and d, we obtain

$$x = (a - b) \cos \theta + d \cos \frac{a-b}{b} \theta,$$
$$y = (a - b) \sin \theta - d \sin \frac{a-b}{b} \theta. \qquad (20)$$

In the particular case in which $a = 2b$, i.e. when a circle rolls inside another of double its diameter, equations (20) become

$$x = (b + d) \cos \theta, \quad y = (b - d) \sin \theta;$$

and accordingly the equation of the roulette is

$$\frac{x^2}{(b+d)^2} + \frac{y^2}{(b-d)^2} = 1;$$

which represents an ellipse whose semi-axes are the sum and the difference of b and d.

This result can also be established geometrically in the following manner:—

285. **Circle rolling inside another of double its Diameter.**—Join C_1 and O to any point L on the circumference of the rolling circle, and let C_1L meet the fixed circumference in A; then since $\angle OCL = 2 OC_1 A$, and $OC_1 = 2OC$, we have arc OA = arc OL; and, accordingly, as the inner circle rolls on the outer the point L moves along $C_1 A$. In like manner any other point on the circumference of the rolling circle describes, during the motion, a diameter of the fixed circle.

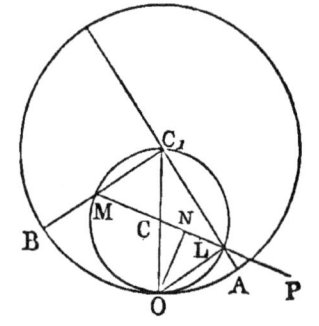

Fig. 59.

Again, *any point P, invariably connected with the rolling circle, describes an ellipse*. For, if L and M be the points in which CP cuts the rolling circle, by what has been just shown, these points move along two fixed right lines C_1A and C_1B, at right angles to each other. Accordingly, by a well-known property of the ellipse, any other point in the line LM describes an ellipse.

The case in which the outer circle rolls on the inner is also worthy of separate consideration.

286. **Circle rolling on another inside it and of half its Diameter.**—In this case, *any diameter of the rolling circle always passes through a fixed point*, which lies on the circumference of the inner circle.

For, let C_1L and C_2L be any two positions of the moving diameter, C_1 and C_2 being the corresponding positions of the centre of the rolling circle: O and O_2 the corresponding positions of the point of contact of the circles. Now, if the outer circle roll from the former to the latter position, the right lines C_1O_2 and CO_2 will coincide in direction, and accordingly the outer circle will have turned through the angle $C_2O_2C_1$; consequently, the moving diameter will have turned through the same angle; and hence $\angle C_2LC_1 = \angle C_2O_2C_1$; therefore the point L lies on the fixed circle, and the diameter always passes through the same point on this circle.

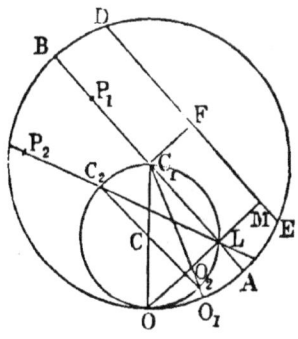

Fig. 60.

Again, *any right line connected with the rolling circle will always touch a fixed circle*.

For, let DE be the moving line in any position, and draw the parallel diameter AB; let fall C_1F and LM perpendicular to DE. Then, by the preceding, AB always passes through a fixed point L; also $LM = C_1F =$ constant; hence DE always touches a circle having its centre at L.

Again, to find the roulette described by any carried point P_1. The right line P_1C_1, as has been shown, always passes through a fixed point L; consequently, since C_1P_1 is a constant length, *the locus of P_1 is a Limaçon* (Art. 269). In like manner, any other point invariably connected with the outer circle describes a Limaçon; unless the point be situated on the circumference of the rolling circle, in which case the locus becomes a cardioid.

Examples—Roulettes.

1. When the radii of the fixed and the rolling circles become equal, prove geometrically that the epicycloid becomes a cardioid, and the epitrochoid a Limaçon (Art. 269).

2. Prove that the equation of the reciprocal polar of an epicycloid, with respect to the fixed circle, is of the form

$$r \sin m\omega = \text{const.}$$

3. Prove that the radius of curvature of an epicycloid varies as the perpendicular on the tangent from the centre of the fixed circle.

4. If $a = 4b$, prove that the equation of the hypocycloid becomes

$$x^{\frac{2}{3}} + y^{\frac{2}{3}} = a^{\frac{2}{3}}.$$

5. Find the equation, in terms of r and p, of the three-cusped hypocycloid; i.e. when $a = 3b$. *Ans.* $r^2 = a^2 - 8p^2$.

6. Find the equation of the pedal in the same curve.
Ans. $p = b \sin 3\omega$.

7. In the case of a curve rolling on another which is equal to it in every respect, corresponding points being in contact, prove that the determination of the roulette of any point P is immediately reduced to finding the pedal of the rolling curve with respect to the point P.

8. Hence, if the curves be equal parabolas, show that the path of the focus is a right line, and that of the vertex a cissoid.

9. In like manner, if the curves be equal ellipses, show that the path of the focus is a circle, and that of any point is a *bicircular quartic*.

10. In Art. 285, prove that the locus of the foci of the ellipses described by the different points on any right line is an equilateral hyperbola.

11. A is a fixed point on the circumference of a circle; the points L and M are taken such that arc $AL = m$ arc AM, where m is a constant; prove that the envelope of LM is an epicycloid or a hypocycloid, according as the arcs AL and AM are measured in the same or opposite directions from the point A.

12. Prove that LM, in the case of an epicycloid, is divided internally in the ratio $m : 1$, at its point of contact with the envelope; and, in the hypocycloid, externally in the same ratio.

13. Show also that the given circle is circumscribed to, or inscribed in, the envelope, according as it is an epicycloid or hypocycloid.

14. Prove, from equation (14), that the intrinsic equation of an epicycloid is

$$s = \frac{4b(a+b)}{a} \sin \frac{a\phi}{a+2b},$$

where s is measured from the vertex of the curve.

15. Hence the equation $s = l \sin n\phi$ represents an epicycloid or a hypocycloid, according as n is less or greater than unity.

16. In an epitrochoid, if the distance, d, of the moving point from the centre of the rolling circle be equal to the distance between the centres of the circles, prove that the polar equation of the locus becomes

$$r = 2(a + b) \cos \frac{a\theta}{a + 2b}.$$

17. Hence show that the curve

$$r = a \sin m\theta$$

is an epitrochoid when $m < 1$, and a hypotrochoid when $m > 1$.

This class of curves was elaborately treated of by the Abbé Grandi in the *Philosophical Transactions for* 1723. He gave them the name of "Rhodoneæ," from a fancied resemblance to the petals of roses. See also Gregory's *Examples on the Differential and Integral Calculus*, p. 183.

For illustrations of the beauty and variety of form of these curves, as well as of epitrochoids and hypotrochoids in general, the student is referred to the admirable figures in Mr. Proctor's *Geometry of Cycloids*.

287. **Centre of Curvature of an Epitrochoid or Hypotrochoid.**—The position of the centre of curvature for any point of an epitrochoid can be easily found from geometrical considerations. For, let C_1 and C_2 be the centres of the rolling and the fixed circles, P_2 the centre of curvature of the roulette described by P_1; and, as before, let O_1 and O_2 be two points on the circles, infinitely near to O, such that $OO_1 = OO_2$. Now, suppose the circle to roll until O_1 and O_2 coincide; then the lines C_1O_1 and C_2O_2 will lie *in directum*, as also the lines P_1O_1 and P_2O_2 (since P_2 is the point of intersection of two consecutive normals to the roulette).

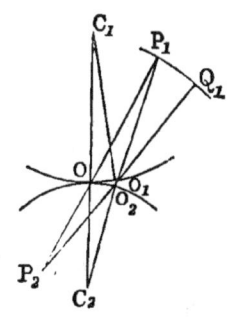

Fig. 61.

Hence $\quad \angle OC_1O_1 + \angle OC_2O_2 = \angle OP_1O_1 + \angle OP_2O_2,$

since each of these sums represents the angle through which the circle has turned.

Again, let $\quad \angle C_1OP_1 = \phi, \quad OO_1 = OO_2 = ds;$

then $\quad \angle OC_1O_1 = \dfrac{ds}{OC_1}, \quad \angle OC_2O_2 = \dfrac{ds}{OC_2},$

$$\angle OP_1O_1 = \frac{ds \cos \phi}{OP_1}, \quad \angle OP_2O_2 = \frac{ds \cos \phi}{OP_2}:$$

consequently we have

$$\frac{1}{OC_1} + \frac{1}{OC_2} = \cos\phi \left(\frac{1}{OP_1} + \frac{1}{OP_2}\right). \tag{21}$$

Or, if $OP_1 = r_1$, $OP_2 = r_2$,

$$\frac{1}{a} + \frac{1}{b} = \cos\phi \left(\frac{1}{r_1} + \frac{1}{r_2}\right).$$

From this, equation r_2, and consequently the radius of curvature of the roulette, can be obtained for any position of the generating point P_1.

If we suppose P_1 to be on the circumference of the rolling circle, we get $\cos\phi = \dfrac{OP_1}{2OC_1}$; whence it follows that

$$OP_2 = \frac{a}{a + 2b} OP_1,$$

which agrees with the result arrived at in Art. 279.

288. **Centre of Curvature of any Roulette.**—The preceding formula can be readily extended to any roulette: for if C_1 and C_2 be respectively the *centres of curvature of the rolling and fixed curves*, corresponding to the point of contact O, we may regard OO_1 and OO_2 as elementary arcs of the circles of curvature, and the preceding demonstration will still hold.

Hence, denoting the radii of curvature OC_1 and OC_2 by ρ_1 and ρ_2, we shall have

$$\frac{1}{\rho_1} + \frac{1}{\rho_2} = \cos\phi \left(\frac{1}{r_1} + \frac{1}{r_2}\right). \tag{22}$$

It can be easily seen, without drawing a separate figure, that we must change the sign of ρ_2 in this formula when the centres of curvature lie at the same side of O.

It may be noted that P_1 is the centre of curvature of the roulette described by the point P_2, if the lower curve be supposed to roll on the upper regarded as fixed.

289. **Geometrical Construction* for the Centre of**

---

* This beautiful construction, and also the formula (22) on which it is based, were given by M. Savary, in his *Leçons des Machines à l'Ecole Polytechnique*. See also Leroy's *Géométrie Descriptive*, Quatrième Edition, p. 347.

Curvature of a Roulette.—The formula (22) leads to a simple and elegant construction for the centre of curvature P_2.

We commence with the case when the base is a right line, as represented in the accompanying figure.

Join P_1 to C_1, the centre of curvature of the rolling curve, and draw ON perpendicular to OP_1, meeting P_1C_1 in N; through N draw NM parallel to OC_1, and the point P_2 in which it meets OP_1 is the centre of curvature required.

For, equation (22) becomes in this case

$$\frac{1}{OC_1} = \cos\phi \left(\frac{1}{OP_1} + \frac{1}{OP_2} \right),$$

whence we get

$$\frac{P_1P_2}{OP_1 \cdot OP_2} = \frac{1}{OC_1 \sin C_1 ON} = \frac{1}{NC_1 \sin C_1 NO} = \frac{NP_1}{NC_1 \cdot OP_1};$$

$$\therefore \frac{P_1P_2}{OP_2} = \frac{NP_1}{NC_1};$$

and, accordingly, the line NP_2 is parallel to OC_1.   Q. E. D.

The construction in the general case is as follows:—

Determine the point N as in the former case, and join it to C_2, the centre of curvature of the fixed curve, then the point of intersection of NC_2 and P_1O is the required centre of curvature.

This is readily established; for, from the equation

$$\frac{1}{OC_1} + \frac{1}{OC_2} = \cos\phi \left(\frac{1}{OP_1} + \frac{1}{OP_2} \right)$$

we get

$$\frac{C_1C_2}{OC_1 \cdot OC_2} = \frac{\cos\phi \, P_1P_2}{OP_1 \cdot OP_2};$$

$$\therefore \frac{C_1C_2}{OC_2} \cdot \frac{OP_2}{P_1P_2} = \frac{OC_1 \cos\phi}{OP_1}.$$

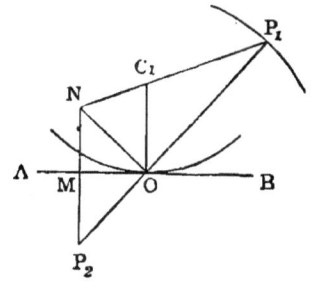

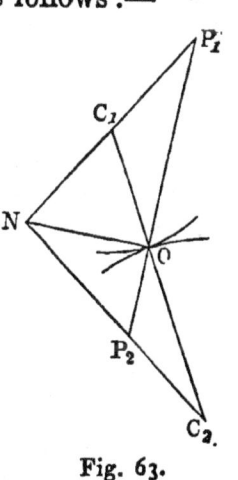

Fig. 62.

Fig. 63.

But, as before,

$$OC_1 \cos \phi = \frac{C_1 N \cdot OP_1}{NP_1}; \quad \therefore \quad \frac{OC_1 \cos \phi}{OP_1} = \frac{NC_1}{NP_1}:$$

hence

$$\frac{C_1 C_2}{OC_2} \cdot \frac{OP_2}{P_1 P_2} = \frac{NC_1}{NP_1}.$$

Consequently, by the well-known property of a transversal cutting the sides of a triangle, the points C_2, P_2, and N are in directum.

The modification in the construction when the rolling curve is a right line can be readily supplied by the student.

290. **Circle of Inflexions.**—The following geometrical construction is in many cases more useful than the preceding.

On the line OC_1 take OD_1 such that

$$\frac{1}{OD_1} = \frac{1}{OC_1} + \frac{1}{OC_2};$$

and on OD_1 as diameter describe a circle. Let E_1 be its point of intersection with OP_1, then we have

$$\cos \phi = \frac{OE_1}{OD_1},$$

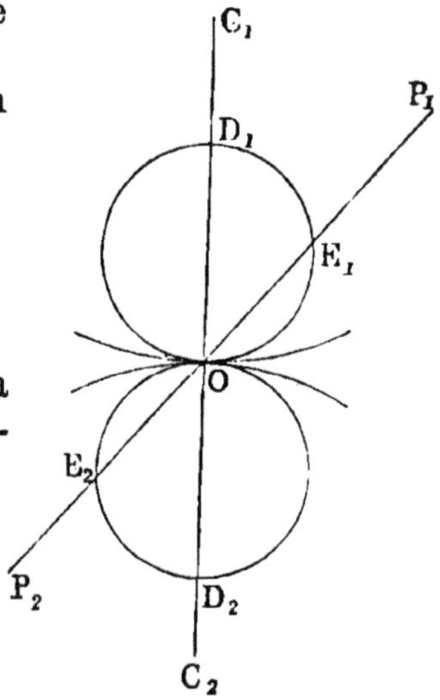

and formula (22) becomes

Fig. 64.

$$\frac{1}{OP_1} + \frac{1}{OP_2} = \frac{1}{OD_1 \cos \phi} = \frac{1}{OE_1}. \qquad (23)$$

Hence, if the tracing point P_1 lie on the circle* $OE_1 D_1$,

---

* This theorem is due to La Hire, who showed that the element of the roulette traced by any point is convex or concave with respect to the point of contact, O, according as the tracing point is inside or outside this circle. (See

the corresponding value of OP_2 is infinite, and consequently P_1 is a *point of inflexion on the roulette*.

In consequence of this property, the circle in question is called the *circle of inflexions*, as each point on it is a point of inflexion on the roulette which it describes.

Again, it can be shown that the lines P_1P_2, P_1O and P_1E_1 are in *continued proportion*; as also C_1C_2, C_1O, and C_1D_1. For, from (23) we have

$$\frac{P_1P_2}{OP_1 \cdot OP_2} = \frac{1}{OE_1}.$$

Hence $\qquad P_1P_2 : P_1O = OP_2 : OE_1;$

$\therefore P_1P_2 : P_1O = P_1P_2 - OP_2 : P_1O - OE_1 = P_1O : P_1E_1.$ (24)

In the same manner it can be shown that

$$C_1C_2 : C_1O = C_1O : C_1D_1. \qquad (25)$$

In the particular case where the base is a right line, the circle of inflexions becomes the circle described on the radius of curvature of the rolling curve as diameter.

Again, if we take $OD_2 = OD_1$, we shall have, by describing a circle on OD_2 as diameter,

$$C_2C_1 : C_2O = C_2O : C_2D_2;$$

and also $\qquad P_2P_1 : P_2O = P_2O : P_2E_2.$ (26)

The importance of these results will be shown further on.

291. **Envelope of a Carried Curve.**—We shall next consider the *envelope of a curve invariably connected with the rolling curve, and carried with it in its motion*.

Since the moving curve touches its envelope in each of its

Memoires de l'Académie des Sciences, 1706.) It is strange that this remarkable result remained almost unnoticed until recent years, when it was found to contain a key to the theory of curvature for roulettes, as well as for the envelopes of any carried curves. How little it is even as yet appreciated in this country will be apparent to any one who studies the most recent productions on roulettes, even by distinguished British Mathematicians.

positions, the path of its point of contact at any instant must be tangential to the envelope; hence the normal at their common point must pass through O, the point of contact of the fixed and rolling curves.

In the particular case in which the carried curve is a right line, its point of contact with its envelope is found by dropping a perpendicular on it from the point of contact O.

For example, suppose a circle to roll on any curve: to find *the envelope** *of any diameter PQ*:—

From O draw ON perpendicular to PQ, then N, by the preceding, is a point on the envelope.

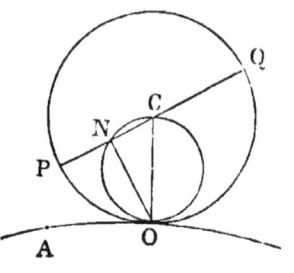

Fig. 65.

On OC describe a semicircle; it will pass through N, and, as in Art. 286, the arc ON = arc OP = OA, if A be the point in which P was originally in contact with the fixed curve. Consequently, the envelope in question is the roulette traced by a point on the circumference of a circle of half the radius of the rolling circle, having the fixed curve AO for its base.

For instance, if *a circle roll on a right line, the envelope of any diameter is a cycloid*, the radius of whose generating circle is half that of the rolling circle.

Again, if *a circle roll on another, the envelope of any diameter of the rolling circle is an epicycloid, or a hypocycloid.*

Moreover, it is obvious that if two carried right lines be parallel, their envelopes will be parallel curves. For example, the envelope of any right line, carried by a circle which rolls on a right line, is a parallel to a cycloid, *i.e.* the *involute of a cycloid*.

These results admit of being stated in a somewhat different form, as follows:

If one point, A, in a plane area move uniformly along a right line, while the area turns uniformly in its own plane, then the envelope of any carried right line is an involute to a cycloid. If the carried line passes through the moving point

---

* The theorems of this Article are, I believe, due to Chasles: *see* his *Histoire de La Géométrie*, p. 69.

A, its envelope is a cycloid. Again, if the point A move uniformly on the circumference of a fixed circle, while the area revolves uniformly, the envelope of any carried right line is an involute to either an epi- or hypo-cycloid. If the carried right line passes through A, its envelope is either an epi- or hypo-cycloid.

292. Centre of Curvature of the Envelope of a Carried Curve.—Let a_1b_1 represent a portion of the carried curve, to which Om is normal at the point m; then, by the preceding, m is the point of contact of a_1b_1 with its envelope.

Now, suppose a_2b_2 to represent a portion of the envelope, and let P_1 be the centre of curvature of a_1b_1, for the point m, and P_2 the corresponding centre of curvature of a_2b_2.

As before, take O_1 and O_2 such that $OO_1 = OO_2$, and join P_1O_1 and P_2O_2. Again, suppose the curve to roll until O_1 and O_2 coincide; then the lines P_1O_1 and P_2O_2 will come *in directum*, as also the lines O_1C_1 and O_2C_2; and, as in Art. 288, we shall have

$$\angle C_1 + \angle C_2 = \angle P_1 + \angle P_2;$$

and consequently

$$\frac{1}{OC_1} + \frac{1}{OC_2} = \cos \phi \left(\frac{1}{OP_1} + \frac{1}{OP_2} \right). \quad (27)$$

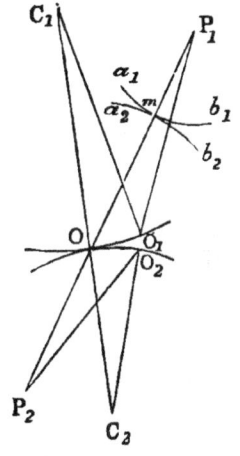

Fig. 66.

From this equation the centre of curvature of the envelope, for any position, can be found. Moreover, it is obvious that the geometrical constructions of Arts. 289, 290, equally apply in this case. It may be remarked that these constructions hold in all cases, whatever be the directions of curvature of the curves.

The case where the moving curve a_1b_1 is a right line is worthy of especial notice.

In this case the normal Om is perpendicular to the moving line; and, since the point P_1 is infinitely distant, we have

$$\frac{\cos \phi}{OP_2} = \frac{1}{OC_1} + \frac{1}{OC_2} = \frac{1}{OD_2} \text{ (Art. 290)};$$

whence, P_2 is situated on the lower circle of inflexions. Hence we infer that *the different centres of curvature of the curves enveloped by all carried right lines, at any instant, lie on the circumference of a circle.*

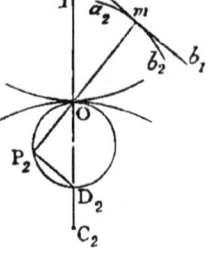

Fig. 67.

As an example, suppose the right line OM to roll on a fixed circle, whose centre is C_2, to find the envelope of any carried right line, LM.

In this case the centre of curvature, P_2, of the envelope of LM, lies, by the preceding, on the circle described on OC as diameter; and, accordingly, CP_2 is perpendicular to the normal $P_1 P_2$.

Hence, since $\angle OLP_1$ remains constant during the motion, the line CP_2 is of constant length; and, if we describe a circle with C as centre, and CP_2 as radius, the envelope of the moving line LM will, in all positions, be an *involute of a circle*. The same reasoning applies to any other moving right line.

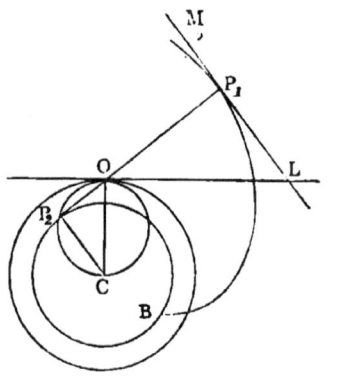

Fig. 68.

We shall conclude with the statement of one or two other important particular cases of the general principle of this Article.

(1). *If the envelope $a_2 b_2$ of the moving curve $a_1 b_1$ be a right line, the centre of curvature P_1 lies on the corresponding circle of inflexions.*

(2). *If the moving right line always passes through a fixed point, that point lies on the circle $OD_2 E_2$.*

292 (*a*). **Expression for Radius of Curvature of Envelope of a Right Line.**—The following expression for the radius of curvature of the envelope of a moving right

line is sometimes useful. Let p be the perpendicular distance of the moving line, in any position, from a fixed point in the plane, and ω the angle that this perpendicular makes with a fixed line in the plane, and ρ the radius of curvature of the envelope at the point of contact; then, by Art. 206, we have

$$\rho = p + \frac{d^2p}{d\omega^2}. \qquad (28)$$

Whenever the conditions of the problem give p in terms of ω (the angle through which the figure has turned), the value of ρ can be found from this equation. For example, the result established in last Article (*see* Fig. 68) can be easily deduced from (28). This is left as an exercise for the student.

293. **On the Motion of a Plane Figure in its Plane.**—We shall now proceed to the consideration of a general method, due to Chasles, which is of fundamental importance in the treatment of roulettes, as also in the general investigation of the motion of a rigid body.

We shall commence with the following theorem :—

When an invariable plane figure moves in its plane, it can be brought from any one position to any other by a single rotation round a fixed point in its plane.

For, let A and B be two points of the figure in its first position, and A_1, B_1 their new positions after a displacement. Join AA_1 and BB_1, and suppose the perpendiculars drawn at the middle points of AA_1 and BB_1 to intersect at O; then we have $AO = A_1O$, and $BO = B_1O$. Also, since the triangles AOB and A_1OB_1 have their sides respectively equal, we have $\angle AOB = \angle A_1OB_1$; $\therefore \angle AOA_1 = \angle BOB_1$.

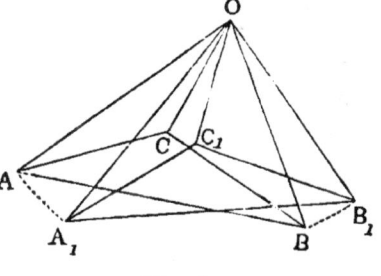

Fig. 69.

Accordingly, AB will be brought to the position A_1B_1 by a rotation through the angle AOA_1 round O. Consequently, any point C in the plane, which is rigidly connected with AB, will be brought from its original to its new position, C_1, by the same rotation.

This latter result can also be proved otherwise thus :—Join OC and OC_1; then the triangles OAC and OA_1C_1 are equal, because $OA = OA_1$, $AC = A_1C_1$, and the angle OAC, being the difference between OAB and BAC, is equal to OA_1C_1, the difference between OA_1B_1 and $B_1A_1C_1$; therefore $OC = OC_1$, and $\angle AOC = \angle A_1OC_1$; and hence $\angle AOA_1 = \angle COC_1$. Consequently the point C is brought to C_1 by a rotation round O through the same angle AOA_1. The same reasoning applies to any other point invariably connected with A and B.

The preceding construction requires modification when the lines AA_1 and BB_1 are parallel. In this case the point, O, of intersection of the lines BA and B_1A_1 is easily seen to be the point of instantaneous rotation.

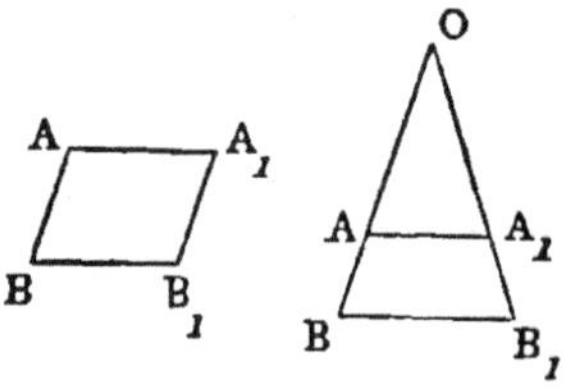

Fig. 70.

For, since $AB = A_1B_1$, and AA_1, BB_1, are parallel, we have $OA = OA_1$, and $OB = OB_1$. Hence, the figure will be brought from its old to its new position by a rotation around O through the angle AOA_1.

Next, let AA_1, and BB_1 be both equal and parallel. In this case the point O is at an infinite distance; but it is obvious that each point in the plane moves through the same distance, equal and parallel to AA_1; and the motion is one of simple *translation*, without any rotation.

In general if we suppose the two positions of the moving figure to be indefinitely near each other, then the line AA_1, joining two infinitely near positions of the same point of the figure, becomes an element of the curve described by that point, and the line OA becomes at the same time a *normal* to the curve. Hence, *the normals to the paths described by all the points of the moving figure pass through O,* which point is called the *instantaneous centre of rotation.*

The position of O is determined whenever the directions of motion of any two points of the moving figure are known; for it is the intersection of the normals to the curves described by those points.

This furnishes a geometrical method of drawing tangents to many curves, as was observed by Chasles.*

---

* This method is given by Chasles as a generalization of the method of Descartes (Art. 273, note). It is itself a particular case of a more general principle concerning homologous figures. See Chasles, *Histoire de la Géométrie*, pp. 548-9: also *Bulletin Universel des Sciences*, 1830.

The following case is deserving of special consideration:—
A right line always passes through a fixed point, while one of its points moves along a fixed line: to find the instantaneous centre of rotation. Let A be the fixed point, and AB any position of the moving line, and take $B'A' = BA$; then the centre of rotation, O, is found as before, and is such that $OA = OA'$, and $OB = OB'$. Accordingly, in the limit the centre of instantaneous rotation is the intersection of BO drawn perpendicular to the fixed line, and AO drawn perpendicular to the moving line at the fixed point.

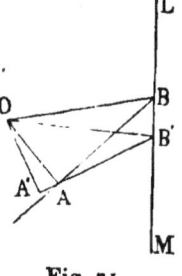

Fig. 71.

In general, if AB be any moving curve, and LM any fixed curve, the *instantaneous centre of rotation is the point of intersection of the normals to the fixed and to the moving curves, for any position.*

Also the normal to the curve described by any point invariably connected with AB is obtained by joining the point to O, the instantaneous centre.

More generally, if a moving curve always touches a fixed curve A, while one point on the moving curve moves along a second fixed curve B, the instantaneous centre is the point of intersection of the normals to A and B at the corresponding points; and the line joining this centre to any describing point is normal to the path which it describes.

We shall illustrate this method of drawing tangents by applying it to the conchoid and the limaçon.

294. **Application to Curves.**—In the Conchoid (Fig. 49, page 332), regarding AP as a moving right line, the instantaneous centre O is the point of intersection of AO drawn perpendicular to AP, with RO drawn perpendicular to LM; and consequently, OP and OP_1 are the normals at P and P_1, respectively.

For the same reason, the normal to the Limaçon (Fig. 48, page 331) at any point P is got by drawing OQ perpendicular to OP to meet the circle in Q, and joining PQ.

Examples.

1. If the radius vector, OP, drawn from the origin to any point P on a curve, be produced to P_1, until PP_1 be a constant length; prove that the normal at P_1 to the locus of P_1, the normal at P to the original curve, and the perpendicular at the origin to the line OP, all pass through the same point.

2. If a constant length measured from the curve be taken on the normals along a given curve, prove that these lines are also normals to the new curve which is the locus of their extremities.

3. An angle of constant magnitude moves in such a manner that its sides constantly touch a given plane curve; prove that the normal to the curve described by its vertex, P, is got by joining P to the centre of the circle passing through P and the points in which the sides of the moveable angle touch the given curve.

4. If on the tangent at each point on a curve a constant length measured from the point of contact be taken, prove that the normal to the locus of the points so found passes through the centre of curvature of the proposed curve.

5. In general, if through each point of a curve a line of given length be drawn making a constant angle with the normal, the normal to the curve locus of the extremities of this line passes through the centre of curvature of the proposed.

295. Motion of any Plane Figure reduced to Roulettes.—Again, the most general motion of any figure in its plane may be regarded as consisting of a number of infinitely small rotations about the different instantaneous centres taken in succession.

Let $O, O', O'', O''',$ &c., represent the successive centres of rotation, and consider the instant when the figure turns through the angle O_1OO' round the point O. This rotation will bring a certain point O_1 of the figure to coincide with the next centre O'. The next rotation takes place around O'; and suppose the point O_2 brought to coincide with the centre of rotation O''. In like manner, by a third rotation the point O_3 is brought to coincide with O''', and so on. By this means the motion of the moveable figure is equivalent to the rolling of the polygon $OO_1O_2O_3\ldots$ invariably connected with the figure, on the polygon $OO'O''O'''\ldots$ fixed in the plane. In the limit, the polygons change into curves, of which one rolls, without

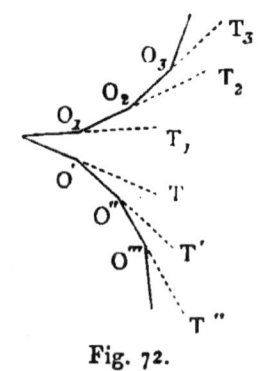

Fig. 72.

sliding, on the other; and hence we conclude that *the general movement of any plane figure in its own plane is equivalent to the rolling of one curve on another fixed curve.*

These curves are called by Reuleaux* the "centrodes" of the moving figures.

For example, suppose two points A and B of the moving figure to slide along two fixed right lines CX and CY; then the instantaneous centre O is the point of intersection of AO and BO, drawn perpendicular to the fixed lines. Moreover, as AB is a constant length, and the angle ACB is fixed, the length CO is constant; consequently the locus of the instantaneous centre is the circle described with C as centre, and CO as radius.

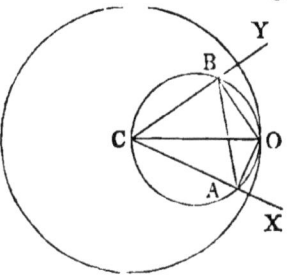

Fig. 73.

Again, if we describe a circle round $CBOA$, this circle is invariably connected with the line AB, and moves with it. Hence the motion of any figure invariably connected with AB is equivalent to the *rolling of a circle inside another of double its radius* (see Art. 285).

Again, if we consider the angle XCY to move so that its legs pass through the fixed points A and B, respectively; then the instantaneous centre O is determined as before. Moreover, the circle BCA becomes a *fixed* circle, along which the instantaneous centre O moves. Also, since CO is of constant length, the outer circle becomes in this case the rolling curve. Hence the motion of any figure invariably connected with the moving lines CX and CY is equivalent to the *rolling of the outer circle on the inner* (compare Art. 286).

295 (a). **Epicyclics.**—As a further example, suppose one point in a plane area to move uniformly along the circumference of a fixed circle, while the area revolves with a uniform angular motion around the point, to find the position of the "centrodes."

The directions of motion are indicated by the arrow heads. Let C be the centre of the fixed circle, P the position

---

* *See* Kennedy's translation of Reuleaux's *Kinematics of Machinery*, pp. 65, &c.

of the moving point at any instant, Q a point in the moving figure such that $CP = PQ$. Now, to find the position of the instantaneous centre of rotations it is necessary to get the *direction of motion* of the point Q.

Let P_1 represent a consecutive position of P, then the simultaneous position of Q is got by first supposing it to move through the infinitely small length QR, equal and parallel to PP_1, and then to turn round P_1, through the angle RP_1Q_1, which the area turns through while P moves to P_1. Moreover, by hypothesis, the angles PCP_1 and RP_1Q_1 are in a constant ratio: if this ratio be denoted by m, we have (since $PQ = PC$)

$$RQ_1 = mPP_1 = mQR.$$

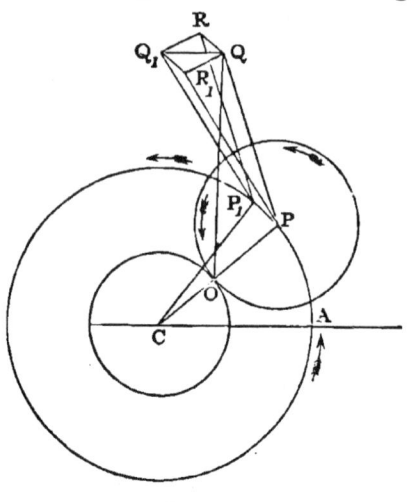

Fig. 74.

Join Q and Q_1, then QQ_1 represents the direction of motion of Q. Hence the right line QO, drawn perpendicular to QQ_1, intersects CP in the *instantaneous centre of rotation*.

Again, since the directions of PO, PQ, and QO are, respectively, perpendicular to QR, RQ_1, and QQ_1, the triangles QPO and Q_1RQ are similar;

$$\therefore PQ = mPO, \text{ i.e. } CP = mPO.$$

Accordingly, the instantaneous centre of rotation is got by cutting off

$$PO = \frac{CP}{m}. \tag{29}$$

Hence, if we describe two circles, one with centre C and radius CO, the other with centre P and radius PO; these circles are the required *centrodes;* and the motion is *equivalent to the rolling of the outer circle on the inner.*

Epicyclics.

Accordingly, any point on the circumference of the outer circle describes an epicycloid, and any point not on this circumference describes an epitrochoid. When the angular motion of PQ is less than that of CP, i.e. when $m < 1$, the point O lies in PC *produced*. Accordingly, in this case, the fixed circle lies *inside* the rolling circle; and the curves traced by any point are still either epitrochoids or epicycloids.

In the preceding we have supposed that the angular rotations take place in the *same direction*. If we suppose them to be in *opposite directions*, the construction has to be modified, as in the accompanying figure.

In this case, the angle RP_1Q_1 must be measured in an opposite direction to that of PCP_1; and, proceeding as in the former case, the direction of motion of Q is represented by QQ_1; accordingly, the perpendicular QO will intersect CP produced, and, as before, we have

$$PO = \frac{PC}{m}.$$

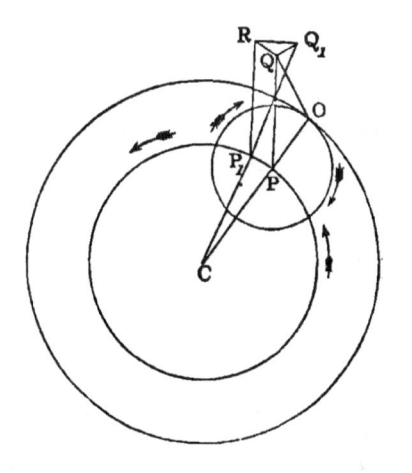

Fig. 75.

Hence the motion is equivalent to the rolling of a circle of radius PO on the *inside of a fixed circle*, whose radius is CO. Accordingly, in this case, the path described by any point in the moving area not on the circumference of the rolling circle is a hypotrochoid.

Also, from Art. 291, it is plain that the envelope of any right line which passes through the point P in the moving area is an epicycloid in the former case, and a hypocycloid in the latter.

Again, if we suppose the point P, instead of moving in a circle, to move uniformly in a right line, the path of any point in the moving area becomes either a trochoid or a cycloid.

Curves traced as above, that is, *by a point which moves*

uniformly round the circumference of a circle, whose centre moves uniformly on the circumference of a fixed circle in the same plane, are called *epicyclics*, and were invented by Ptolemy (about A.D. 140) for the purpose of explaining the planetary motions. In this system* the fixed circle is called the *deferent*, and that in which the tracing point moves is called the *epicycle*. The motion in the fixed circle may be supposed in all cases to take place in the same direction around C, that indicated by the arrows in our figures. Such motion is called *direct*. The case for which the motion in the epicycle is direct is exhibited in Fig. 74.

Angular motion in the reverse direction is called *retrograde*. This case is exhibited in Fig. 75. The corresponding epicyclics are called by Ptolemy *direct* and *retrograde* epicyclics.

The preceding investigation shows that every direct epicyclic is an epitrochoid, and every retrograde epicyclic a hypotrochoid.

It is obvious that the greatest distance in an epicyclic from the centre C is equal to the sum of the radii of the circles, and the least to their difference. Such points on the epicyclic are called *apocentres* and *pericentres*, respectively.

Again, if a represent the radius of the fixed circle or deferent, and β the radius of the revolving circle or epicycle; then, if the curve be referred to rectangular axes, that of x passing through an apocentre, it is easily seen that we have for a *direct epicyclic*

$$\left.\begin{array}{l} x = a \cos \theta + \beta \cos m\theta, \\ y = a \sin \theta + \beta \sin m\theta. \end{array}\right\} \quad (30)$$

---

* The importance of the epicyclic method of Ptolemy, in representing approximately the planetary paths relative to the earth at rest, has recently been brought prominently forward by Mr. Proctor, to whose work on the *Geometry of Cycloids* the student is referred for fuller information on the subject.

We owe also to Mr. Proctor the remark that the invention of cycloids, epicycloids, and epitrochoids, is properly attributable to Ptolemy and the ancient astronomers, who, in their treatment of epicyclics, first investigated some of the properties of such curves. It may, however, be doubted if Ptolemy had any idea of the shape of an epicyclic, as no trace of such is to be found in the entire of his great work, *The Almagest*.

The formulæ for a *retrograde* epicyclic are obtained by changing the sign of m (compare Art. 284).

It is easily seen that *every epicyclic admits of a twofold generation.*

For, if we make $m\theta = \phi$, equation (30) may be written

$$x = \beta \cos\phi + a \cos\frac{\phi}{m},$$

$$y = \beta \sin\phi + a \sin\frac{\phi}{m},$$

which is equivalent to an interchange of the radii of the deferent circle and of the epicycle, and an alteration of m into $\frac{1}{m}$. This result can also be seen immediately geometrically.

It may be remarked that this contains Euler's theorem (Art. 280) under it as a particular case.

296. **Properties of the Circle of Inflexions.**—It should be especially observed that the results established in Art. 290, relative to the circle of inflexions, hold in all cases of the motion of a figure in its plane, and hence we infer that the *distances of any moving point from the centre of curvature of its path, from the instantaneous centre of rotation, and from the circle of inflexions, are in continued proportion.*

Again, from Art. 292, we infer that if a moveable curve slide on a fixed curve, the *distances of the centre of curvature of the moving, from that of the fixed curve, from the centre of instantaneous rotation, and from the circle of inflexions, are in continued proportion.*

The particular cases mentioned in these Articles obviously hold also in this case, and admit of similar enunciations.

These principles are the key to the theory of the curvature of the paths of points carried by moving curves, as also to the curvature of the envelopes of carried curves.

We shall illustrate this statement by a few applications.

297. **Example on the Construction of Circle of Inflexions.**—*Suppose two curves a_1b_1 and c_1d_1, invariably connected with a moving plane figure, always to touch two fixed curves a_2b_2 and c_2d_2, to find the centre of curvature of the roulette described by any point R_1 of the moving figure.*

The instantaneous circle of inflexions is easily constructed in the following manner:—Let P_1 and P_2 be the centres of curvature for the point of contact m for the curves $a_1 b_1$ and $a_2 b_2$, respectively: and let Q_1, Q_2, be the corresponding points for the curves $c_1 d_1$ and $c_2 d_2$. Take

$$P_1 E_1 = \frac{P_1 O^2}{P_1 P_2}, \text{ and } Q_1 F_1 = \frac{Q_1 O^2}{Q_1 Q_2};$$

then, by Art. 290, the points E_1 and F_1 lie on the circle of inflexions. Accordingly, the circle which passes through O, E_1 and F_1, is the circle of inflexions.

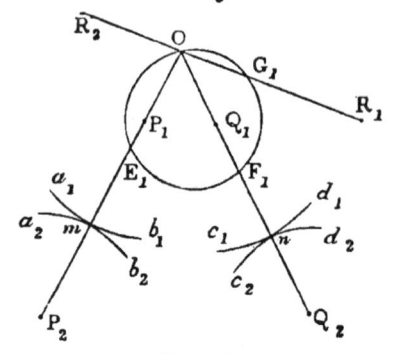

Fig. 76.

Hence, if $R_1 O$ meet this circle in G_1, and we take $R_1 R_2 = \dfrac{R_1 O^2}{R_1 G_1}$, the point R_2 (by the same theorem) is *the centre of curvature of the roulette described by* R_1.

In the same case, by a like construction, the centre of curvature of the envelope of any carried curve can be found.

The modifications when any of the curves $a_1 b_1$, $a_2 b_2$, &c., becomes a right line, or reduces to a single point, can also be readily seen by aid of the principles already established for such cases.

298. **Theorem of Bobillier.***—*If two sides of a moving triangle always touch two fixed circles, the third side also always touches a fixed circle.*

Let ABC be the moving triangle; the side AB touching at c a fixed circle whose centre is γ, and AC touching at b a circle with centre β. Then the instantaneous centre O is the point of intersection of $b\beta$ and $c\gamma$.

Again, the angle $\beta O \gamma$, being the supplement of the constant angle BAC, is given; and consequently the instantaneous centre O always lies on a fixed circle.

---

* *Cours de géométrie pour les écoles des arts et métiers.* See also Collignon, *Traité de Mécanique Cinématique*, p. 306.

This theorem admits of a simple proof by elementary geometry. The investigation above has however the advantage of connecting it with the general theory given in the preceding Articles, as well as of leading to the more general theorem stated at the end of this Article.

Also if Oa be drawn perpendicular to the third side BC, a is the point in which the side touches its envelope (Art. 291). Produce aO to meet the circle in a; and since the angle $aO\beta$ is equal to the angle ACB, it is constant; and consequently the point a is a fixed point on the circle. Again, by (4) Art. 292, the circle $\beta O\gamma$ passes through the centre of curvature of the envelope of any carried right line; and accordingly a is the centre of curvature of the envelope of BC; but a has already been proved to be a fixed point; consequently BC in all positions touches a fixed circle whose centre is a. (Compare Art. 286.)

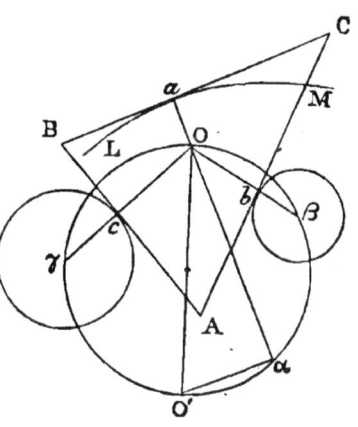

Fig. 77.

This result can be readily extended to the case where the sides AB and AC slide on any curves; for we can, for an infinitely small motion, substitute for the curves the osculating circles at the points b and c, and the construction for the point a will give the centre of curvature of the envelope of the third side BC.

298 (a). **Analytical Demonstration.**—The result of the preceding Article can also be established analytically, as was shown by Mr. Ferrers, in the following manner:—

Let a, b, c represent the lengths of the sides of the moving triangle, and p_1, p_2, p_3 the perpendiculars from any point on the sides a, b, c, respectively; then, by elementary geometry, we have

$$ap_1 + bp_2 + cp_3 = 2 \text{ (area of triangle)} = 2\Delta.$$

Again, if ρ_1, ρ_2, ρ_3 be the radii of curvature of the envelopes of the three sides, and ω the angle through which each of the perpendiculars has turned, we have by (28),

$$a\rho_1 + b\rho_2 + c\rho_3 = 2\Delta. \tag{31}$$

Hence, if two of the radii of curvature be given the third can be determined.

2 B

We next proceed to consider the conchoid of Nicomedes.

299. Centre of Curvature for a Conchoid.—Let A be the pole, and LM the directrix of a conchoid. Construct the instantaneous centre O, as before: and produce AO until $OA_1 = AO$.

It is easily seen that the circle circumscribing A_1OR_1 is the instantaneous circle of inflexions: for the instantaneous centre O always lies on this circle; also R_1 lies on the circle by Art. 290, since it moves along a right line: again, A lies on the lower circle of inflexions of same Article, and consequently A_1 lies on the circle of inflexions.

Hence, to find the centre of curvature of the conchoid described by the moving point P_1, produce P_1O to meet the circle of inflexions in F_1, and take

$$P_1P_2 = \frac{P_1O^2}{P_1F_1};$$

then, by (22), P_2 is the centre of curvature belonging to the point P_1 on the conchoid.

In the same case, the centre of curvature of the curve described by any other point Q_1, which is invariably connected with the moving line, can be found. For, if we produce Q_1O to meet the circle of inflexions in E_1, and take $Q_1Q_2 = \frac{Q_1O^2}{Q_1E_1}$; then, by the same theorem, Q_2 is the centre of curvature required.

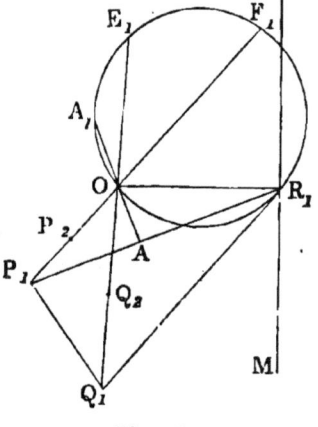

Fig. 78.

A similar construction holds in all other cases.

300. Spherical Roulettes.—The method of reasoning adopted respecting the motion of a plane figure in its plane is applicable identically to the motion of a curve on the surface of a sphere, and leads to the following results, amongst others:—

(1). A spherical curve can be brought from any one position on a sphere to any other by means of a single rotation around a diameter of the sphere.

(2). The elementary motion of a moveable figure on a sphere may be regarded as an infinitely small rotation around a certain diameter of the sphere. This diameter is called the instantaneous axis of rotation, and its points of intersection with the sphere are called the poles of rotation.

(3). The great circles drawn, for any position, from the pole to each of the points of the moving curve are normals to the curves described by these points.

(4). When the instantaneous paths of any two points are given, the instantaneous poles are the points of intersection of the great circles drawn normal to the paths.

(5). The continuous movement of a figure on a sphere may be reduced to the rolling of a curve fixed relatively to the moving figure on another curve fixed on the sphere. By aid of these principles the properties of spherical roulettes* can be discussed.

301. **Motion of a Rigid Body about a Fixed Point.**—We shall next consider the motion of any rigid body around a fixed point. Suppose a sphere described having its centre at the fixed point; its surface will intersect the rigid body in a spherical curve A, which will be carried with the body during its motion. The elementary motion of this curve, by the preceding Article, is an infinitely small rotation around a diameter of the sphere; and hence the motion of the solid consists in a rotation around an instantaneous axis passing through the fixed point.

Again, the continuous motion of A on the sphere by (5) (preceding Article) is reducible to the rolling of a curve L, connected with the figure A, on a curve λ, traced on the sphere. But the rolling of L on λ is equivalent to the rolling of the cone with vertex O standing on L, on the cone with the same vertex standing on λ. Hence the most general motion of a rigid body having a fixed point is equivalent to the rolling of a conical surface, having the fixed point for its summit, and appertaining to the solid, on a cone fixed in space, having the same vertex.

These results are of fundamental importance in the general theory of rotation.

---

* On the Curvature of Spherical Epicycloids, see Resal; *Journal de l'École Polytechnique*, 1858, pp. 235, &c.

Examples.

1. If the radius of the generating circle be one-fourth that of the fixed, prove immediately that the hypocycloid becomes the envelope of a right line of constant length whose extremities move on two rectangular lines.

2. Prove that the evolute of a cardioid is another cardioid in which the radius of the generating circle is one-third of that for the original circle.

3. Prove that the entire length of the cardioid is eight times the diameter of its generating circle.

4. Show that the points of inflexion in the trochoid are given by the equation $\cos\theta + \dfrac{d}{a} = 0$; hence find when they are real and when imaginary.

5. One leg of a right angle passes through a fixed point, whilst its vertex slides along a given curve; show that the problem of finding the envelope of the other leg of the right angle is reducible to the investigation of a locus.

6. Show that the equation of the pedal of an epicycloid with respect to any origin is of the form

$$r = (a + 2b)\cos\frac{a\theta}{a + 2b} - c\cos(\theta + a).$$

7. In figure 57, Art. 281, show that the points C, P' and Q are *in directum*.

8. Prove that the locus of the vertex of an angle of given magnitude, whose sides touch two given circles, is composed of two limaçons.

9. The legs of a given angle slide on two given circles: show that the locus of any carried point is a limaçon, and the envelope of any carried right line is a circle.

10. Find the equation to the tangent to the hypocycloid when the radius of the fixed circle is three times that of the rolling.

Ans. $x\cos\omega + y\sin\omega = b\sin 3\omega$.

This is called the three-cusped hypocycloid. See Ex. 5, Art. 286.

11. Apply the method of envelopes to deduce the equation of the three-cusped hypocycloid.

Substituting for $\sin 3\omega$ its value, and making $t = \cot\omega$, the equation of the tangent becomes

$$xt^3 + (y - 3b)t^2 + xt + b + y = 0,$$

in which t is an arbitrary parameter. If t be eliminated between this and its derived equation taken with respect to t, we shall get for the equation of the hypocycloid,

$$(x^2 + y^2)^2 + 18b^2(x^2 + y^2) + 24bx^2y - 8by^3 = 27b^4.$$

12. If two tangents to a cycloid intersect at a constant angle, prove that the length of the portion which they intercept on the tangent at the vertex of the cycloid is constant.

13. If two tangents to a hypocycloid intersect at a constant angle, prove that the arc which they intercept on the circle inscribed in the hypocycloid is of constant length.

14. The vertex of a right angle moves along a right line, and one of its legs passes through a fixed point: show geometrically that the other leg envelopes a parabola, having the fixed point for focus.

15. One angle of a given triangle moves along a fixed curve, while the opposite side passes through a fixed point: find, for any position, the centre of curvature of the envelope of either of the other sides, and also that of the curve described by any carried point.

16. If a right line move in any manner in a plane, prove that the locus of the centres of curvature of the paths of the different points on the line, at any instant, is a conic.—(Resal, *Journal de l'Ecole Polytechnique*, 1858, p. 112).

This, as well as the following, can be proved without difficulty from equation (22), p. 352.

17. When a conic rolls on any curve, the locus of the centres of curvature of the elements described simultaneously by all the points on the conic is a new conic, touching the other at the instantaneous centre of rotation.—(Mannheim, *same Journal*, p. 179.)

18. An ellipse rolls on a right line: prove that ρ, the radius of curvature of the path described by either focus, is given by the equation $\dfrac{1}{\rho} = \dfrac{1}{a} - \dfrac{1}{r}$; where r is the distance of the focus from the point of contact, and a is the semi-axis major.—(Mannheim, *Ibid.*)

19. The extremities of a right line of given length move along two fixed right lines: give a geometrical construction for the centre of curvature of the envelope in any position.

20. Prove that the locus of the intersection of tangents to a cycloid which intersect at a constant angle is a prolate trochoid (La Hire, *Mém. de l'Acad. des Sciences*, 1704).

21. More generally, prove that the corresponding locus for an epicycloid is an epitrochoid, and for a hypocycloid is a hypotrochoid. (Chasles, *Hist. de la Géom.*, p. 125).

22. If a variable circle touch a given cycloid, and also touch the tangent at the vertex, the locus of its centre is a cycloid. (Professor Casey, *Phil. Trans.*, 1877.)

23. Being given three fixed tangents to a variable cycloid, the envelope of the tangent at its vertex is a parabola. (*Ibid.*)

24. If two tangents to a cycloid cut at a constant angle, the locus of the centre of the circle described about the triangle, formed by the two tangents and the chord of contact, is a right line. (*Ibid.*)

25. If a curve (A) be such that the radius of curvature at each point is n times the normal intercepted between the point and a fixed straight line (B),

then when the curve rolls along another straight line, (B) will envelope a curve in which the radius of curvature is $n + 1$ times the normal.

Thus, when $n = -2$, (A) is a parabola, and (B) the directrix; and when the parabola rolls along a straight line, its directrix envelopes a catenary (for which $n = -1$), to which the straight line is directrix.

When the catenary rolls along a straight line, its directrix passes through a fixed point, for which $n = 0$.

When the point moves along a straight line, the straight line which it carries with it envelopes a circle ($n = 1$), and (B) is a diameter.

When the circle rolls along a straight line, its diameter envelopes a cycloid ($n = 2$), to which (B) is the base. When the cycloid rolls along a straight line its base envelopes a curve which is the involute of the four-cusped hypocycloid, passing through two cusps, and is in figure like an ellipse whose major axis is twice the minor. (Professor Wolstenholme.)

The fundamental theorem given above follows immediately from equation (27), p. 357.

26. Prove the following extension of Bobillier's theorem:—If two sides of a moving triangle always touch the involutes to two circles, the third side will always touch the involute to a circle.

27. Investigate the conditions of equilibrium of a heavy body which rests on a fixed rough surface.

In this case it is plain that, in the position of equilibrium, the centre of gravity G of the body must be vertically over the point of contact of the body with the fixed surface.

Again, if we suppose the body to receive a slight displacement by rolling on the fixed surface, the equilibrium will be stable or unstable, from elementary mechanical considerations, according as the new position of G is higher or lower than its former position, *i.e.* according as G is situated *inside or outside the circle of inflexions* (Art. 290).

Hence, if ρ_1 and ρ_2 be the radii of curvature for the corresponding fixed and rolling curve, and h the distance of G from the point of contact of the surfaces, the equilibrium is stable or unstable according as h is $<$ or $> \dfrac{\rho_1 \rho_2}{\rho_1 + \rho_2}$. See Walton's *Problems*, p. 190; also, for a complete investigation of the case where $h = \dfrac{\rho_1 \rho_2}{\rho_1 + \rho_2}$, Minchin's *Statics*, pp. 320-2, 2nd Edition.

28. Apply the method of Art. 285 to prove the following construction for the axes of an ellipse, being given a pair of its conjugate semi-diameters OP, OQ, in magnitude and position. From P draw a perpendicular to OQ, and on it take $PD = PQ$; join P to the centre of the circle described on OD as diameter by a right line, and let it cut the circumference in the points E and F; then the right lines OE and OF are the axes of the ellipse, in position, and the segments PE and PF are the lengths of its semi-axes (Mannheim, *Nouv. An. de Math.* 1857, p. 188).

29. An involute to a circle rolls on a right line: prove that its centre describes a parabola.

30. A cycloid rolls on an equal cycloid, corresponding points being in contact: show that the locus of the centre of curvature of the rolling curve at the point of contact is a trochoid, whose generating circle is equal to that of either cycloid.

CHAPTER XX.

ON THE CARTESIAN OVAL.

302. Equation of Cartesian Oval.—In this Chapter* it is proposed to give a short discussion of the principal properties of the Cartesian Oval, treated geometrically.

We commence by writing the equation of the curve in its usual form, viz.,

$$r_1 \pm \mu r_2 = a,$$

where r_1 and r_2 represent the distances of any point on the curve from two fixed points, or foci, F_1 and F_2, while μ and a are constants, of which we may assume that μ is less than unity. We also assume that a is greater than $F_1 F_2$, the distance between the fixed points.

It is easily seen that the curve consists of two ovals, one lying inside the other; the former corresponding to the equation $r_1 + \mu r_2 = a$, and the latter to $r_1 - \mu r_2 = a$. Now, with F_1 as centre, and a as radius, describe a circle. Through F_2 draw any chord DE, join $F_1 D$ and $F_1 E$; then, if P be the point in which $F_1 D$ meets the inner oval, we have

$$PD = a - r_1 = \mu r_2 = \mu P F_2.$$

From this relation the point P can be readily found.

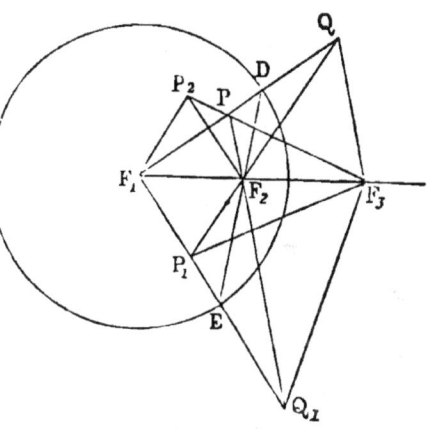

Fig. 79.

---

* This Chapter is taken, with slight modifications, from a Paper published by me in *Hermathena*, No. IV., p. 509.

Again, let Q be the corresponding point for the outer oval $r_1 - \mu r_2 = a$; and we have, in like manner, $DQ = \mu F_2 Q$;

$$\therefore F_2 Q : F_2 P = QD : DP;$$

consequently, $F_2 D$ bisects the angle $PF_2 Q$.

Produce QF_2 and PF_2 to intersect $F_1 E$, and let P_1 and Q_1 be the points of intersection.

Then, since the triangles $PF_2 D$ and $P_1 F_2 E$ are equiangular, we have $P_1 E = \mu P_1 F_2$; and consequently the point P_1 lies on the inner oval. In like manner it is plain that Q_1 lies on the outer.

Again, by an elementary theorem in geometry, we have

$$F_2 P \cdot F_2 Q = PD \cdot DQ + F_2 D^2;$$

$$\therefore (1 - \mu^2) F_2 P \cdot F_2 Q = F_2 D^2.$$

Also, by similar triangles, we get

$$F_2 P : F_2 P_1 = F_2 D : F_2 E;$$

consequently

$$(1 - \mu^2) F_2 Q \cdot F_2 P_1 = F_2 D \cdot F_2 E = \text{const.} \qquad (2)$$

Therefore the rectangle under $F_2 Q$ and $F_2 P_1$ is constant; a theorem due to M. Quetelet.

303. Construction for Third Focus.—Next, draw QF_3, making $\angle F_2 Q F_3 = \angle F_2 F_1 P_1$; then, since the points P_1, F_1, Q, F_3 lie on the circumference of a circle, we get

$$F_1 F_2 \cdot F_2 F_3 = F_2 Q \cdot F_2 P_1 = \text{const.} \qquad (3)$$

Hence the point F_3 is determined.

We proceed to show that F_3 possesses the same properties relative to the curve as F_1 and F_2; in other words, that F_3 is a *third focus*.*

For this purpose it is convenient to write the equation of the curve in the form

$$mr_1 \pm lr_2 = nc_3, \qquad (4)$$

in which c_3 represents $F_1 F_2$, and l, m, n are constants.

It may be observed that in this case we have $n > m > l$.

---

* This fundamental property of the curve was discovered by Chasles. See *Histoire de la Géométrie*, note xxi., p. 352.

Now, since $\angle F_1F_3Q = \angle F_1P_1F_2 = \angle F_1PF_2$, the triangles F_1PF_2 and F_1F_3Q are equiangular; but, by (4), we have

$$mF_1P + lF_2P = nF_1F_2;$$

accordingly we have

$$mF_1F_3 + lF_3Q = nF_1Q,$$

or
$$nF_1Q - lF_3Q = mF_1F_3;$$

i.e. denoting the distance from F_3 by r_3 and F_1F_3 by c_2,

$$nr_1 - lr_3 = mc_2.$$

This shows that the distances of any point on the outer oval from F_1 and F_3 are connected by an equation similar in form to (4); and, consequently, F_3 is a *third focus of the curve*.

304. **Equations of Curve, relative to each pair of Foci.**—In like manner, since the triangles F_1QF_2 and F_1F_3P are equiangular, the equation

$$mF_1Q - lF_2Q = nF_1F_2$$

gives
$$mF_1F_3 - lF_3P = nF_1P.$$

Hence, for the inner oval, we have

$$nr_1 + lr_3 = mc_2.$$

This, combined with the preceding result, shows that the conjugate ovals of a Cartesian, referred to its two extreme foci, are represented by the equation

$$nr_1 \pm lr_3 = mc_2. \tag{5}$$

In like manner, it is easily seen that the conjugate ovals referred to the foci F_2 and F_3 are comprised under the equation

$$nr_2 - mr_3 = \pm lc_1, \tag{6}$$

where
$$c_1 = F_2F_3.$$

305. **Relation between the Constants.**—The equation connecting the constants l, m, n in a Cartesian, which has three points F_1, F_2, F_3 for its foci, can be readily found.

For, if we substitute in (3), c_3 for F_1F_2, &c., the equation is easily reduced to the form

$$l^2 c_1 + n^2 c_3 = m^2 c_2,$$

or
$$l^2 F_2F_3 + m^2 F_3F_1 + n^2 F_1F_2 = 0, \qquad (7)$$

in which the lengths F_2F_3, &c., are taken with their proper signs, viz., $F_3F_1 = -F_1F_3$, &c.

306. Conjugate Ovals are Inverse Curves.—Next, since the four points F_2, P, Q, F_3, lie in a circle, we have

$$F_1P \cdot F_1Q = F_1F_2 \cdot F_1F_3 = \text{const.} \qquad (8)$$

Consequently *the two conjugate ovals are inverse to each other with respect to a circle** whose centre is F_1, and whose radius is a mean proportional between F_1F_2 and F_1F_3.

It follows immediately from this, since F_2 lies inside both ovals, that F_3 lies outside both. It hence may be called the external focus. This is on the supposition that the constants† are connected by the relations $n > m > l$.

Also we have

$$\angle PF_3F_2 = \angle PQF_2 = \angle F_2Q_1P_1 = \angle F_2F_3P_1;$$

hence the lines F_3P and F_3P_1 are equally inclined to the axis F_1F_3. Consequently, if P_2 be the second point in which the line F_3P meets the inner oval, it follows, from the symmetry of the curve, that the points P_2 and P_1 are the

---

* It is easily seen that when $l = 0$ the Cartesian whose foci are F_1, F_2, F_3, reduces to this circle. Again, if $n = 0$, the Cartesian becomes another circle, whose centre is F_3, and which, as shall be presently seen, cuts orthogonally the system of Cartesians which have F_1, F_2, F_3 for their foci. These circles are called by Prof. Crofton (*Transactions, London Mathematical Society*, 1866), the *Confocal Circles* of the Cartesian system.

† From the above discussion it will appear, that if the general equation of a Cartesian be written $\lambda r + \mu r' = \nu c$, where c represents the distance between the foci; then (1) if, of the constants, λ, μ, ν, the greatest be ν, the curve is referred to its two internal foci; (2) if ν be intermediate between λ and μ, the curve is referred to the two extreme foci; (3) if ν be the least of the three, the curve is referred to the external and middle focus; (4) if $\lambda = \mu$, the curve is a conic; (5) if $\nu = \lambda$, or $\nu = \mu$, the curve is a limaçon; (6) if one of the constants λ, μ, ν vanish, the curve is a circle.

reflexions of each other with respect to the axis F_1F_2, and the triangles $F_1P_2F_2$ and $F_1P_1F_2$ are equal in every respect.

Again, since

$$\angle F_2PF_3 = \angle F_2QF_3 = \angle F_2F_1P_1 = \angle F_2F_1P_2,$$

the four points P, P_2, F_1 and F_2 lie on the circumference of a circle.

From this we have

$$F_3P \cdot F_3P_2 = F_3F_1 \cdot F_3F_2 = \text{constant}.$$

Hence, *the rectangle under the segments, made by the inner oval, on any transversal from the external focus, is constant.*

In like manner it can be shown that the same property holds for the segments made by the outer oval.

If we suppose P and P_2 to coincide, the line F_3P becomes a tangent to the oval, and the length of this tangent becomes constant, being a mean proportional between F_3F_1 and F_3F_2.

Accordingly, the tangents drawn from the external focus to a system of triconfocal Cartesians are of equal length.

This result may be otherwise stated, as follows:—*A system of triconfocal Cartesians is cut orthogonally by the confocal circle whose centre is the external focus of the system* (Prof. Crofton).

This theorem is a particular case of another—also due, I believe, to Prof. Crofton—which shall be proved subsequently, viz., that if two *triconfocal* Cartesians intersect, they cut each other orthogonally.

307. Construction for Tangent at any Point.— We next proceed to give a geometrical method of drawing the tangent and the normal at any point on a Cartesian.

Retaining the same notation as before, let R be the point in which the line F_2D meets the circle which passes through the points P, F_2, F_3, Q; then it can be shown that the lines PR and RQ are the normals at P and Q to the Cartesian oval which has F_1 and F_2 for its internal foci, and F_3 for its external. For, from equation (4), we have for the outer oval

$$m\frac{dr_1}{ds} - l\frac{dr_2}{ds} = 0.$$

Hence, if ω_1 and ω_2 be the angles which the normal at Q makes with QF_1 and QF_2 respectively, we have

$$m \sin \omega_1 = l \sin \omega_2; \text{ or } \sin \omega_1 : \sin \omega_2 = l : m. \tag{9}$$

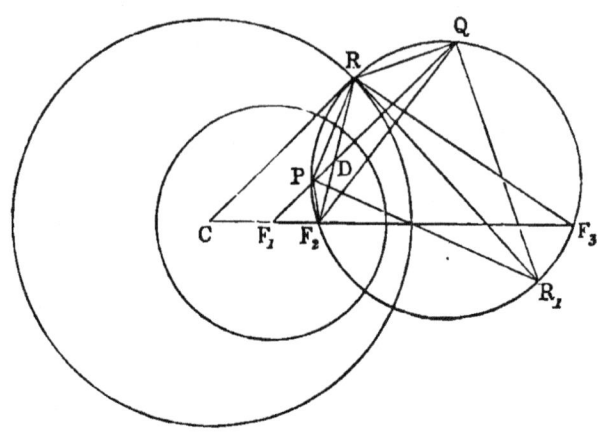

Fig. 80.

Again, we have seen at the commencement that

$$l : m = DQ : F_2Q;$$

also, by similar triangles,

$$RQ : RF_2 = DQ : F_2Q = l : m; \tag{10}$$

but

$$RQ : RF_2 = \sin RQP : \sin RQF_2;$$

hence

$$\sin RQF_1 : \sin RQF_2 = l : m.$$

Consequently, by (9), the line RQ is the normal at Q to the outer oval. In like manner it follows immediately that PR is normal to the inner oval.

This theorem is given by Prof. Crofton in the following form :—*The arc of a Cartesian oval makes equal angles with the right line drawn from the point to any focus and the circular arc drawn from it through the two other foci.*

This result furnishes an easy method of drawing the tangent at any point on a Cartesian whose three foci are given.

The construction may be exhibited in the following form:—

Let F_1, F_2, F_3 be the three foci, and P the point in question. Describe a circle through P and two foci F_2 and F_3, and let Q be the second point in which F_1P meets this circle; then the line joining P to R, the middle point of the arc cut off by PQ, is the normal.

308. **Confocal Cartesians intersect Orthogonally.** —It is plain, for the same reason, that the line drawn from P to R_1, the middle point of the other segment standing on PQ, is normal to a second Cartesian passing through P, and having the same three points as foci—F_2 and F_3 for its internal foci, and F_1 for its external.

Hence it follows that *through any point two Cartesian ovals can be drawn having three given points—which are in directum—for foci.*

Also *the two curves so described cut orthogonally.*

Again, if RC be drawn touching the circle PRQ, it is parallel to PQ, and hence

$$F_2C : F_1C = F_2R : RD = F_2R^2 : F_2R \cdot RD;$$

but
$$F_2R \cdot RD = RP^2;$$

$$\therefore F_2C : F_1C = F_2R^2 : PR^2 = m^2 : l^2. \tag{11}$$

Hence the point C is fixed.

Again
$$CR : F_1D = RF_2 : DF_2 = m^2 : m^2 - l^2;$$

$$\therefore CR = \frac{m^2 a}{m^2 - l^2}, \tag{12}$$

which determines the length of CR.

Next, since $RP = RQ$, if with R as centre and RP as radius a circle be described, it will touch each of the ovals, from what has been shown above.

Also, since C is a fixed point by (11), and CR a constant length by (12), it follows that the *locus of the centre of a circle which touches both branches of a Cartesian is a circle* (Quetelet, *Nouv. Mém. de l'Acad. Roy. de Brux.* 1827).

This construction is shown in the following figure, in which the form of two conjugate ovals, having the points F_1, F_2, F_3, for foci, is exhibited.

Again, since the ratio of F_2R to RP is constant, we get the following theorem, which is also due to M. Quetelet:—

A Cartesian oval is the envelope of a circle, whose centre moves on the circumference of a given circle, while its radius is in a constant ratio to the distance of its centre from a given point.

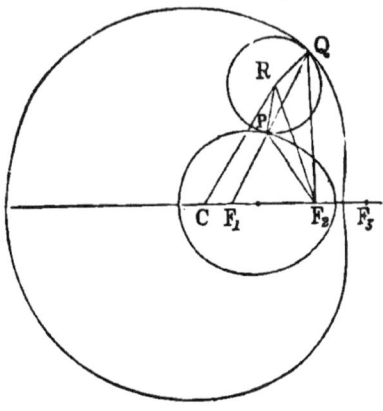

Fig. 81.

310. **Cartesian Oval as an Envelope.**—This construction has been given in a different form by Professor Casey, *Transactions Royal Irish Academy*, 1869.

If a circle cut a given circle orthogonally, while its centre moves along another given* circle, its envelope is a Cartesian oval.

This follows immediately; for the rectangle under F_1P and F_1Q is constant (8), and therefore the length of the tangent from F_1 to the circle is constant.

This result is given by Prof. Casey as a particular case of the general and elegant property of bicircular quartics, viz.: if in the preceding construction the centre of the moving circle describe any conic, instead of a circle, its envelope is a bicircular quartic.

---

* It is easily seen that the three foci of the Cartesian oval are: the centre of the orthogonal centre, and the limiting points of this and the other fixed circle.

Examples.

1. Find the polar equation of a Cartesian oval referred to a focus as pole. If the focus F_1 be taken as pole, and the line F_1F_2 as prime vector, we easily obtain, for the polar equation of the curve,

$$(m^2 - l^2)r^2 - 2c_3(mn - l^2 \cos \theta)r + c_3{}^2(n^2 - l^2) = 0.$$

The equations with respect to the other foci, taken as poles, are obtained by a change of letters.

2. Hence any equation of the form

$$r^2 - 2(a + b \cos \theta)r + c^2 = 0$$

represents a Cartesian oval.

3. Hence deduce Quetelet's theorem of Art. 302.

4. If any chord meet a Cartesian in four points, the sum of their distances from any focus is constant?

For, if we eliminate θ between the equation of the curve and the equation of an arbitrary line, we get a biquadratic in r, of which $-4a$ is the coefficient of the second term.

5. Show that the equation of a Cartesian may in general be brought to the form

$$S^2 = k^3 L,$$

where S represents a circle, and L a right line, and k is a constant.

6. Hence show that the curve is the envelope of the variable circle

$$\lambda^2 k L + 2\lambda S + k^2 = 0.$$

Compare Art. 309.

7. From this show that the curve has three foci; *i.e.* three evanescent circles having double contact with the curve.

8. The base angles of a variable triangle move on two fixed circles, while the two sides pass through the centres of the circles, and the base passes through a fixed point on the line joining the centres; prove that the locus of the vertex is a Cartesian.

9. Prove that the inverse of a Cartesian with respect to any point is a bicircular quartic. (*See* Salmon, *Higher Plane Curves*, Arts. 280, 281.)

10. Prove that the Cartesian

$$r^2 - 2(a + b \cos \theta)r + c^2 = 0$$

has three real foci, or only one according as

$$a - b \text{ is} > \text{ or } < c.$$

CHAPTER XXI.

ELIMINATION OF CONSTANTS AND FUNCTIONS.

311. Elimination of Constants.—The process of differentiation is often applied for the elimination of constants and functions from an equation, so as to form *differential equations* independent of the particular constants and functions employed.

We commence with the simple example $y^2 = ax + b$. By differentiation we get $2y\dfrac{dy}{dx} = a$, a result independent of b. A second differentiation gives

$$\left(\frac{dy}{dx}\right)^2 + y\frac{d^2y}{dx^2} = 0\,;$$

a differential equation containing neither a nor b, and which accordingly is satisfied by each of the individual equations which result from giving all possible values to a and b in the proposed.

In general, let the proposed equation be of the form $f(x, y, a) = 0$. By differentiation with respect to x, we get

$$\frac{df}{dx} + \frac{df}{dy}\frac{dy}{dx} = 0.$$

The elimination of a between this and the equation $f(x, y, a) = 0$ leads to a differential equation involving x, y and $\dfrac{dy}{dx}$, which holds for all the equations got by varying a in the proposed.

Again, if the given equation in x and y contain two constants, a and b; by two differentiations with respect to x, we obtain two differential equations, between which and the original, when the constants a and b are eliminated, we get a differential equation containing x, y, $\dfrac{dy}{dx}$ and $\dfrac{d^2y}{dx^2}$.

In general, for an equation containing n constants, the resulting differential equation contains x, y, $\dfrac{dy}{dx}$, $\dfrac{d^2y}{dx^2} \ldots \dfrac{d^n y}{dx^n}$; arising from the elimination of the n constants between the given equation and the n equations derived from it by successive differentiation.

EXAMPLES.

1. Eliminate a from the equation
$$y^2 - 2ay + x^2 = a^2. \qquad \textit{Ans. } (x^2 - 2y^2)\left(\dfrac{dy}{dx}\right)^2 - 4xy\dfrac{dy}{dx} - x^2 = 0.$$

2. Eliminate a and β from the equation
$$(y - a)^2 = p(x - \beta). \qquad \textit{Ans. } 2\left(\dfrac{dy}{dx}\right)^3 + p\dfrac{d^2y}{dx^2} = 0.$$

3. Eliminate the constants a and β from the equation
$$y = a \cos nx + \beta \sin nx. \qquad \textit{Ans. } \dfrac{d^2y}{dx^2} + n^2 y = 0.$$

4. Eliminate a and b from the equation
$$(x - a)^2 + (y - b)^2 = c^2. \qquad \textit{Ans. } c^2 = \dfrac{\left\{1 + \left(\dfrac{dy}{dx}\right)^2\right\}^3}{\left(\dfrac{d^2y}{dx^2}\right)^2}.$$

This agrees with the formula for the radius of curvature in Art. 226.

5. Eliminate a and β from the equation
$$y = ax \cos\left(\dfrac{n}{x} + \beta\right). \qquad \textit{Ans. } \dfrac{d^2y}{dx^2} + \dfrac{n^2 y}{x^4} = 0.$$

6. Eliminate the constants $a_0, a_1, \ldots a_n$ from the equation
$$y = \phi(x) + a_0 x^n + a_1 x^{n-1} + \ldots a_n. \qquad \textit{Ans. } \dfrac{d^{n+1} y}{dx^{n+1}} = \phi^{(n+1)}(x).$$

7. Eliminate the constants a and β from the equation
$$y = \alpha e^{ax} + \beta e^{bx}. \qquad \textit{Ans. } \dfrac{d^2y}{dx^2} - (a + b)\dfrac{dy}{dx} + aby = 0.$$

8. Eliminate a and b from the equation
$$xy = ae^x + be^{-x}. \qquad \textit{Ans. } x\dfrac{d^2y}{dx^2} + 2\dfrac{dy}{dx} - xy = 0.$$

9. Eliminate, by differentiation, c, c' from the equation
$$y = cxe^{\frac{1}{x}} + c'xe^{-\frac{1}{x}}. \qquad \textit{Ans. } x^4 \dfrac{d^2y}{dx^2} = y.$$

2 C

312. Elimination of Transcendental Functions.—

The process of differentiation can also be employed for the elimination of transcendental functions from equations of given form; for example, the logarithmic function can be eliminated by differentiation from the equation $y = \log \phi(x)$, which gives $\dfrac{dy}{dx} = \dfrac{\phi'(x)}{\phi(x)}$. We have met several instances of this process already; thus, in Art. 86, we found that the elimination of the symbolic functions, sin and $\sin^{-1}$, from the equation $y = \sin(m \sin^{-1} x)$ leads to the differential equation

$$(1 - x^2)\frac{d^2y}{dx^2} - x\frac{dy}{dx} + m^2 y = 0.$$

The principles involved in this process are of great importance in connexion with the converse problem—viz., the procedure from the differential equation to the primitive from which it is derived. This part of the subject belongs to the Integral Calculus in connexion with the *solution of differential equations*.

EXAMPLES.

1. $y = \tan^{-1} x$.   *Ans.* $\dfrac{dy}{dx} = \dfrac{1}{1 + x^2}$.

2. $y = \cos\left(\dfrac{y}{x}\right)$.   *Ans.* $x^2 \dfrac{dy}{dx} = \sqrt{1 - y^2}\left(y - x\dfrac{dy}{dx}\right)$.

3. Eliminate the exponential and logarithmic functions from the equation

$$y = \log(e^x + e^{-x}). \qquad \textit{Ans. } \frac{d^2y}{dx^2} + \left(\frac{dy}{dx}\right)^2 = 1.$$

4. Eliminate the circular and exponential functions from $y = e^x \sin x$.

Here $\dfrac{dy}{dx} = e^x \sin x + e^x \cos x = y + e^x \cos x$;

therefore $\dfrac{d^2y}{dx^2} = \dfrac{dy}{dx} + e^x \cos x - e^x \sin x = 2\dfrac{dy}{dx} - 2y$.

5. $y = \dfrac{e^x + e^{-x}}{e^x - e^{-x}}$.   *Ans.* $\dfrac{dy}{dx} = 1 - y^2$.

6. $y = \sin(\log x)$.   *Ans.* $x^2 \dfrac{d^2y}{dx^2} + x\dfrac{dy}{dx} + y = 0$.

In the preceding examples we considered only the case of a single independent variable; the differential equations arrived at in such cases are called *ordinary* differential equations.

When our equations are of such a nature as to admit of two or more independent variables, the equations derived from them by differentiation are called *partial* differential equations. We proceed to consider some cases of elimination which introduce differential equations of this class.

313. **Elimination of Arbitrary Functions.**—The equations hitherto considered contained only two variables; we now proceed to the more general case of an equation involving three variables, two of which accordingly can be regarded as *independent*. We shall denote the independent variables by the letters x and y, and the dependent variable by z. It will also be found convenient to adopt the usual notation, and to represent the partial differential coefficients

$$\frac{dz}{dx}, \quad \frac{dz}{dy}, \quad \frac{d^2z}{dx^2}, \quad \frac{d^2z}{dxdy}, \text{ and } \frac{d^2z}{dy^2},$$

by the letters p, q, r, s and t, respectively.

We proceed to show that in this case we are enabled by differentiation to eliminate functions whose forms are altogether arbitrary. In fact we have already met with examples of this process; for instance, if $z = x^n \phi \left(\dfrac{y}{x}\right)$ we have seen, in Art. 102, that in all cases we have

$$x \frac{dz}{dx} + y \frac{dz}{dy} = nz,$$

whatever be the form of the function ϕ; this function accordingly may be regarded as completely *arbitrary* in its form, and the preceding differential equation holds whatever form is assigned to it. This can also be shown immediately by differentiation. Conversely, it can be established without difficulty that $x^n \phi \left(\dfrac{y}{x}\right)$ is the most general form of z which satisfies the preceding partial differential equation. This process, as in the case of ordinary differential equations, comes under the province of the Integral Calculus, and is mentioned here merely for the purpose of showing the connexion between the integration of differential equations and the formation of such equations by the method of elimination.

As another simple example, let it be proposed to eliminate the arbitrary function from the equation $z = f(x^2 + y^2)$.

Here $p = \dfrac{dz}{dx} = 2xf'(x^2 + y^2)$, $q = \dfrac{dz}{dy} = 2yf'(x^2 + y^2)$;

hence we get $yp - xq = 0$;

an equation which holds for all values of z whatever the form of the function (f) may be.

Examples.

1. $z = \phi(ax + by)$.   Ans. $aq = bp$.
2. $y - bz = \phi(x - az)$.   ,, $ap + bq = 1$.
3. $x - a = (z - \gamma)\,\phi\!\left(\dfrac{y - \beta}{z - \gamma}\right)$.   ,, $(x - a)p + (y - \beta)q = z - \gamma$.
4. $\phi(x^n + y^m) = z^r$.   ,, $nx^{n-1}q = my^{m-1}p$.
5. $z^2 = xy + \phi\!\left(\dfrac{x}{y}\right)$.   ,, $xzp + yzq = xy$.
6. $x + \sqrt{x^2 + y^2 + z^2} = x^{1-n}\,\phi\!\left(\dfrac{y}{x}\right)$.   ,, $z = px + qy + n\sqrt{x^2 + y^2 + z^2}$.

314. Condition that one Expression is a Function of another.—Let $z = \phi(v)$, where v is a known function of x and y.

Here $\dfrac{dz}{dx} = \phi'(v)\dfrac{dv}{dx}$, $\dfrac{dz}{dy} = \phi'(v)\dfrac{dv}{dy}$;

therefore $\dfrac{dz}{dx}\dfrac{dv}{dy} - \dfrac{dz}{dy}\dfrac{dv}{dx} = 0$, or $p\dfrac{dv}{dy} - q\dfrac{dv}{dx} = 0$.

This furnishes the condition that z should be a function of the quantity represented by v. Also, denoting z by V, and supposing V and v to be two given *explicit* functions of x and y, the *condition that V is a function of v* is that the equation

$$\dfrac{dV}{dx}\dfrac{dv}{dy} - \dfrac{dV}{dy}\dfrac{dv}{dx} = 0 \qquad (1)$$

Condition that one Expression is a Function of another.

shall hold for all values of x and y, i.e. shall be identically satisfied. For instance, if

$$V = \frac{\sqrt{1-x^2} - \sqrt{1-y^2}}{x+y}, \text{ and } v = x\sqrt{1-y^2} + y\sqrt{1-x^2},$$

we get
$$\frac{dV}{dx}\frac{dv}{dy} - \frac{dV}{dy}\frac{dv}{dx} = 0, \text{ identically};$$

hence V is a function of v in this case.

This can also be independently verified; for, if $x = \sin\theta$, and $y = \sin\phi$, we get

$$V = \frac{\cos\theta - \cos\phi}{\sin\theta - \sin\phi} = -\tan\frac{\theta+\phi}{2}:$$

$$v = \sin\theta\cos\phi + \cos\theta\sin\phi = \sin(\theta+\phi);$$

which establishes the result required.

We have here assumed that whenever equation (1) is satisfied identically, V is expressible as a function of v: this can be easily shown as follows:—

Since V and v are supposed to be given functions of x and y, if one of these variables, y, be eliminated between them we can represent V as a function of v and x.

Accordingly, let
$$V = f(x, v);$$

then
$$\frac{dV}{dx} = \frac{df}{dx} + \frac{df}{dv}\frac{dv}{dx}, \quad \frac{dV}{dy} = \frac{df}{dv}\frac{dv}{dy};$$

therefore
$$\frac{dV}{dx}\frac{dv}{dy} - \frac{dV}{dy}\frac{dv}{dx} = \frac{df}{dx}\frac{dv}{dy}.$$

Hence, since the left-hand side is zero by hypothesis, we must have $\frac{df}{dx} = 0$; i.e. the function $f(x, v)$ or V reduces to a function of v simply; which establishes the proposition.

315. More generally, let it be proposed to eliminate the arbitrary function ϕ from the equation

$$V = \phi(v),$$

where V and v are given functions of three variables, x, y, and z.

Regarding x and y as independent variables, we get by differentiation

$$\frac{dV}{dx} + p\frac{dV}{dz} = \phi'(v)\left(\frac{dv}{dx} + p\frac{dv}{dz}\right),$$

$$\frac{dV}{dy} + q\frac{dV}{dz} = \phi'(v)\left(\frac{dv}{dy} + q\frac{dv}{dz}\right):$$

eliminating $\phi'(v)$ we obtain

$$\frac{dV}{dx}\frac{dv}{dy} - \frac{dV}{dy}\frac{dv}{dx} + p\left(\frac{dV}{dz}\frac{dv}{dy} - \frac{dv}{dz}\frac{dV}{dy}\right)$$

$$+ q\left(\frac{dV}{dx}\frac{dv}{dz} - \frac{dv}{dx}\frac{dV}{dz}\right) = 0; \qquad (2)$$

a result independent of the arbitrary function ϕ.

This equation can also be established as follows :—

Differentiating the equation $V = \phi(v)$, considering x, y, z as all variables, we get

$$\frac{dV}{dx}dx + \frac{dV}{dy}dy + \frac{dV}{dz}dz = \phi'(v)\left(\frac{dv}{dx}dx + \frac{dv}{dy}dy + \frac{dv}{dz}dz\right).$$

Then, since the form of $\phi(v)$ is perfectly arbitrary, this equation must hold whatever be the form of the function $\phi'(v)$, and hence we must have

$$\left.\begin{array}{l} \dfrac{dV}{dx}dx + \dfrac{dV}{dy}dy + \dfrac{dV}{dz}dz = 0, \\[2mm] \dfrac{dv}{dx}dx + \dfrac{dv}{dy}dy + \dfrac{dv}{dz}dz = 0. \end{array}\right\} \qquad (3)$$

Condition that one Expression is a Function of another.

Moreover, introducing the condition that z depends on x and y, we have

$$dz = p\,dx + q\,dy;$$

consequently, eliminating dx, dy, dz between this and the equations in (3), we get

$$\begin{vmatrix} \dfrac{dV}{dx}, & \dfrac{dV}{dy}, & \dfrac{dV}{dz} \\ \dfrac{dv}{dx}, & \dfrac{dv}{dy}, & \dfrac{dv}{dz} \\ p, & q, & -1 \end{vmatrix} = 0; \qquad (4)$$

which agrees with the result in (2).

EXAMPLES.

Eliminate the arbitrary functions in the following cases:—

1. $z = \phi(a \sin x + b \sin y)$.
   Ans. $b \cos y \dfrac{dz}{dx} - a \cos x \dfrac{dz}{dy} = 0$.

2. $z = e^{\frac{y}{a}} \phi(x - y)$.
   $\dfrac{dz}{dx} + \dfrac{dz}{dy} = \dfrac{z}{a}$.

3. $z^2 = xy + \phi\left(\dfrac{x}{y}\right)$.
   $x \dfrac{dz}{dx} + y \dfrac{dz}{dy} = \dfrac{xy}{z}$.

4. $\dfrac{1}{z} - \dfrac{1}{x} = \phi\left(\dfrac{1}{y} - \dfrac{1}{x}\right)$.
   $x^2 \dfrac{dz}{dx} + y^2 \dfrac{dz}{dy} = z^2$.

5. $z = \dfrac{y^2 \phi(y) + x}{1 - x\phi(y)}$.
   $(x^2 + y^2) \dfrac{dz}{dx} = y^2 + z^2$.

6. $z = a\sqrt{x^2 + y^2} + \phi\left(\dfrac{y}{x}\right)$.
   $x \dfrac{dz}{dx} + y \dfrac{dz}{dy} = a\sqrt{x^2 + y^2}$.

7. $z = (x + y)^n \phi(x^2 - y^2)$.
   $y \dfrac{dz}{dx} + x \dfrac{dz}{dy}$.

8. $x^2 + y^2 + z^2 = \phi(ax + by + cz)$.
   Ans. $(bz - cy)\dfrac{dz}{dx} + (cx - az)\dfrac{dz}{dy} = ay - bx$.

316. Next, let it be required to eliminate the arbitrary function ϕ from the equation

$$F\{x, y, z, \phi(u)\} = 0,$$

where u is a given explicit function of x, y, and z.

Regarding x and y as the independent variables, we may differentiate the equation with respect to x, and also with respect to y; then, since z is a function of x and y, we have

$$\frac{d \cdot \phi(u)}{dx} = \phi'(u)\left(\frac{du}{dx} + \frac{du}{dz}p\right),$$

and

$$\frac{d \cdot \phi(u)}{dy} = \phi'(u)\left(\frac{du}{dy} + \frac{du}{dz}q\right);$$

hence we obtain two partial differential equations involving x, y, z, p, q, $\phi(u)$, and $\phi'(u)$. Accordingly, if $\phi(u)$ and $\phi'(u)$ be eliminated between these and the original equation, we shall have a resulting equation containing only x, y, z, p, and q.

317. **Case of two or more Arbitrary Functions.**—If the given equation contain more than one arbitrary function, we have to proceed to partial differentiations of a higher degree in order to eliminate the functions: thus, in the case of two arbitrary functions, $\phi(u)$ and $\psi(v)$, the first differentiations with respect to x and y introduce the functions $\phi'(u)$ and $\psi'(v)$. It is plainly impossible, in general, to eliminate the four arbitrary functions between three equations; we accordingly must proceed to form the three partial differentials of the second order, introducing two new arbitrary functions $\phi''(u)$ and $\psi''(v)$. Here, again, it is in general impossible to eliminate the *six* functions between *six* equations, so that it is necessary to proceed to differentials of the third order: in doing so we obtain four new equations, containing two additional functions, $\phi'''(u)$ and $\psi'''(v)$. After the elimination of the eight arbitrary functions there would remain, in general, *two* resulting partial differential equations of the third order.

318. There is one case, however, in which we can always obtain a resulting partial differential equation of the second order—viz., where the arbitrary functions are functions of the same quantity, u.

Case of Two or more Arbitrary Functions.

Thus, suppose the given equation of the form

$$F\{x, y, z, \phi(u), \psi(u)\} = 0, \qquad (5)$$

where u is a known function of x, y, and z.

By differentiation we get

$$\frac{dF}{dx} + p\frac{dF}{dz} + \frac{dF}{du}\left(\frac{du}{dx} + p\frac{du}{dz}\right) = 0,$$

$$\frac{dF}{dy} + q\frac{dF}{dz} + \frac{dF}{du}\left(\frac{du}{dy} + q\frac{du}{dz}\right) = 0.$$

Eliminating $\dfrac{dF}{du}$ between these equations, we obtain

$$\frac{dF}{dx}\frac{du}{dy} - \frac{dF}{dy}\frac{du}{dx} + p\left(\frac{dF}{dz}\frac{du}{dy} - \frac{dF}{dy}\frac{du}{dz}\right)$$

$$+ q\left(\frac{dF}{dx}\frac{du}{dz} - \frac{dF}{dz}\frac{du}{dx}\right) = 0. \qquad (6)$$

This equation contains only the original functions $\phi(u)$, $\psi(u)$, along with x, y, z, p, and q. Again, if we apply the same method to it, we can form a new partial differential equation, involving the same functions $\phi(u)$ and $\psi(u)$, along with x, y, z, p, q, r, s, t.

The elimination of the unknown functions, $\phi(u)$ and $\psi(u)$, between this last equation and equations (5) and (6), leads to the required partial differential equation of the second order. The result in (6) admits also of being arrived at by the method adopted in the second proof of Art. 315. For regarding x, y, z, as all variables, we get from (5), on differentiation,

$$\frac{dF}{dx}dx + \frac{dF}{dy}dy + \frac{dF}{dz}dz + \frac{dF}{du}\left(\frac{du}{dx}dx + \frac{du}{dy}dy + \frac{du}{dz}dz\right) = 0. \qquad (7)$$

But
$$\frac{dF}{du} = \frac{dF}{d\phi(u)}\phi'(u) + \frac{dF}{d\psi(u)}\psi'(u),$$

and accordingly, since (7) must hold for all values of $\phi'(u)$ and $\psi'(u)$, we have

and
$$\left.\begin{array}{r}\dfrac{dF}{dx}dx + \dfrac{dF}{dy}dy + \dfrac{dF}{dz}dz = 0,\\[1em] \dfrac{du}{dx}dx + \dfrac{du}{dy}dy + \dfrac{du}{dz}dz = 0.\end{array}\right\} \qquad (8)$$

Eliminating between these equations and

$$dz = pdx + qdy,$$

we get the following determinant:

$$\begin{vmatrix} \dfrac{dF}{dx}, & \dfrac{dF}{dy}, & \dfrac{dF}{dz} \\[1em] \dfrac{du}{dx}, & \dfrac{du}{dy}, & \dfrac{du}{dz} \\[1em] p, & q, & -1 \end{vmatrix} = 0 ; \qquad (9)$$

which plainly is identical with (6).

This admits also of the following statement: substitute c instead of u in the proposed equation: then regarding c as constant, differentiate the resulting equation, as also the equation $u = c$ (on the same hypothesis): on combining the resulting equations with

$$dz = pdx + qdy,$$

we get another equation connecting $\phi(c)$ and $\psi(c)$; and applying the same method to it, we obtain the result, on eliminating the arbitrary functions $\phi(c)$ and $\psi(c)$ between the original equation and the two others thus arrived at.

These methods will be illustrated in the following examples.

EXAMPLES.

1. $$z = x\phi(z) + y\psi(z).$$

Here
$$p = \phi(z) + \{x\phi'(z) + y\psi'(z)\}\, p,$$
$$q = \psi(z) + \{x\phi'(z) + y\psi'(z)\}\, q.$$

Hence
$$\frac{p}{q} = \frac{\phi(z)}{\psi(z)} = f(z), \text{ suppose.}$$

Applying the principle of Art. 314, we have
$$q\frac{d}{dx}\left(\frac{p}{q}\right) - p\frac{d}{dy}\left(\frac{p}{q}\right) = 0,$$

or
$$q^2 r - 2pqs + p^2 t = 0.$$

Otherwise thus: let $z = c$, and we get $dz = 0$, and $\phi(c)dx + \psi(c)dy = 0$; also $pdx + qdy = 0$;

therefore
$$\frac{p}{q} = \frac{\phi(c)}{\psi(c)}.$$

Differentiating again, we have
$$qdp - pdq = 0,$$

or
$$q\{rdx + sdy\} - p(sdx + tdy) = 0,$$

which, combined with
$$pdx + qdy = 0,$$

leads to the same result as before.

2. $$z = x\phi(ax + by) + y\psi(ax + by).$$

Here
$$p = \phi(ax + by) + a\{x\phi'(ax + by) + y\psi'(ax + by)\},$$
$$q = \psi(ax + by) + b\{x\phi'(ax + by) + y\psi'(ax + by)\};$$

therefore
$$bp - aq = b\phi(ax + by) - a\psi(ax + by);$$

hence
$$br - as = a\{b\phi'(ax + by) - a\psi'(ax + by)\},$$
$$bs - at = b\{b\phi'(ax + by) - a\psi'(ax + by)\};$$

therefore
$$b^2 r - 2abs + a^2 t = 0.$$

Otherwise thus: let $ax + by = c$, then $adx + bdy = 0$; also, $dz = \phi(c)\,dx + \psi(c)\,dy$, and $dz = pdx + qdy$; hence

$$bp - aq = b\phi(c) - a\psi(c).$$

Differentiating again, we get

$$bdp - adq = 0, \text{ or } b(rdx + sdy) - a(sdx + tdy) = 0.$$

Combining this with the equation $adx + bdy = 0$, we get

$$b^2 r - 2abs + a^2 t = 0,$$

as before.

319. **Case of n Arbitrary Functions of same Function.**—It can be readily seen that the preceding method is capable of extension to the elimination of any number n of arbitrary functions from an equation, provided that they are all functions of the same quantity u.

For the equation (6) plainly holds in this case, and, proceeding as in the last Article, we obtain a series of equations (the last being of the nth order of differentiation), each containing the n arbitrary functions along with the variables and their derived functions. If the n functions be eliminated between the n differential equations and the original equation, we obtain a differential equation of the nth order which is independent of the arbitrary functions in question.

EXAMPLES.

1. Given $y = e^{ax}(C + C'x)$, prove that
$$\frac{d^2y}{dx^2} - 2a\frac{dy}{dx} + a^2y = 0.$$

2. Eliminate the constants from the equation

$y = C_1 e^{2x} \cos 3x + C_2 e^{2x} \sin 3x.$ Ans. $\frac{d^2y}{dx^2} - 4\frac{dy}{dx} + 13y = 0.$

3. Eliminate C and C' from the equations

(a). $y = \frac{\cos mx}{n^2 - m^2} + C \cos nx + C' \sin nx,$

(b) $y = x \sin nx + C \cos nx + C' \sin nx.$

Ans. (a) $\frac{d^2y}{dx^2} + n^2 y = \cos mx.$ (b) $\frac{d^2y}{dx^2} + n^2 y = 2n \cos nx.$

4. Eliminate the arbitrary functions from the equation

$z = \frac{x^3 y}{6} + \phi(y + ax) + \psi(y - ax).$ Ans. $r - a^2 t = xy.$

5. Eliminate the functions from the equation

$y = A \cos(a \sin^{-1}\frac{x}{a} + a).$ Ans. $(c^2 - x^2)\frac{d^2y}{dx^2} - x\frac{dy}{dx} + a^2 y = 0.$

6. Eliminate A and a from

$y = A \cos(nx + a).$ Ans. $\frac{d^2y}{dx^2} - n \cot nx \frac{dy}{dx} + n^2 y \sin^2 nx = 0.$

7. If $z = \cos ax \, \phi\left(\frac{y}{x}\right) + \sin ax \, \psi\left(\frac{y}{x}\right)$, prove that
$$rx^2 + 2sxy + ty^2 + a^2x^2 z = 0.$$

8. If a_1, a_2, a_3 be the roots of the equation
$$z^3 + p_1 z^2 + p_2 z + p_3 = 0,$$

prove that the result of eliminating the exponentials from the equation
$$y = C_1 e^{a_1 z} + C_2 e^{a_2 z} + C_3 e^{a_3 z}$$

is
$$\frac{d^3 y}{dx^3} + p_1 \frac{d^2 y}{dx^2} + p_2 \frac{dy}{dx} + p_3 y = 0.$$

9. Find the result of the elimination of the arbitrary functions from
$$z = \phi(x + ay) + \psi(x - ay). \qquad Ans. \ a^2 r - t = 0.$$

10. If $z = f\left(\dfrac{y}{x}\right) + \phi(xy)$, prove that
$$x^2 r - y^2 t + xp - yq = 0.$$

11. If $ae^y + be^{-y} = ce^z + de^{-z}$, prove that
$$\left[\frac{d^2 y}{dx^2} + \left(\frac{dy}{dx}\right)^2 - \frac{dy}{dx}\right]\left[\left(\frac{dy}{dx}\right)^2 - 1\right] = 3 \frac{dy}{dx} \left(\frac{d^3 y}{dx^3}\right)^3.$$

12.
$$z = x^n \phi\left(\frac{y}{x}\right) + x^m \psi\left(\frac{y}{x}\right).$$

$Ans. \ x^2 r + 2xys + y^2 t - (m + n - 1)(px + qy) + mnz = 0.$

13. Eliminate the arbitrary functions from the equation
$$z = \phi\{x + f(y)\}. \qquad Ans. \ ps - qr = 0.$$

14. If the substitution of Ae^{ax} for y satisfies the differential equation with constant coefficients,
$$\frac{d^n y}{dx^n} + p_1 \frac{d^{n-1} y}{dx^{n-1}} + \ldots + p_{n-1} \frac{dy}{dx} + p_n y = 0,$$

prove that a must be a root of the equation
$$z^n + p_1 z^{n-1} + \ldots + p_{n-1} z + p_n = 0.$$

15. Eliminate the constants from the equation
$$ax^2 + 2bxy + cy^2 + 2dx + 2ey + f = 0.$$

$Ans. \ 40 r^3 - 45 q r^2 s + 9 q^2 t = 0,$

where $\quad p = \dfrac{dy}{dx}, \quad q = \dfrac{d^2 y}{dx^2}, \quad r = \dfrac{d^3 y}{dx^3},$ &c.

CHAPTER XXII.

CHANGE OF THE INDEPENDENT VARIABLE.

320. Case of a Single Independent Variable.—We have already pointed out the distinction between independent and dependent variables in the formation of differential coefficients.

In applications of the Differential Calculus it is sometimes necessary to make our differential equations depend on new independent variables instead of those which had been originally selected.

To show how this transformation is effected we commence with the case of one independent variable, and suppose V to represent any function of x, y, $\dfrac{dy}{dx}$, $\dfrac{d^2y}{dx^2}$, &c. We proceed to show how the expressions for $\dfrac{dy}{dx}$, $\dfrac{d^2y}{dx^2}$, &c., are transformed, when, instead of x, any function of x is taken as the independent variable.

Let this new function be denoted by t, and suppose that $\dfrac{dx}{dt}$, $\dfrac{d^2x}{dt^2}$, &c., are represented by $\dot{x}$, $\ddot{x}$, &c., then in all cases we have

$$\frac{du}{dt} = \frac{du}{dx}\frac{dx}{dt} = \dot{x}\,\frac{du}{dx},$$

where u is any function of x;

or
$$\frac{d}{dx}(u) = \frac{1}{\dot{x}}\frac{d}{dt}(u). \qquad (1)$$

Hence
$$\frac{dy}{dx} = \frac{1}{\dot{x}}\frac{dy}{dt}, \qquad (2)$$

also $\quad \dfrac{d^2y}{dx^2} = \dfrac{d}{dx}\left(\dfrac{dy}{dx}\right) = \dfrac{d}{dx}\left(\dfrac{1}{\dot{x}}\dfrac{dy}{dt}\right) = \dfrac{1}{\dot{x}}\dfrac{d}{dt}\left(\dfrac{1}{\dot{x}}\dfrac{dy}{dt}\right),$

{substituting $\dfrac{1}{\dot{x}}\dfrac{dy}{dt}$ instead of u in (1)};

hence $\quad \dfrac{d^2y}{dx^2} = \dfrac{\dot{x}\dfrac{d^2y}{dt^2} - \ddot{x}\dfrac{dy}{dt}}{\dot{x}^3}.$ \hfill (3)

Again, $\quad \dfrac{d^3y}{dx^3} = \dfrac{d}{dx}\left(\dfrac{\dot{x}\dfrac{d^2y}{dt^2} - \ddot{x}\dfrac{dy}{dt}}{(\dot{x})^3}\right) = \dfrac{1}{\dot{x}}\dfrac{d}{dt}\left(\dfrac{\dot{x}\dfrac{d^2y}{dt^2} - \ddot{x}\dfrac{dy}{dt}}{\dot{x}^3}\right)$

$= \dfrac{\dot{x}^2\dfrac{d^3y}{dt^3} - 3\dot{x}\ddot{x}\dfrac{d^2y}{dt^2} + \dfrac{dy}{dt}\{3(\ddot{x})^2 - \dot{x}\,\dddot{x}\}}{(\dot{x})^5};$ \hfill (4)

and so on for differentiations of higher degrees.

If y be taken as the independent variable, we obtain the corresponding values by making

$$\dfrac{dy}{dt} = 1, \qquad \dfrac{d^2y}{dt^2} = 0, \&c.$$

Hence $\quad \dfrac{dy}{dx} = \dfrac{1}{\dfrac{dx}{dy}}, \qquad \dfrac{d^2y}{dx^2} = -\dfrac{\dfrac{d^2x}{dy^2}}{\left(\dfrac{dx}{dy}\right)^3};$ \hfill (5)

$\dfrac{d^3y}{dx^3} = \dfrac{3\left(\dfrac{d^2x}{dy^2}\right)^2 - \dfrac{dx}{dy}\dfrac{d^3x}{dy^3}}{\left(\dfrac{dx}{dy}\right)^5};$ \hfill (6)

and so on.

The preceding results can also be arrived at otherwise, as follows. The essential distinction of an independent variable is, that its differential is regarded as constant; accordingly, in differentiating $\dfrac{dy}{dx}$ when x is the independent variable we have $d\left(\dfrac{dy}{dx}\right) = \dfrac{d^2y}{dx}$. However, when x is no longer regarded as the independent variable, we must consider the numerator and the denominator of the fraction $\dfrac{dy}{dx}$ as both variables, and by Art. 15, we get

$$d\left(\frac{dy}{dx}\right) = \frac{dx\,d^2y - dy\,d^2x}{dx^2}, \text{ or } \frac{d}{dx}\left(\frac{dy}{dx}\right) = \frac{dx\,d^2y - dy\,d^2x}{dx^3}.$$

Differentiating again on the same hypothesis, we get

$$\frac{d}{dx}\left(\frac{d^2y}{dx^2}\right) = \frac{dx^2\,d^3y - dx\,dy\,d^3y - 3dx\,d^2x\,d^2y + 3(d^2x)^2\,dy}{dx^5}.$$

These results are perfectly general whatever function of x be taken as the independent variable. Their identity with the equations previously arrived at is manifest.

EXAMPLES.

1. Being given that $x = a(\theta - \sin\theta)$, $y = a(1 - \cos\theta)$, find the value of $\dfrac{d^2y}{dx^2}$.

   Ans. $\dfrac{-1}{a(1 - \cos\theta)^2}$.

2. Hence deduce the expression for the radius of curvature in a cycloid.

3. If $x = (a + b)\cos\theta - b\cos\dfrac{a+b}{b}\theta$, $y = (a + b)\sin\theta - b\sin\dfrac{a+b}{b}\theta$, find the value of $\dfrac{d^2y}{dx^2}$.

   Here $\dfrac{dy}{dx} = \dfrac{\cos\theta - \cos\dfrac{a+b}{b}\theta}{\sin\dfrac{a+b}{b}\theta - \sin\theta} = \tan\left(\dfrac{a}{2b} + 1\right)\theta,$

   $\dfrac{d^2y}{dx^2} = \dfrac{a + 2b}{4b(a+b)\sin\dfrac{a\theta}{2b}\cos^3\left(\dfrac{a}{2b} + 1\right)\theta}.$

4. Change the independent variable from x to θ in the expression $\dfrac{d^2y}{dx^2}$, supposing $x = \sin\theta$.

   Here $\dfrac{dy}{dx} = \dfrac{1}{\cos\theta}\dfrac{dy}{d\theta}$, $\therefore \dfrac{d^2y}{dx^2} = \dfrac{1}{\cos\theta}\dfrac{d}{d\theta}\left(\dfrac{1}{\cos\theta}\dfrac{dy}{d\theta}\right) = \dfrac{1}{\cos^2\theta}\dfrac{d^2y}{d\theta^2} + \dfrac{\sin\theta\,\dfrac{dy}{d\theta}}{\cos^3\theta}.$

2 D

5. Transform the equation

$$x^2 \frac{d^2y}{dx^2} + ax \frac{dy}{dx} + by = 0$$

into another in which θ is the independent variable, being given $x = e^\theta$.

Here
$$\frac{dy}{d\theta} = \frac{dy}{dx}\frac{dx}{d\theta} = x\frac{dy}{dx};$$

hence
$$\frac{d}{d\theta}\left(\frac{dy}{d\theta}\right) = x\frac{d}{dx}\left(x\frac{dy}{dx}\right), \text{ or } \frac{d^2y}{d\theta^2} = x^2\frac{d^2y}{dx^2} + x\frac{dy}{dx};$$

therefore
$$x^2 \frac{d^2y}{dx^2} = \frac{d^2y}{d\theta^2} - \frac{dy}{d\theta},$$

and the transformed equation is

$$\frac{d^2y}{d\theta^2} + (a - 1)\frac{dy}{d\theta} + by = 0.$$

6. Transform the equation

$$x^2 \frac{d^2y}{dx^2} + 2x \frac{dy}{dx} + \frac{a^2}{x^2} y = 0$$

into another where z is the independent variable, being given $x = \frac{1}{z}$.

It is evident that in this case $x\dfrac{dy}{dx} = -z\dfrac{dy}{dz}$, and

$$x\frac{d}{dx}\left(x\frac{dy}{dx}\right) = z\frac{d}{dz}\left(z\frac{dy}{dz}\right),$$

or
$$x^2\frac{d^2y}{dx^2} + x\frac{dy}{dx} = z^2\frac{d^2y}{dz^2} + z\frac{dy}{dz};$$

therefore
$$x^2\frac{d^2y}{dx^2} + 2x\frac{dy}{dx} = z^2\frac{d^2y}{dz^2},$$

and the transformed equation is

$$\frac{d^2y}{dz^2} + a^2 y = 0.$$

7. Change the independent variable from x to z in the equation

$$x^4 \frac{d^2y}{dx^2} + a^2 y = 0, \text{ where } x = \frac{1}{z}.$$

$$\text{Ans. } \frac{d^2y}{dz^2} + \frac{2}{z}\frac{dy}{dz} + a^2 y = 0.$$

321. Two Independent Variables.—We will next consider the process of transformation for two independent variables, and commence with the transformations introduced by changing from rectangular to polar coordinates in analytic geometry. In this case we have

$$x = r \cos \theta, \quad y = r \sin \theta; \tag{7}$$

and therefore
$$r^2 = x^2 + y^2, \quad \tan \theta = \frac{y}{x}. \tag{8}$$

Accordingly, any function, V, of x and y may be regarded as a function of r and θ, and by Art. 98 we have

$$\left. \begin{array}{l} \dfrac{dV}{d\theta} = \dfrac{dV}{dx}\dfrac{dx}{d\theta} + \dfrac{dV}{dy}\dfrac{dy}{d\theta} \\[6pt] \dfrac{dV}{dr} = \dfrac{dV}{dx}\dfrac{dx}{dr} + \dfrac{dV}{dy}\dfrac{dy}{dr} \end{array} \right\}. \tag{9}$$

But, from (7),

$$\frac{dx}{dr} = \cos \theta, \quad \frac{dx}{d\theta} = -r \sin \theta = -y, \quad \frac{dy}{dr} = \sin \theta, \quad \frac{dy}{d\theta} = x; \tag{10}$$

hence we obtain

$$\frac{dV}{d\theta} = x \frac{dV}{dy} - y \frac{dV}{dx}, \tag{11}$$

$$r \frac{dV}{dr} = x \frac{dV}{dx} + y \frac{dV}{dy}. \tag{12}$$

These transformations are useful in the Planetary Theory. Again, we have

$$\left. \begin{array}{l} \dfrac{dV}{dx} = \dfrac{dV}{dr}\dfrac{dr}{dx} + \dfrac{dV}{d\theta}\dfrac{d\theta}{dx} \\[6pt] \dfrac{dV}{dy} = \dfrac{dV}{dr}\dfrac{dr}{dy} + \dfrac{dV}{d\theta}\dfrac{d\theta}{dy} \end{array} \right\}. \tag{13}$$

But from (8) we have

$$\frac{dr}{dx} = \frac{x}{r} = \cos\theta, \quad \frac{dr}{dy} = \sin\theta, \tag{14}$$

$$\frac{d\theta}{dx} = -\cos^2\theta \frac{y}{x^2} = -\frac{\sin\theta}{r}, \quad \frac{d\theta}{dy} = \frac{\cos\theta}{r}; \tag{15}$$

therefore

$$\frac{dV}{dx} = \cos\theta \frac{dV}{dr} - \frac{\sin\theta}{r}\frac{dV}{d\theta}, \tag{16}$$

$$\frac{dV}{dy} = \sin\theta \frac{dV}{dr} + \frac{\cos\theta}{r}\frac{dV}{d\theta}. \tag{17}$$

The two latter equations can also be derived by solving for $\frac{dV}{dx}$ and $\frac{dV}{dy}$ from the equations (11) and (12).

322. Transformation of $\frac{d^2V}{dx^2}$ and $\frac{d^2V}{dy^2}$.—Since formula (16) holds, whatever be the form of the function V, we have

$$\frac{d}{dx}(\phi) = \cos\theta \frac{d}{dr}(\phi) - \frac{\sin\theta}{r}\frac{d}{d\theta}(\phi),$$

where ϕ stands for any function of x and y. On substituting $\frac{dV}{dx}$ instead of ϕ, this equation becomes

$$\frac{d}{dx}\left(\frac{dV}{dx}\right) = \cos\theta \frac{d}{dr}\left[\cos\theta \frac{dV}{dr} - \frac{\sin\theta}{r}\frac{dV}{d\theta}\right]$$

$$- \frac{\sin\theta}{r}\frac{d}{d\theta}\left[\cos\theta \frac{dV}{dr} - \frac{\sin\theta}{r}\frac{dV}{d\theta}\right]$$

$$= \cos^2\theta \frac{d^2V}{dr^2} - \frac{\cos\theta \sin\theta}{r}\frac{d^2V}{drd\theta} + \frac{\cos\theta \sin\theta}{r^2}\frac{dV}{d\theta}$$

$$- \frac{\sin\theta}{r}\left[\cos\theta \frac{d^2V}{drd\theta} - \sin\theta \frac{dV}{dr}\right]$$

$$+ \frac{\sin\theta}{r}\left[\frac{\cos\theta}{r}\frac{dV}{d\theta} + \frac{\sin\theta}{r}\frac{d^2V}{d\theta^2}\right],$$

Transformation of $\dfrac{d^2V}{dx^2}$ and $\dfrac{d^2V}{dy^2}$. 405

or $\dfrac{d^2V}{dx^2} = \cos^2\theta \dfrac{d^2V}{dr^2} + \dfrac{2\sin\theta\cos\theta}{r}\left[\dfrac{1}{r}\dfrac{dV}{d\theta} - \dfrac{d^2V}{dr\,d\theta}\right]$

$\qquad + \dfrac{\sin^2\theta}{r}\dfrac{dV}{dr} + \dfrac{\sin^2\theta}{r^2}\dfrac{d^2V}{d\theta^2}.$

In like manner we get

$\dfrac{d^2V}{dy^2} = \sin^2\theta \dfrac{d^2V}{dr^2} - \dfrac{2\sin\theta\cos\theta}{r}\left[\dfrac{1}{r}\dfrac{dV}{d\theta} - \dfrac{d^2V}{dr\,d\theta}\right]$

$\qquad + \dfrac{\cos^2\theta}{r}\dfrac{dV}{dr} + \dfrac{\cos^2\theta}{r^2}\dfrac{d^2V}{d\theta^2}.$

This result can be also readily deduced from the preceding by substituting in it $\dfrac{\pi}{2} - \theta$ for θ.

If these equations be added we have

$$\dfrac{d^2V}{dx^2} + \dfrac{d^2V}{dy^2} = \dfrac{d^2V}{dr^2} + \dfrac{1}{r}\dfrac{dV}{dr} + \dfrac{1}{r^2}\dfrac{d^2V}{d\theta^2}. \qquad (18)$$

323. **Transformation of** $\dfrac{d^2V}{dx^2} + \dfrac{d^2V}{dy^2} + \dfrac{d^2V}{dz^2}$ **to polar Coordinates.**

Let the polar transformation be represented by the equations

$\qquad x = r\sin\phi\cos\theta, \quad y = r\sin\phi\sin\theta, \quad z = r\cos\phi;$

also, assume $\rho = r\sin\phi$, and we have

$\qquad x = \rho\cos\theta, \quad y = \rho\sin\theta;$

hence, by (18), $\dfrac{d^2V}{dx^2} + \dfrac{d^2V}{dy^2} = \dfrac{d^2V}{d\rho^2} + \dfrac{1}{\rho}\dfrac{dV}{d\rho} + \dfrac{1}{\rho^2}\dfrac{d^2V}{d\theta^2}.$

Again, from the equations

$$\rho = r \sin \phi, \quad z = r \cos \phi,$$

we have in like manner

$$\frac{d^2V}{d\rho^2} + \frac{d^2V}{dz^2} = \frac{d^2V}{dr^2} + \frac{1}{r}\frac{dV}{dr} + \frac{1}{r^2}\frac{d^2V}{d\phi^2}.$$

Accordingly

$$\frac{d^2V}{dx^2} + \frac{d^2V}{dy^2} + \frac{d^2V}{dz^2} = \frac{d^2V}{dr^2} + \frac{1}{\rho}\frac{dV}{d\rho} + \frac{1}{\rho^2}\frac{d^2V}{d\theta^2} + \frac{1}{\rho}\frac{dV}{dr} + \frac{1}{r^2}\frac{d^2V}{d\phi^2}.$$

But by (17) we have

$$\frac{dV}{d\rho} = \sin \phi \frac{dV}{dr} + \frac{\cos \phi}{r}\frac{dV}{d\phi};$$

therefore

$$\frac{1}{\rho}\frac{dV}{d\rho} = \frac{1}{r}\frac{dV}{dr} + \frac{\cot \phi}{r^2}\frac{dV}{d\phi}.$$

Hence we get finally

$$\frac{d^2V}{dx^2} + \frac{d^2V}{dy^2} + \frac{d^2V}{dz^2} = \frac{d^2V}{dr^2} + \frac{1}{r^2 \sin^2 \phi}\frac{d^2V}{d\theta^2}$$

$$+ \frac{1}{\rho^2}\frac{d^2V}{d\phi^2} + \frac{2}{r}\frac{dV}{dr} + \frac{\cot \phi}{r^2}\frac{dV}{d\phi}. \quad (19)$$

324. Remarks on Partial Differentials.—As already stated in Art. 113, the student must be careful to attach the correct meaning to the partial differential coefficients in each case.

Thus in finding $\frac{dx}{dr}$ in (10) we regard x as a function of r and θ, and differentiate on the supposition that θ *is constant;* in like manner the value of $\frac{dr}{dx}$ in (14) is found on the supposition that y *is constant.*

The beginner, accordingly, must not fall into the confusion of supposing that in this case we have $\dfrac{dr}{dx} \times \dfrac{dx}{dr} = 1$. This caution is necessary, as even advanced students, from not paying proper attention to the meanings of partial derived functions, sometimes fall into the error referred to.

325. Geometrical Illustration.—The following geometrical method of determining the proper values of $\dfrac{dr}{dx}$ and $\dfrac{dx}{dr}$ under the preceding hypotheses may assist the beginner towards forming correct ideas on this important subject.

Let P be the point whose coordinates are x and y; then $OM = x$, $PM = y$, $OP = r$, $POX = \theta$. Now, in finding $\dfrac{dx}{dr}$, regarding θ as constant, we take on the radius vector OP produced a portion $PQ = \Delta r$, and draw QN perpendicular to OX; then Δx, the corresponding increment in x, is represented by MN or PL;

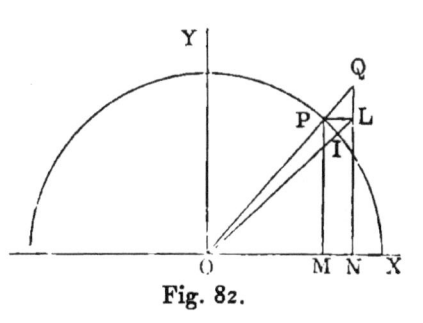

Fig. 82.

therefore $\quad \dfrac{\Delta x}{\Delta r} = \dfrac{PL}{PQ} = \cos \theta, \quad \text{or} \quad \dfrac{dx}{dr} = \cos \theta.$

Again, to find $\dfrac{dr}{dx}$ on the supposition that y is constant: let MN be Δx, the increment in x, and draw the parallelogram $PLMN$, and join OL, meeting in I a circle described with radius r and centre O; then LI represents the corresponding increment in r, and we have

$$\dfrac{dr}{dx} = \text{limit of } \dfrac{\Delta r}{\Delta x} = \text{limit of } \dfrac{IL}{PL} = \cos \theta,$$

so that in this case the values of $\dfrac{dr}{dx}$ and $\dfrac{dx}{dr}$ are each equal to $\cos \theta$ or $\dfrac{x}{r}$, as before.

The values of $\dfrac{dr}{d\theta}$, $\dfrac{d\theta}{dx}$, &c., can be also readily represented geometrically in a similar manner.

326. Linear Transformations.—If we are given

$$x = aX + bY + cZ, \ y = a'X + b'Y + c'Z, \ z = a''X + b''Y + c''Z, \quad (20)$$

then any function V, of x, y and z, is transformed into a function of X, Y, Z; and, as in Ex. 2, Art. 98, we have

$$\frac{dV}{dX} = a\frac{dV}{dx} + a'\frac{dV}{dy} + a''\frac{dV}{dz},$$

$$\frac{dV}{dY} = b\frac{dV}{dx} + b'\frac{dV}{dy} + b''\frac{dV}{dz},$$

$$\frac{dV}{dZ} = c\frac{dV}{dx} + c'\frac{dV}{dy} + c''\frac{dV}{dz}.$$

Again, proceeding to second differentiation, we get

$$\frac{d^2V}{dX^2} = a\frac{d}{dx}\left(a\frac{dV}{dx} + a'\frac{dV}{dy} + a''\frac{dV}{dz}\right) + a'\frac{d}{dy}\left(a\frac{dV}{dx} + a'\frac{dV}{dy} + a''\frac{dV}{dz}\right)$$

$$+ a''\frac{d}{dz}\left(a\frac{dV}{dx} + a'\frac{dV}{dy} + a''\frac{dV}{dz}\right)$$

$$= a^2\frac{d^2V}{dx^2} + 2aa'\frac{d^2V}{dx\,dy} + 2aa''\frac{d^2V}{dx\,dz} + 2a'a''\frac{d^2V}{dz^2}$$

$$+ a'^2\frac{d^2V}{dy^2} + a''^2\frac{d^2V}{dz^2}.$$

Similarly we have

$$\frac{d^2V}{dY^2} = b^2\frac{d^2V}{dx^2} + b'^2\frac{d^2V}{dy^2} + b''^2\frac{d^2V}{dz^2} + 2bb'\frac{d^2V}{dx\,dy}$$

$$+ 2bb''\frac{d^2V}{dx\,dz} + 2b'b''\frac{d^2V}{dy\,dz};$$

$$\frac{d^2V}{dZ^2} = c^2\frac{d^2V}{dx^2} + c'^2\frac{d^2V}{dy^2} + c''^2\frac{d^2V}{dz^2} + 2cc'\frac{d^2V}{dx\,dy}$$

$$+ 2cc''\frac{d^2V}{dx\,dz} + 2c'c''\frac{d^2V}{dy\,dz}.$$

327. Orthogonal Transformations.—If the transformation be such that

$$x^2 + y^2 + z^2 = X^2 + Y^2 + Z^2,$$

we have

$$a^2 + a'^2 + a''^2 = 1, \quad b^2 + b'^2 + b''^2 = 1, \quad c^2 + c'^2 + c''^2 = 1. \quad (21)$$

$$ab + a'b' + a''b'' = 0, \quad ac + a'c' + a''c'' = 0, \quad bc + b'c' + b''c'' = 0. \quad (22)$$

Again, multiplying the first of equations (20) by a, the second by a', and the third by a'', we get on addition, by aid of (21) and (22),

$$X = ax + a'y + a''z.$$

In like manner, if the equations (20) be respectively multiplied by b, b', b'', we get

$$Y = bx + b'y + b''z;$$

similarly

$$Z = cx + c'y + c''z.$$

If these equations be squared and added, we obtain

$$a^2 + b^2 + c^2 = 1, \quad a'^2 + b'^2 + c'^2 = 1, \quad a''^2 + b''^2 + c''^2 = 1. \quad (23)$$

$$aa' + bb' + cc' = 0, \quad aa'' + bb'' + cc'' = 0, \quad a'a'' + b'b'' + c'c'' = 0. \quad (24)$$

Hence in this case, if the equations of the last Article be added, we shall have

$$\frac{d^2V}{dx^2} + \frac{d^2V}{dy^2} + \frac{d^2V}{dz^2} = \frac{d^2V}{dX^2} + \frac{d^2V}{dY^2} + \frac{d^2V}{dZ^2}. \quad (25)$$

The transformations in this and the preceding Article are necessary when the axes of co-ordinates are changed in Analytic Geometry of three dimensions; and equation (25) shows that, in transforming from one rectangular system to another, the function $\dfrac{d^2V}{dx^2} + \dfrac{d^2V}{dy^2} + \dfrac{d^2V}{dz^2}$ is unaltered.

328. General Case of Transformation for Two Independent Variables.—Suppose that we are given the equations

$$x = \phi(r, \theta), \quad y = \psi(r, \theta), \qquad (26)$$

then any function V of x and y may be regarded as a function of r and θ, and we have, from (9),

$$\frac{dV}{d\theta} = \frac{dV}{dx}\frac{dx}{d\theta} + \frac{dV}{dy}\frac{dy}{d\theta},$$

$$\frac{dV}{dr} = \frac{dV}{dx}\frac{dx}{dr} + \frac{dV}{dy}\frac{dy}{dr},$$

where the values of $\dfrac{dx}{d\theta}, \dfrac{dy}{d\theta}, \dfrac{dx}{dr}, \dfrac{dy}{dr}$ can be determined from equations (26).

Whenever these equations can be solved for r and θ, separately, we can determine, by direct differentiation, the values of $\dfrac{dr}{dx}, \dfrac{dr}{dy}, \dfrac{d\theta}{dx}, \dfrac{d\theta}{dy}$, and hence by substituting in (13) we can obtain the values of $\dfrac{dV}{dx}$ and $\dfrac{dV}{dy}$.

When, however, this process is impracticable we can obtain the values of $\dfrac{dr}{dx}, \dfrac{dr}{dy}$, &c., by solving for $\dfrac{dV}{dx}$ and $\dfrac{dV}{dy}$ from the preceding equations.

Thus, we obtain

$$\frac{dV}{dx} = \frac{\dfrac{dV}{d\theta}\dfrac{dy}{dr} - \dfrac{dV}{dr}\dfrac{dy}{d\theta}}{\dfrac{dx}{d\theta}\dfrac{dy}{dr} - \dfrac{dx}{dr}\dfrac{dy}{d\theta}}; \qquad (27)$$

$$\frac{dV}{dy} = \frac{\dfrac{dV}{d\theta}\dfrac{dx}{dr} - \dfrac{dV}{dr}\dfrac{dx}{d\theta}}{\dfrac{dx}{dr}\dfrac{dy}{d\theta} - \dfrac{dx}{d\theta}\dfrac{dy}{dr}}. \qquad (28)$$

The values of $\dfrac{d^2V}{dx^2}$, $\dfrac{d^2V}{dy^2}$, &c., can be deduced from these: but the general formulæ are too complicated to be of much interest or utility.

329. **Concomitant Functions.**—We add one or two results in connexion with linear transformations, commencing with the case of two variables. We suppose x and y changed into $aX + bY$ and $a'X + b'Y$, respectively, so that any function $\phi(x, y)$ is transformed into a function of X and Y; let the latter be denoted by $\phi_1(X, Y)$, and we have

$$\phi(x, y) = \phi_1(X, Y).$$

Again, let x' and y' be transformed by the same substitutions, *i.e.*,

$$x' = aX' + bY', \quad y' = a'X' + b'Y';$$

then since $x + kx' = a(X + kX') + b(Y + kY'),$

and $y + ky' = a'(X + kX') + b'(Y + kY'),$

it is evident that

$$\phi(x + kx', y + ky') = \phi_1(X + kX', Y + kY').$$

Hence, expanding by the theorem of Art. 127, and equating like powers of k, we get

$$x'\frac{d\phi}{dx} + y'\frac{d\phi}{dy} = X'\frac{d\phi_1}{dX} + Y'\frac{d\phi_1}{dY}, \qquad (29)$$

$$x'^2\frac{d^2\phi}{dx^2} + 2x'y'\frac{d^2\phi}{dx\,dy} + y'^2\frac{d^2\phi}{dy^2} = X'^2\frac{d^2\phi_1}{dX^2} + 2X'Y'\frac{d^2\phi_1}{dX\,dY} + Y'^2\frac{d^2\phi_1}{dY^2},$$

&c. &c. (30)

Accordingly, if u represent any function of x and y, the expressions denoted by

$$\left(x'\frac{d}{dx} + y'\frac{d}{dy}\right)u, \quad \left(x'\frac{d}{dx} + y'\frac{d}{dy}\right)^2 u, \ \&c.,$$

are unaltered by linear transformation.

Similar results obviously hold for linear transformations whatever be the number of variables (Salmon's *Higher Algebra*, Art. 125).

Functions, such as the above, whose relations to a quantic are unaltered by linear transformation, have been called *concomitants* by Professor Sylvester.

330. **Transformation of Coordinate Axes.**—When applied to transformation from one system of coordinate axes to another, the preceding leads to some important results, by applying Boole's method* (Salmon's *Conics*, Art. 159).

For in the case of two dimensions when the origin is unaltered we have

$$x'^2 + 2x'y' \cos \omega + y'^2 = X'^2 + 2X'Y' \cos \Omega + Y'^2, \quad (31)$$

where ω and Ω denote the angle between the original axes and that between the transformed axes, respectively.

Multiply (31) by λ, and add to (30): then denoting $\phi(x, y)$ by u, and $\phi_1(X, Y)$ by U, we get

$$x'^2\left(\frac{d^2u}{dx^2} + \lambda\right) + 2x'y'\left(\frac{d^2u}{dxdy} + \lambda \cos \omega\right) + y'^2\left(\frac{d^2u}{dy^2} + \lambda\right)$$

$$= X'^2\left(\frac{d^2U}{dX^2} + \lambda\right) + 2X'Y'\left(\frac{d^2U}{dXdY} + \lambda \cos \Omega\right) + Y'^2\left(\frac{d^2U}{dY^2} + \lambda\right).$$

Now, suppose λ assumed so as to make the first side of this equation a perfect square, it is obvious that the other side will be a perfect square also. The former condition gives

$$\left(\frac{d^2u}{dx^2} + \lambda\right)\left(\frac{d^2u}{dy^2} + \lambda\right) = \left(\frac{d^2u}{dxdy} + \lambda \cos \omega\right)^2,$$

---

* I am indebted to Prof. Burnside for the suggestion that the equations of this Article are immediately obtained by Boole's method.

or
$$\lambda^2 \sin^2 \omega + \lambda \left(\frac{d^2u}{dx^2} + \frac{d^2u}{dy^2} - 2 \frac{d^2u}{dx\,dy} \cos \omega \right)$$
$$+ \frac{d^2u}{dx^2} \frac{d^2u}{dy^2} - \left(\frac{d^2u}{dx\,dy} \right)^2 = 0.$$

Accordingly, we must have at the same time

$$\lambda^2 \sin^2 \Omega + \lambda \left(\frac{d^2U}{dX^2} + \frac{d^2U}{dY^2} - 2 \frac{d^2U}{dX\,dY} \cos \Omega \right)$$
$$+ \frac{d^2U}{dX^2} \frac{d^2U}{dY^2} - \left(\frac{d^2U}{dX\,dY} \right)^2 = 0.$$

Hence, comparing coefficients, we get

$$\frac{\dfrac{d^2u}{dx^2} \dfrac{d^2u}{dy^2} - \left(\dfrac{d^2u}{dx\,dy} \right)^2}{\sin^2 \omega} = \frac{\dfrac{d^2U}{dX^2} \dfrac{d^2U}{dY^2} - \left(\dfrac{d^2U}{dX\,dY} \right)^2}{\sin^2 \Omega}, \quad (32)$$

and

$$\frac{\dfrac{d^2u}{dx^2} + \dfrac{d^2u}{dy^2} - 2 \dfrac{d^2u}{dx\,dy} \cos \omega}{\sin^2 \omega} = \frac{\dfrac{d^2U}{dX^2} + \dfrac{d^2U}{dY^2} - 2 \dfrac{d^2U}{dX\,dY} \cos \Omega}{\sin^2 \Omega}. \quad (33)$$

Consequently, if u be any function of the coordinates of a point, the expressions

$$\frac{\dfrac{d^2u}{dx^2} \dfrac{d^2u}{dy^2} - \left(\dfrac{d^2u}{dx\,dy} \right)^2}{\sin^2 \omega} \quad \text{and} \quad \frac{\dfrac{d^2u}{dx^2} + \dfrac{d^2u}{dy^2} - 2 \dfrac{d^2u}{dx\,dy} \cos \omega}{\sin^2 \omega}$$

are unaltered when the axes of coordinates are changed in any manner, the origin remaining the same.

In the particular case of rectangular axes, it follows that

$$\frac{d^2u}{dx^2} + \frac{d^2u}{dy^2} \quad \text{and} \quad \frac{d^2u}{dx^2} \frac{d^2u}{dy^2} - \left(\frac{d^2u}{dx\,dy} \right)^2$$

preserve the same values when the axes are turned round through any angle.

331. Application to Orthogonal Transformation.
—It is easy to extend the preceding results to three or more variables when the transformations are orthogonal (Art. 327).

Thus, in the case of three variables we have

$$x'^2 + y'^2 + z'^2 = X'^2 + Y'^2 + Z'^2.$$

Multiplying this by λ and adding the result to the equation that corresponds to (30), it follows that the expression

$$x'^2\left(\frac{d^2u}{dx^2} + \lambda\right) + y'^2\left(\frac{d^2u}{dy^2} + \lambda\right) + z'^2\left(\frac{d^2u}{dz^2} + \lambda\right) + 2y'z'\frac{d^2u}{dy\,dz}$$
$$+ 2z'x'\frac{d^2u}{dz\,dx} + 2x'y'\frac{d^2u}{dx\,dy}$$

is unaltered by orthogonal transformation.

Next, suppose that λ is such that the quadratic function in x', y' and z' is the product of two linear factors; then, by Art. 107, we have

$$\begin{vmatrix} \dfrac{d^2u}{dx^2} + \lambda, & \dfrac{d^2u}{dx\,dy}, & \dfrac{d^2u}{dx\,dz} \\ \dfrac{d^2u}{dx\,dy}, & \dfrac{d^2u}{dy^2} + \lambda, & \dfrac{d^2u}{dy\,dz} \\ \dfrac{d^2u}{dx\,dz}, & \dfrac{d^2u}{dy\,dz}, & \dfrac{d^2u}{dz^2} + \lambda \end{vmatrix} = 0. \qquad (34)$$

But, as the transformed expression must also be the product of two linear factors, we have

$$\begin{vmatrix} \dfrac{d^2u}{dx^2} + \lambda, & \dfrac{d^2u}{dx\,dy}, & \dfrac{d^2u}{dx\,dz} \\ \dfrac{d^2u}{dy\,dx}, & \dfrac{d^2u}{dy^2} + \lambda, & \dfrac{d^2u}{dy\,dz} \\ \dfrac{d^2u}{dx\,dz}, & \dfrac{d^2u}{dy\,dz}, & \dfrac{d^2u}{dz^2} + \lambda \end{vmatrix} = \begin{vmatrix} \dfrac{d^2U}{dX^2} + \lambda, & \dfrac{d^2U}{dX\,dY}, & \dfrac{d^2U}{dX\,dZ} \\ \dfrac{d^2U}{dX\,dY}, & \dfrac{d^2U}{dY^2} + \lambda, & \dfrac{d^2U}{dY\,dZ} \\ \dfrac{d^2U}{dX\,dZ}, & \dfrac{d^2U}{dY\,dZ}, & \dfrac{d^2U}{dZ^2} + \lambda \end{vmatrix}. \qquad (35)$$

Equating the coefficients of like powers of λ, we see that the expressions

$$\frac{d^2u}{dx^2} + \frac{d^2u}{dy^2} + \frac{d^2u}{dz^2},$$

$$\frac{d^2u}{dx^2}\frac{d^2u}{dy^2} - \left(\frac{d^2u}{dx\,dy}\right)^2 + \frac{d^2u}{dx^2}\frac{d^2u}{dz^2} - \left(\frac{d^2u}{dx\,dz}\right)^2 + \frac{d^2u}{dy^2}\frac{d^2u}{dz^2} - \left(\frac{d^2u}{dy\,dz}\right)^2,$$

and

$$\begin{vmatrix} \dfrac{d^2u}{dx^2}, & \dfrac{d^2u}{dx\,dy}, & \dfrac{d^2u}{dx\,dz} \\ \dfrac{d^2u}{dx\,dy}, & \dfrac{d^2u}{dy^2}, & \dfrac{d^2u}{dy\,dz} \\ \dfrac{d^2u}{dx\,dz}, & \dfrac{d^2u}{dy\,dz}, & \dfrac{d^2u}{dz^2} \end{vmatrix}$$

are unaltered by orthogonal transformation.

The first of these results has been already arrived at by direct substitution (Art. 327).

JACOBIANS.

332. The results in the preceding Article are particular cases of a class of general theorems in determinants, which were first developed by Jacobi (Crelle's *Journal*, 1841).

Thus, if u, v, w be functions of x, y, z, the determinant

$$J = \begin{vmatrix} \dfrac{du}{dx}, & \dfrac{du}{dy}, & \dfrac{du}{dz} \\ \dfrac{dv}{dx}, & \dfrac{dv}{dy}, & \dfrac{dv}{dz} \\ \dfrac{dw}{dx}, & \dfrac{dw}{dy}, & \dfrac{dw}{dz} \end{vmatrix}, \qquad (36)$$

was styled by Jacobi a functional determinant   Such a

determinant is now usually represented by the notation

$$\frac{d(u, v, w)}{d(x, y, z)},$$

and is called the Jacobian of the system u, v, w with respect to the variables x, y, z.

In the particular case where u, v, w are the partial differential coefficients of the same function of the variables x, y, z, their Jacobian becomes of the form (35), and is called the Hessian of the primitive function. Thus the determinant in (35) is called the Hessian of u, after Hesse, who first introduced such functions into analysis, and pointed out their importance in the general theory of curves and surfaces.

More generally, if $y_1, y_2, y_3 \ldots y_n$ be functions of $x_1, x_2, x_3, \ldots x_n$, the determinant

$$\begin{vmatrix} \dfrac{dy_1}{dx_1}, & \dfrac{dy_1}{dx_2}, & \cdots & \dfrac{dy_1}{dx_n} \\ \dfrac{dy_2}{dx_1}, & \dfrac{dy_2}{dx_2}, & \cdots & \dfrac{dy_2}{dx_n} \\ \cdot & \cdot & \cdot & \cdot \\ \dfrac{dy_n}{dx_1}, & \dfrac{dy_n}{dx_2}, & \cdots & \dfrac{dy_n}{dx_n} \end{vmatrix}$$

is called the Jacobian of the system of functions $y_1, y_2, \ldots y_n$ with respect to the variables $x_1, x_2, \ldots x_n$; and is denoted by

$$\frac{d(y_1, y_2, \ldots y_n)}{d(x_1, x_2, \ldots x_n)}. \qquad (37)$$

Again, if $y_1, y_2, \ldots y_n$ be differential coefficients of the same function, the Jacobian is styled, as above, the Hessian of the function. A Jacobian is frequently represented by the notation

$$J(y_1, y_2, \ldots y_n),$$

the variables $x_1, x_2, \ldots x_n$ being understood.

If the equations for $y_1, y_2, \ldots y_n$ be of the following form:

$$y_1 = f_1(x_1),$$
$$y_2 = f_2(x_1, x_2),$$
$$y_3 = f_3(x_1, x_2, x_3),$$
$$\cdot \quad \cdot \quad \cdot \quad \cdot$$
$$y_n = f_n(x_1, x_2, \ldots x_n),$$

it is obvious that their Jacobian reduces to its leading term, viz.,

$$J = \frac{dy_1}{dx_1} \cdot \frac{dy_2}{dx_2} \ldots \frac{dy_n}{dx_n}. \qquad (38)$$

This is a case of a more general theorem, which will be given subsequently (Art. 336).

EXAMPLES.

1. Find the Jacobian of $y_1, y_2, \ldots y_n$, being given

$$y_1 = 1 - x_1, \quad y_2 = x_1(1 - x_2), \quad y_3 = x_1 x_2(1 - x_3) \ldots$$

$$y_n = x_1 x_2 \ldots x_{n-1}(1 - x_n). \quad Ans. \; J = (-1)^n x_1^{n-1} x_2^{n-2} \ldots x_{n-1}.$$

2. Find the Jacobian of $x_1, x_2, \ldots x_n$ with respect to $\theta_1, \theta_2, \ldots \theta_n$, being given

$$x_1 = \cos\theta_1, \quad x_2 = \sin\theta_1 \cos\theta_2, \quad x_3 = \sin\theta_1 \sin\theta_2 \cos\theta_3, \ldots$$

$$x_n = \sin\theta_1 \sin\theta_2 \sin\theta_3 \ldots \sin\theta_{n-1} \cos\theta_n.$$

$$Ans. \; \frac{d(x_1, x_2, \ldots x_n)}{d(\theta_1, \theta_2, \ldots \theta_n)} = (-1)^n \sin^n\theta_1 \cdot \sin^{n-1}\theta_2 \ldots \sin\theta_n.$$

333. Case of the Functions not being Independent.—If the system $y_1, y_2, \ldots y_n$ be connected by a relation, it is easily seen that their Jacobian is always zero.

For, suppose the equation of connexion represented by

$$F(y_1, y_2, \ldots y_n) = 0;$$

2 E

then, differentiating with respect to the variables $x_1, x_2 \ldots x_n$, we get the following system of equations:—

$$\frac{dF}{dy_1}\frac{dy_1}{dx_1} + \frac{dF}{dy_2}\frac{dy_2}{dx_1} + \ldots + \frac{dF}{dy_n}\frac{dy_n}{dx_1} = 0,$$

$$\frac{dF}{dy_1}\frac{dy_1}{dx_2} + \frac{dF}{dy_2}\frac{dy_2}{dx_2} + \ldots + \frac{dF}{dy_n}\frac{dy_n}{dx_2} = 0,$$

$$\cdots \cdots \cdots \cdots \cdots \cdots \cdots$$

$$\frac{dF}{dy_1}\frac{dy_1}{dx_n} + \frac{dF}{dy_2}\frac{dy_2}{dx_n} + \ldots + \frac{dF}{dy_n}\frac{dy_n}{dx_n} = 0;$$

whence, eliminating $\dfrac{dF}{dy_1}$, $\dfrac{dF}{dy_2}$, $\ldots$ $\dfrac{dF}{dy_n}$, we get

$$\frac{d(y_1, y_2, \ldots y_n)}{d(x_1, x_2, \ldots x_n)} = 0. \qquad (39)$$

The converse of this result will be established in Art. 337; and we infer that whenever the Jacobian of a system of functions vanishes identically, the functions are not independent. This is an extension of the result arrived at in Art. 314.

334. Case of Functions of Functions.—If we suppose u_1, u_2, u_3 to be functions of y_1, y_2, y_3, where y_1, y_2, y_3 are given functions of x_1, x_2, x_3; then we have

$$\frac{du_1}{dx_1} = \frac{du_1}{dy_1}\frac{dy_1}{dx_1} + \frac{du_1}{dy_2}\frac{dy_2}{dx_1} + \frac{du_1}{dy_3}\frac{dy_3}{dx_1},$$

$$\frac{du_1}{dx_2} = \frac{du_1}{dy_1}\frac{dy_1}{dx_2} + \frac{du_1}{dy_2}\frac{dy_2}{dx_2} + \frac{du_1}{dy_3}\frac{dy_3}{dx_2},$$

$$\frac{du_1}{dx_3} = \frac{du_1}{dy_1}\frac{dy_1}{dx_3} + \frac{du_1}{dy_2}\frac{dy_2}{dx_3} + \frac{du_1}{dy_3}\frac{dy_3}{dx_3},$$

$$\cdots \cdots \cdots \cdots \cdots \cdots \text{&c.}$$

Hence, by the ordinary rule for the multiplication of determinants, we get

$$\begin{vmatrix} \dfrac{du_1}{dx_1}, & \dfrac{du_1}{dx_2}, & \dfrac{du_1}{dx_3} \\ \dfrac{du_2}{dx_1}, & \dfrac{du_2}{dx_2}, & \dfrac{du_2}{dx_3} \\ \dfrac{du_3}{dx_1}, & \dfrac{du_3}{dx_2}, & \dfrac{du_3}{dx_3} \end{vmatrix} = \begin{vmatrix} \dfrac{du_1}{dy_1}, & \dfrac{du_1}{dy_2}, & \dfrac{du_1}{dy_3} \\ \dfrac{du_2}{dy_1}, & \dfrac{du_2}{dy_2}, & \dfrac{du_2}{dy_3} \\ \dfrac{du_3}{dy_1}, & \dfrac{du_3}{dy_2}, & \dfrac{du_3}{dy_3} \end{vmatrix} \cdot \begin{vmatrix} \dfrac{dy_1}{dx_1}, & \dfrac{dy_1}{dx_2}, & \dfrac{dy_1}{dx_3} \\ \dfrac{dy_2}{dx_1}, & \dfrac{dy_2}{dx_2}, & \dfrac{dy_2}{dx_3} \\ \dfrac{dy_3}{dx_1}, & \dfrac{dy_3}{dx_2}, & \dfrac{dy_3}{dx_3} \end{vmatrix}, \quad (40)$$

or

$$\frac{d(u_1, u_2, u_3)}{d(x_1, x_2, x_3)} = \frac{d(u_1, u_2, u_3)}{d(y_1, y_2, y_3)} \cdot \frac{d(y_1, y_2, y_3)}{d(x_1, x_2, x_3)}.$$

It follows as a particular case, that

$$\frac{d(y_1, y_2, y_3)}{d(x_1, x_2, x_3)} \times \frac{d(x_1, x_2, x_3)}{d(y_1, y_2, y_3)} = 1. \quad (41)$$

These results are readily generalized, and it can be shown by the method given above, that

$$\frac{d(u_1, u_2, \ldots u_n)}{d(x_1, x_2, \ldots x_n)} = \frac{d(u_1, u_2, \ldots u_n)}{d(y_1, y_2, \ldots y_n)} \cdot \frac{d(y_1, y_2, \ldots y_n)}{d(x_1, x_2, \ldots x_n)}. \quad (42)$$

This is a generalization of the elementary theorem (Art. 19),

$$\frac{du}{dx} = \frac{du}{dy} \frac{dy}{dx}.$$

Again,

$$\frac{d(y_1, y_2, \ldots y_n)}{d(x_1, x_2, \ldots x_n)} \frac{d(x_1, x_2, \ldots x_n)}{d(y_1, y_2, \ldots y_n)} = 1. \quad (43)$$

This may be regarded as a generalization of the result

$$\frac{dx}{dy} = \frac{1}{\dfrac{dy}{dx}}.$$

335. Jacobian of Implicit Functions.—Next, if u, v, w, instead of being given explicitly in terms of x, y, z, be connected with them by equations such as

$$F_1(x, y, z, u, v, w) = 0, \; F_2(x, y, z, u, v, w) = 0, \; F_3(x, y, z, u, v, w) = 0,$$

then u, v, w may be regarded as implicit functions of x, y, z.

In this case we have, by differentiation,

$$\frac{dF_1}{dx} + \frac{dF_1}{du}\frac{du}{dx} + \frac{dF_1}{dv}\frac{dv}{dx} + \frac{dF_1}{dw}\frac{dw}{dx} = 0,$$

$$\frac{dF_1}{dy} + \frac{dF_1}{du}\frac{du}{dy} + \frac{dF_1}{dv}\frac{dv}{dy} + \frac{dF_1}{dw}\frac{dw}{dy} = 0,$$

$$\cdot \quad \cdot \quad \cdot \quad \cdot \quad \cdot \quad \cdot \quad \cdot$$

$$\frac{dF_2}{dx} + \frac{dF_2}{du}\frac{du}{dx} + \frac{dF_2}{dv}\frac{dv}{dx} + \frac{dF_2}{dw}\frac{dw}{dx} = 0,$$

$$\cdot \quad \cdot \quad \cdot \quad \cdot \quad \cdot \quad \cdot \quad \cdot$$

Hence we observe, from the ordinary rule for multiplication of determinants, that

$$\begin{vmatrix} \frac{dF_1}{du}, & \frac{dF_1}{dv}, & \frac{dF_1}{dw} \\ \frac{dF_2}{du}, & \frac{dF_2}{dv}, & \frac{dF_2}{dw} \\ \frac{dF_3}{du}, & \frac{dF_3}{dv}, & \frac{dF_3}{dw} \end{vmatrix} \cdot \begin{vmatrix} \frac{du}{dx}, & \frac{dv}{dx}, & \frac{dw}{dx} \\ \frac{du}{dy}, & \frac{dv}{dy}, & \frac{dw}{dy} \\ \frac{du}{dz}, & \frac{dv}{dz}, & \frac{dw}{dz} \end{vmatrix} = - \begin{vmatrix} \frac{dF_1}{dx}, & \frac{dF_1}{dy}, & \frac{dF_1}{dz} \\ \frac{dF_2}{dx}, & \frac{dF_2}{dy}, & \frac{dF_2}{dz} \\ \frac{dF_3}{dx}, & \frac{dF_3}{dy}, & \frac{dF_3}{dz} \end{vmatrix}. \quad (44)$$

This result may be writtten

$$\frac{d(F_1, F_2, F_3)}{d(u, v, w)} \cdot \frac{d(u, v, w)}{d(x, y, z)} = - \frac{d(F_1, F_2, F_3)}{d(x, y, z)}.$$

The preceding can be generalized, and it can be readily shown by a like demonstration that if y_1, y_2, y_3, $\ldots$ y_n

are connected with $x_1, x_2, x_3 \ldots x_n$ by n equations of the form

$$F_1(x_1, x_2 \ldots x_n, y_1, y_2 \ldots y_n) = 0,$$
$$F_2(x_1, x_2 \ldots x_n, y_1, y_2 \ldots y_n) = 0,$$
$$\cdot \quad \cdot \quad \cdot \quad \cdot \quad \cdot \quad \cdot \quad \cdot$$
$$F_n(x_1, x_2 \ldots x_n, y_1, y_2 \ldots y_n) = 0,$$

we shall have the following relation between the Jacobians:

$$\frac{d(F_1, F_2, \ldots F_n)}{d(y_1, y_2, \ldots y_n)} \cdot \frac{d(y_1, y_2, \ldots y_n)}{d(x, x_2, \ldots x_n)} = (-1)^n \frac{d(F_1, F_2, \ldots F_n)}{d(x_1, x_2, \ldots x_n)}.$$

Accordingly

$$\frac{d(y_1, y_2, \ldots y_n)}{d(x_1, x_2, \ldots x_n)} = (-1)^n \frac{\dfrac{d(F_1, F_2, \ldots F_n)}{d(x_1, x_2, \ldots x_n)}}{\dfrac{d(F_1, F_2, \ldots F_n)}{d(y_1, y_2, \ldots y_n)}}. \quad (45)$$

336. Again, if we suppose that the equations connecting the variables are transformed, by elimination or otherwise, to the following shape—

$$\phi_1(x_1, x_2, \ldots x_n, y_1) = 0,$$
$$\phi_2(x_2, x_3, \ldots x_n, y_1, y_2) = 0,$$
$$\phi_3(x_3, x_4, \ldots x_n, y_1, y_2, y_3) = 0,$$
$$\cdot \quad \cdot \quad \cdot \quad \cdot \quad \cdot \quad \cdot$$
$$\phi_n(x_n, y_1, y_2, \ldots y_n) = 0,$$

then the Jacobian determinant

$$\frac{d(\phi_1, \phi_2, \ldots \phi_n)}{d(y_1, y_2, \ldots y_n)},$$

as in Art. 332, reduces to its leading term

$$\frac{d\phi_1}{dy_1} \frac{d\phi_2}{dy_2} \frac{d\phi_3}{dy_3} \cdots \frac{d\phi_n}{dy_n}.$$

In like manner

$$\frac{d(\phi_1, \phi_2, \ldots \phi_n)}{d(x_1, x_2, \ldots x_n)}$$

reduces to

$$\frac{d\phi_1}{dx_1} \frac{d\phi_2}{dx_2} \ldots \frac{d\phi_n}{dx_n}.$$

Accordingly, in this case, the Jacobian

$$\frac{d(y_1, y_2, \ldots y_n)}{d(x_1, x_2, \ldots x_n)} = (-1)^n \frac{\dfrac{d\phi_1}{dx_1} \dfrac{d\phi_2}{dx_2} \ldots \dfrac{d\phi_n}{dx_n}}{\dfrac{d\phi_1}{dy_1} \dfrac{d\phi_2}{dy_2} \ldots \dfrac{d\phi_n}{dy_n}} \qquad (46)$$

337. Case where $J = 0$.—We can now prove that if the Jacobian vanishes, the functions $y_1, y_2, \ldots y_n$ are not independent of one another.

For, if $J(y_1, y_2, \ldots y_n) = 0$, we must have

$$\frac{d\phi_1}{dx_1} \frac{d\phi_2}{dx_2} \ldots \frac{d\phi_n}{dx_n} = 0\,;$$

that is, we have $\dfrac{d\phi_i}{dx_i} = 0$ for some value of i between 1 and n.

Hence ϕ_i must not contain x_i; and accordingly the corresponding equation is of the form

$$\phi_i(x_{i+1}, \ldots x_n, \; y_1, y_2, \ldots y_i) = 0.$$

Hence between this and the remaining equations,

$$\phi_{i+1} = 0, \quad \phi_{i+2} = 0, \; \ldots \; \phi_n = 0,$$

the variables $x_{i+1}, x_{i+2}, \ldots x_n$ can be eliminated so as to give a final equation between $y_1, y_2, \ldots y_n$ alone. This establishes our theorem.

Jacobian of Implicit Functions.

338. In the particular case where

$$y_1 = F_1(x_1, x_2, \ldots x_n),$$
$$y_2 = F_2(y_1, x_2, \ldots x_n),$$
$$\cdot \quad \cdot \quad \cdot \quad \cdot$$
$$y_n = F_n(y_1, y_2, \ldots y_{n-1}, x_n),$$

we have

$$\frac{d(y_1, y_2, \ldots y_n)}{d(x_1, x_2, \ldots x_n)} = \frac{dy_1}{dx_1} \cdot \frac{dy_2}{dx_2} \cdots \frac{dy_n}{dx_n} \cdots \quad (47)$$

It may be observed that the theory of Jacobians is of fundamental importance in the transformation of Multiple Integrals (see *Int. Calc.*, Art. 225).

EXAMPLES.

1. Find the Jacobian of $y_1, y_2, \ldots y_n$ with respect to $r, \theta_1, \theta_2, \ldots \theta_{n-1}$, being given the system of equations

$$y_1 = r \cos \theta_1, \quad y_2 = r \sin \theta_1 \cos \theta_2, \quad y_3 = r \sin \theta_1 \sin \theta_2 \cos \theta_3, \ldots$$

$$y_n = r \sin \theta_1 \sin \theta_2 \ldots \sin \theta_{n-1}.$$

If we square and add we get

$$y_1^2 + y_2^2 + \ldots y_n^2 = r^2.$$

Assuming this instead of the last of the given equations we readily find

$$J = r^{n-1} \sin^{n-2} \theta_1 \sin^{n-3} \theta_2 \ldots \sin \theta_{n-2}.$$

2. Find the Jacobian of $y_1, y_2, \ldots y_n$, being given

$$y_1 = x_1(1 - x_2), \quad y_2 = x_1 x_2(1 - x_3) \ldots$$

$$y_{n-1} = x_1 x_2 \ldots x_{n-1}(1 - x_n),$$

$$y_n = x_1 x_2 \ldots x_n.$$

Here $y_1 + y_2 + \ldots y_n = x_1$, and we get

$$\frac{d(y_1, y_2, \ldots y_n)}{d(x_1, x_2, \ldots x_n)} = x_1^{n-1} x_2^{n-2} \ldots x_{n-1}.$$

339. If $y_1, y_2, \ldots y_n$, which are given functions of the n variables $x_1, x_2, \ldots x_n$, be connected by an independent relation

$$F(y_1, y_2, \ldots y_n) = 0, \qquad (48)$$

we may, in virtue of this relation, regard one of the variables, x_n suppose, as a function of the remaining variables, and thus consider $y_1, y_2, \ldots y_{n-1}$ as functions of $x_1, x_2, \ldots x_{n-1}$. In this case it can be shown that

$$\frac{d(y_1, y_2, \ldots y_{n-1})}{d(x_1, x_2, \ldots x_{n-1})} = \frac{\dfrac{dF}{dy_n}}{\dfrac{dF}{dx_n}} \frac{d(y_1, y_2, \ldots y_n)}{d(x_1, x_2, \ldots x_n)}.$$

For, if we regard x_n as a function of x_1, we have

$$\frac{d}{dx_1}(y_1) = \frac{dy_1}{dx_1} + \frac{dy_1}{dx_n}\frac{dx_n}{dx_1}, \quad \frac{d}{dx_1}(y_2) = \frac{dy_2}{dx_1} + \frac{dy_2}{dx_n}\frac{dx_n}{dx_1}, \text{ &c.}$$

Also, from equation (48),

$$\frac{dF}{dx_1} + \frac{dF}{dx_n}\frac{dx_n}{dx_1} = 0, \quad \frac{dF}{dx_2} + \frac{dF}{dx_n}\frac{dx_n}{dx_2} = 0, \text{ &c.}$$

Again, let $\lambda_1 = \dfrac{\dfrac{dF}{dx_1}}{\dfrac{dF}{dx_n}}, \quad \lambda_2 = \dfrac{\dfrac{dF}{dx_2}}{\dfrac{dF}{dx_n}}, \ldots \lambda_{n-1} = \dfrac{\dfrac{dF}{dx_{n-1}}}{\dfrac{dF}{dx_n}};$

then $\quad \dfrac{dx_n}{dx_1} = -\lambda_1, \quad \dfrac{dx_n}{dx_2} = -\lambda_2, \ldots \dfrac{dx_n}{dx_{n-1}} = -\lambda_{n-1}.$

Hence $\quad \dfrac{d}{dx_1}(y_1) = \dfrac{dy_1}{dx_1} - \lambda_1 \dfrac{dy_1}{dx_n}, \quad \dfrac{d}{dx_2}(y_1) = \dfrac{dy_1}{dx_2} - \lambda_2 \dfrac{dy_1}{dx_n}, \text{ &c.}$

$$\cdots \cdots \cdots \cdots \text{ &c.}$$

accordingly, substituting in the Jacobian

$$\frac{d(y_1, y_2, \ldots y_{n-1})}{d(x_1, x_2, \ldots x_{n-1})},$$

Jacobian of Implicit Functions.

it becomes

$$\begin{vmatrix} \dfrac{dy_1}{dx_1} - \lambda_1 \dfrac{dy_1}{dx_n}, & \dfrac{dy_1}{dx_2} - \lambda_2 \dfrac{dy_1}{dx_n}, & \cdots & \dfrac{dy_1}{dx_{n-1}} - \lambda_{n-1} \dfrac{dy_1}{dx_n} \\ \dfrac{dy_2}{dx_1} - \lambda_1 \dfrac{dy_2}{dx_n}, & \dfrac{dy_2}{dx_2} - \lambda_2 \dfrac{dy_2}{dx_n}, & \cdots & \dfrac{dy_2}{dx_{n-1}} - \lambda_{n-1} \dfrac{dy_2}{dx_n} \\ \cdot & \cdot & \cdots & \cdot \\ \dfrac{dy_{n-1}}{dx_1} - \lambda_1 \dfrac{dy_{n-1}}{dx_n}, & \dfrac{dy_{n-1}}{dx_2} - \lambda_2 \dfrac{dy_{n-1}}{dx_n}, & \cdots & \dfrac{dy_{n-1}}{dx_{n-1}} - \lambda_{n-1} \dfrac{dy_{n-1}}{dx_n} \end{vmatrix}$$

If this determinant be bordered by introducing an additional column as in the following determinant, the other terms of the additional row being cyphers, its value is readily seen to be

$$\begin{vmatrix} \dfrac{dy_1}{dx_1}, & \dfrac{dy_1}{dx_2}, & \cdots & \dfrac{dy_1}{dx_n} \\ \dfrac{dy_2}{dx_1}, & \dfrac{dy_2}{dx_2}, & \cdots & \dfrac{dy_2}{dx_n} \\ \cdot & \cdot & \cdots & \cdot \\ \dfrac{dy_{n-1}}{dx_1}, & \dfrac{dy_{n-1}}{dx_2}, & \cdots & \dfrac{dy_{n-1}}{dx_n} \\ \lambda_1, & \lambda_2, & \cdots & 1 \end{vmatrix},$$

or

$$\dfrac{1}{\dfrac{dF}{dx_n}} \begin{vmatrix} \dfrac{dy_1}{dx_1}, & \dfrac{dy_1}{dx_2}, & \cdots & \dfrac{dy_1}{dx_n} \\ \dfrac{dy_2}{dx_1}, & \dfrac{dy_2}{dx_2}, & \cdots & \dfrac{dy_2}{dx_n} \\ \cdot & \cdot & \cdots & \cdot \\ \dfrac{dy_{n-1}}{dx_1}, & \dfrac{dy_{n-1}}{dx_2}, & \cdots & \dfrac{dy_{n-1}}{dx_n} \\ \dfrac{dF}{dx_1}, & \dfrac{dF}{dx_2}, & \cdots & \dfrac{dF}{dx_n} \end{vmatrix}$$

Again, we have

$$\frac{dF}{dx_1} = \frac{dF}{dy_1}\frac{dy_1}{dx_1} + \frac{dF}{dy_2}\frac{dy_2}{dx_1} + \ldots + \frac{dF}{dy_n}\frac{dy_n}{dx_1},$$

$$\frac{dF}{dx_2} = \frac{dF}{dy_1}\frac{dy_1}{dx_2} + \frac{dF}{dy_2}\frac{dy_2}{dx_2} + \ldots + \frac{dF}{dy_n}\frac{dy_n}{dx_2},$$

. . . .

Substituting these values in the last row of the preceding the theorem is established, since we readily find that the determinant is reducible to

$$\frac{\dfrac{dF}{dy_n}}{\dfrac{dF}{dx_n}} \begin{vmatrix} \dfrac{dy_1}{dx_1}, & \dfrac{dy_1}{dx_2}, & \ldots & \dfrac{dy_1}{dx_n} \\ \dfrac{dy_2}{dx_1}, & \dfrac{dy_2}{dx_2}, & \ldots & \dfrac{dy_2}{dx_n} \\ \cdot & \cdot & \cdot & \cdot \\ \dfrac{dy_n}{dx_1}, & \dfrac{dy_n}{dx_2}, & \ldots & \dfrac{dy_n}{dx_n} \end{vmatrix}. \qquad (49)$$

It may be well to guard the student from the supposition that this latter determinant is zero, as in Arts. 333 and 337. The distinction is, that in the former cases the equation $F(y_1, y_2, \ldots y_n) = 0$, connecting the y functions, is deduced by the elimination of the variables $x_1, x_2, \ldots x_n$ from the equations of connexion, whereas in the case here considered it is an additional and independent relation.

Examples.

1. Being given $y = f(u)$, and $u = \phi(x)$, find $\dfrac{d^2y}{dx^2}$.

   Ans. $f'(u)\,\phi''(x) + f''(u)\{\phi'(x)\}^2$.

2. If $y = F(t)$, $t = f(u)$, $u = \phi(x)$, find the value of $\dfrac{d^2y}{dx^2}$.

   Ans. $F'(t)\,f'(u)\,\phi''(x) + \{\phi'(x)\}^2\{f''(u)\,F'(t) + (f'(u))^2\,F''(t)\}$.

3. Change the independent variable from x to z in the equation
$$x^4 \frac{d^2y}{dx^2} - 2nx^3 \frac{dy}{dx} + a^2 y = 0, \text{ where } x = \frac{1}{z}.$$

   Ans. $\dfrac{d^2y}{dz^2} + \dfrac{2(n+1)}{z}\dfrac{dy}{dz} + a^2 y = 0$.

4. Transform $(1-x^2)\dfrac{d^2y}{dx^2} - x\dfrac{dy}{dx} + a^2 y = 0$, being given $x = \sin z$.

   Ans. $\dfrac{d^2y}{dz^2} + a^2 y = 0$.

5. If V be a function of r, where $r^2 = x^2 + y^2$, prove that
$$\frac{d^2V}{dx^2} + \frac{d^2V}{dy^2} = \frac{d^2V}{dr^2} + \frac{1}{r}\frac{dV}{dr}.$$

6. If V be a function of r, where $r^2 = x^2 + y^2 + z^2$, prove that
$$\frac{d^2V}{dx^2} + \frac{d^2V}{dy^2} + \frac{d^2V}{dz^2} = \frac{d^2V}{dr^2} + \frac{2}{r}\frac{dV}{dr}.$$

7. If $x = r\sin\theta\cos\phi$, $y = r\sin\theta\sin\phi$, $z = r\cos\theta$, prove that $\dfrac{dx}{dr} = \dfrac{dr}{dx}$, where in finding $\dfrac{dx}{dr}$, θ and ϕ are regarded as constants; while in finding $\dfrac{dr}{dx}$, y and z are regarded as constants.

8. If z be a function of two independent variables, x and y, which are connected with two other variables, u and v, by the equations
$$f_1(x, y, u, v) = 0, \qquad f_2(x, y, u, v) = 0;$$
show how to express $\dfrac{dz}{dx}$ and $\dfrac{dz}{dy}$ in terms of $\dfrac{dz}{du}$ and $\dfrac{dz}{dv}$.

Examples.

9. Transform the equation

$$\frac{d^2y}{dx^2} + \frac{2x}{1+x^2}\frac{dy}{dx} + \frac{y}{(1+x^2)^2} = 0$$

into another in which θ is the independent variable, supposing $x = \tan\theta$.

$$Ans. \quad \frac{d^2y}{d\theta^2} + y = 0.$$

10. If z be a function of x and y, and $u = px + qy - z$, prove that when p and q are taken as independent variables, we have

$$\frac{du}{dp} = x, \quad \frac{du}{dq} = y, \quad \frac{d^2u}{dp^2} = \frac{t}{rt-s^2}, \quad \frac{d^2u}{dp\,dq} = -\frac{s}{rt-s^2}, \quad \frac{d^2u}{dq^2} = \frac{r}{rt-s^2},$$

where p, q, r, s, t, denote the partial differential coefficients of z, as in Art. 313.

11. If the equation

$$x^n \frac{d^n y}{dx^n} + A_1 x^{n-1} \frac{d^{n-1} y}{dx^{n-1}} + \ldots + A_{n-1} x \frac{dy}{dx} + A_n = 0$$

be transformed to depend on θ, where $x = e^\theta$, prove that the coefficients in the transformed differential equation are all constants.

12. In orthogonal transformations, prove that

$$\frac{dV^2}{dx^2} + \frac{dV^2}{dy^2} + \frac{dV^2}{dz^2} = \frac{dV^2}{dX^2} + \frac{dV^2}{dY^2} + \frac{dV^2}{dZ^2}.$$

13. Given $x = \dfrac{\phi(t)}{F(t)}, \quad y = \dfrac{\psi(t)}{F(t)}$, prove that

$$\frac{d^2y}{dx^2} = \left\{\frac{F(t)}{F(t)\,\phi'(t) - \phi(t)\,F'(t)}\right\}^3 \begin{vmatrix} F(t), & F'(t), & F''(t) \\ \phi(t), & \phi'(t), & \phi''(t) \\ \psi(t), & \psi'(t), & \psi''(t) \end{vmatrix}.$$

14. Being given

$$y_1 = r\sin\theta_1 \sin\theta_2, \quad y_2 = r\sin\theta_1 \cos\theta_2,$$
$$y_3 = r\cos\theta_1 \sin\theta_3, \quad y_4 = r\cos\theta_1 \cos\theta_3,$$

find the value of the Jacobian $\dfrac{d(y_1, y_2, y_3, y_4)}{d(r, \theta_1, \theta_2, \theta_3)}$.

$$Ans. \quad r^3 \sin\theta_1 \cos\theta_1.$$

Examples.

15. Find the Jacobian $\dfrac{d(x, y, z)}{d(r, \theta, \phi)}$, being given

$$x = r \cos \theta \cos \phi, \quad y = r \sin \theta \sqrt{1 - m^2 \sin^2\phi}, \quad z = r \sin \phi \sqrt{1 - n^2 \sin^2\theta},$$

where $m^2 + n^2 = 1$.

$$\text{Ans. } \frac{r^2 (m^2 \cos^2\phi + n^2 \cos^2\theta)}{\sqrt{1 - m^2 \sin^2\phi} \sqrt{1 - n^2 \sin^2 \theta}}.$$

16. Being given

$$y_1 = \frac{x_2 x_3}{x_1}, \quad y_2 = \frac{x_1 x_3}{x_2}, \quad y_3 = \frac{x_1 x_2}{x_3},$$

find the value of the Jacobian of y_1, y_2, y_3. *Ans.* 4.

17. In the Jacobian

$$\frac{d(y_1, y_2, \ldots y_n)}{d(x_1, x_2, \ldots x_n)}$$

if we make

$$y_1 = \frac{u_1}{u}, \quad y_2 = \frac{u_2}{u}, \quad \ldots \quad y_n = \frac{u_n}{u},$$

prove that it becomes

$$\frac{1}{u^{n+1}} \begin{vmatrix} u, & u_1, & u_2, & \cdots & u_n \\ \dfrac{du}{dx_1}, & \dfrac{du_1}{dx_1}, & \dfrac{du_2}{dx_1}, & \cdots & \dfrac{du_n}{dx_1} \\ \dfrac{du}{dx_2}, & \dfrac{du_1}{dx_2}, & \dfrac{du_2}{dx_2}, & \cdots & \dfrac{du_n}{dx_2} \\ \cdot & \cdot & \cdot & & \cdot \\ \dfrac{du}{dx_n}, & \dfrac{du_1}{dx_n}, & \dfrac{du_2}{dx_n}, & \cdots & \dfrac{du_n}{dx_n} \end{vmatrix}.$$

This determinant is represented by the notation $K(u, u_1, \ldots u_n)$.

18. If a homogeneous relation exists between $u, u_1, \ldots u_n$, prove that

$$K(u, u_1, \ldots u_n) = 0.$$

19. In the same case, if $y_1, y_2, \ldots y_n$ possess a common factor, so that $y_i = u_i u$, &c., prove that

$$J(y_1, y_2, \ldots y_n) = 2u^n J(u_1, u_2, \ldots u_n) - u^{n-1} K(u, u_1, \ldots u_n).$$

Miscellaneous Examples.

1. If α, β, γ be the roots of the cubic

$$x^3 + px^2 + qx + r = 0,$$

show that

$$\begin{vmatrix} \dfrac{dp}{d\alpha}, & \dfrac{dq}{d\alpha}, & \dfrac{dr}{d\alpha} \\ \dfrac{dp}{d\beta}, & \dfrac{dq}{d\beta}, & \dfrac{dr}{d\beta} \\ \dfrac{dp}{d\gamma}, & \dfrac{dq}{d\gamma}, & \dfrac{dr}{d\gamma} \end{vmatrix} = (\gamma - \beta)(\beta - \alpha)(\alpha - \gamma).$$

2. Being given the three simultaneous equations

$$\phi_1(x_1, x_2, x_3, x_4) = 0, \quad \phi_2(x_1, x_2, x_3, x_4) = 0, \quad \phi_3(x_1, x_2, x_3, x_4) = 0,$$

determine the values of $\dfrac{dx_2}{dx_1}, \dfrac{dx_3}{dx_1}, \dfrac{dx_4}{dx_1}$.

3. If u be a solution of the differential equation

$$\frac{d^2V}{dx^2} + \frac{d^2V}{dy^2} + \frac{d^2V}{dz^2} = 0,$$

prove that $x \dfrac{du}{dx} + y \dfrac{du}{dy} + z \dfrac{du}{dz}$ will also be a solution of it.

4. If x and y be not independent, prove that the equation $\dfrac{d^2u}{dx\,dy} = \dfrac{d^2u}{dy\,dx}$ does not hold, in general.

5. Prove that the points of intersection of a curve of the fourth degree with its asymptotes lie on a conic; and in general for a curve of the degree n they lie on a curve of the degree $n - 2$.

6. Prove that every curve of the third degree is capable of being projected into a central curve. (Chasles.)

For if the harmonic polar of a point of inflexion be projected to infinity, the point of inflexion will be projected into a centre of the projected curve (see p. 282).

7. Two ellipses having the same foci are described infinitely near one another; how does the interval between them vary?

(a). How will the interval vary if the ellipses be concentric, similar, and similarly placed?

8. Eliminate the arbitrary functions from the equation $z = \phi(x) \cdot \psi(y)$.

9. Show that in order to eliminate n arbitrary functions from an equation containing two independent variables, it is, in general, requisite to proceed to differentials of the order $2n - 1$. How many resulting equations would be obtained in this case?

10. In the Lemniscate $r^2 = a^2 \cos 2\theta$, show that the angle between the tangent and radius vector is $\dfrac{\pi}{2} + 2\theta$.

11. In transforming from rectangular to polar coordinates, prove that

$$\frac{d^2V}{dx^2} + \frac{d^2V}{dy^2} + \frac{d^2V}{dz^2} = \frac{1}{r^2}\left\{\frac{d}{dr}\left(r^2 \frac{dV}{dr}\right) + \frac{1}{\sin\phi}\frac{d}{d\phi}\left(\sin\phi \frac{dV}{d\phi}\right) + \frac{1}{\sin^2\phi}\frac{d^2V}{d\theta^2}\right\}.$$

12. Prove that the ellipses

$$a^2y^2 + b^2x^2 = a^2b^2 \quad (1), \qquad a^2x^2 \sec^4\phi + b^2y^2 \cosec^4\phi = a^4e^4 \quad (2),$$

are so related that the envelope of (2) for different values of ϕ is the evolute of (1); and the point of contact of (2) with its envelope is the centre of curvature at the point of (1) whose excentric angle is ϕ.

13. Being given the equations

$$bx = \lambda\mu, \quad by = \sqrt{(\lambda^2 - b^2)(b^2 - \mu^2)},$$

prove that

$$dx^2 + dy^2 = (\lambda^2 - \mu^2)\left\{\frac{d\lambda^2}{\lambda^2 - b^2} + \frac{d\mu^2}{b^2 - \mu^2}\right\}.$$

14. If $1 - y - ay^m = 0$, develop y^r in terms of a by Lagrange's Theorem.

15. Being given $x = r \cos\theta$, $y = r \sin\theta$, transform

$$\frac{\left\{1 + \left(\dfrac{dy}{dx}\right)^2\right\}^{\frac{3}{2}}}{\dfrac{d^2y}{dx^2}}$$

into a function of r and θ, where θ is taken as the independent variable.

$$Ans. \quad \frac{\left\{r^2 + \left(\dfrac{dr}{d\theta}\right)^2\right\}^{\frac{3}{2}}}{r^2 - r\dfrac{d^2r}{d\theta^2} + 2\left(\dfrac{dr}{d\theta}\right)^2}.$$

16. Apply the method of infinitesimals to find a point such that the sum of its distances from three given points shall be a minimum.

Let ρ_1, ρ_2, ρ_3 denote the three distances, and we have $d\rho_1 + d\rho_2 + d\rho_3 = 0$: suppose $d\rho_1 = 0$, then $d(\rho_2 + \rho_3) = 0$, and it is easily seen that ρ_1 bisects the angle between ρ_2 and ρ_3, and similarly for the others; therefore, &c.

17. Eliminate the circular and exponential function from the equation $y = e^{\sin^{-1}x}$.

18. One leg of a right angle passes through a fixed point, whilst its vertex slides along a given curve; show that the problem of finding the envelope of the other leg of the right angle may be reduced to the investigation of a locus.

19. If two pairs of conjugates, in a system of lines in involution, be given by the equations

$$u = ax^2 + 2bxy + cy^2 = 0, \quad u' = a'x^2 + 2b'xy + c'y^2 = 0,$$

show that the double lines are given by the equation

$$\frac{du}{dx}\frac{du'}{dy} - \frac{du}{dy}\frac{du'}{dx} = 0. \quad \text{(Salmon's \emph{Conics}, Art. 342.)}$$

20. If

$$u_1 = \frac{x_1}{x_n}, \quad u_2 = \frac{x_2}{x_n}, \quad u_{n-1} = \frac{x_{n-1}}{x_n},$$

where $x_1, x_2, \ldots x_n$ are connected by the relation

$$x_1^2 + x_2^2 + x_3^2 + \ldots + x_n^2 = 1,$$

prove that the Jacobian

$$\frac{d(u_1, u_2, \ldots u_{n-1})}{d(x_1, x_2, \ldots x_{n-1})} = \frac{1}{x_n^{n+1}}.$$

21. If the variables $y_1, y_2, \ldots y_n$ are related to $x_1, x_2, \ldots x_n$ by the equations

$$y_1 = a_1 x_1 + a_2 x_2 + \ldots + a_n x_n,$$
$$y_2 = b_1 x_1 + b_2 x_2 + \ldots + b_n x_n,$$
$$\cdot \quad \cdot \quad \cdot \quad \cdot \quad \cdot$$
$$y_n = l_1 x_1 + l_2 x_2 + \ldots + l_n x_n,$$

and we have also

$$x_1^2 + x_2^2 + \ldots + x_n^2 = 1,$$
$$y_1^2 + y_2^2 + \ldots + y_n^2 = 1,$$

prove that the Jacobian

$$\frac{d(y_1, y_2, \ldots y_{n-1})}{d(x_1, x_2, \ldots x_{n-1})} = \frac{y_n}{x_n}.$$

22. Prove that the equation

$$ry^2 - 2sxy + tx^2 = px + qy - z$$

may be reduced to the form $\dfrac{d^2z}{dv^2} + z = 0$ by putting $x = u \cos v, \quad y = u \sin v$.

23. Investigate the nature of the singular point which occurs at the origin of coordinates in the curve

$$x^4 - 2ax^2 y - axy^2 + a^2 y^2 = 0.$$

24. Investigate the form of the curve represented by the equation $y = e^{-\frac{1}{x}}$

Miscellaneous Examples.

25. How would you ascertain whether a proposed expression, V, involving x, y, and z, is a function of two linear functions of these same variables?

Ans. The given function must be homogeneous; and the equations

$$\frac{dV}{dx} = 0, \quad \frac{dV}{dy} = 0, \quad \frac{dV}{dz} = 0,$$

must be capable of being satisfied by the same values of x, y, z: i.e. the result of the elimination of x, y, and z between these equations must vanish identically.

26. If $y = \phi(x^2)$, prove that

$$\frac{d^n y}{dx^n} = (2x^n) \phi^{(n)}(x^2) + n(n-1)(2x)^{n-2} \phi^{(n-1)}(x^2)$$

$$+ \frac{n(n-1)(n-2)(n-3)}{1 \cdot 2} 2x)^{n-4} \phi^{(n-2)}(x^2), \&c.$$

27. If $x + iy = (a + i\beta)^n$, where $i = \sqrt{-1}$, prove that

$$\frac{dx^2 + dy^2}{x^2 + y^2} = n^2 \frac{da^2 + d\beta^2}{a^2 + \beta^2}.$$

28. If $\tan\phi \tan\psi = \dfrac{1}{\sqrt{1-c^2}}$ prove that $\dfrac{d\phi}{d\psi} + \sqrt{\dfrac{1-c^2 \sin^2\phi}{1-c^2 \sin^2\psi}} = 0$.

29. If $x = \dfrac{1}{ky}$, prove that

$$\frac{dx}{\sqrt{(1-x^2)(1-k^2 x^2)}} \quad \text{transforms into} \quad \frac{dy}{\sqrt{(1-y^2)(1-k^2 y^2)}}.$$

30. Prove that $\dfrac{d}{dx}(xu) = \left(1 + x\dfrac{d}{dx}\right)u.$

31. Hence prove that

$$\left(x\frac{d}{dx}\right)\left(x\frac{d}{dx} - 1\right)u = x^2 \frac{d^2 u}{dx^2}.$$

For

$$\left(x\frac{d}{dx}\right)\left(x\frac{du}{dx}\right) = \left(x + x^2 \frac{d}{dx}\right)\frac{du}{dx} = x\frac{du}{dx} + x^2 \frac{d^2 u}{dx^2};$$

therefore

$$\left(x\frac{d}{dx} - 1\right)\left(x\frac{du}{dx}\right) = x^2 \frac{d^2 u}{dx^2}.$$

32. Prove that

$$\left(x\frac{d}{dx}\right)\left(x\frac{d}{dx} - 1\right)\left(x\frac{d}{dx} - 2\right)u = x^3 \frac{d^3 u}{dx^3}.$$

By the preceding example we have

$$\left(x\frac{d}{dx} - 2\right)\left(x\frac{d}{dx} - 1\right)\left(x\frac{d}{dx}\right)u = \left(x\frac{d}{dx} - 2\right)x^2 \frac{d^2 u}{dx^2};$$

2 F

but
$$\frac{d}{dx}\left(x^2 \frac{d^2u}{dx^2}\right) = x^2 \frac{d^3u}{dx^3} + 2x \frac{d^2u}{dx^2};$$

therefore
$$\left(x\frac{d}{dx} - 2\right) x^2 \frac{d^2u}{dx^2} = x^3 \frac{d^3u}{dx^3}.$$

33. Prove, in general, that

$$\left(x\frac{d}{dx}\right)\left(x\frac{d}{dx} - 1\right)\left(x\frac{d}{dx} - 2\right)\ldots\left(x\frac{d}{dx} - n + 1\right)u = x^n \frac{d^nu}{dx^n}.$$

This can be easily arrived at from the preceding by the method of mathematical induction; that is, assuming that the theorem holds for any positive integer n, prove that it holds for the next higher integer $(n + 1)$, &c.

34. Find $\dfrac{1}{r} + \dfrac{d^2\left(\dfrac{1}{r}\right)}{d\theta^2}$ in terms of r when $r^2 = a^2 \cos 2\theta$. *Ans.* $\dfrac{3a^4}{r^5}$.

35. If $u = (x^2 + y^2 + z^2)^{\frac{1}{2}}$, prove that

$$\frac{d^4u}{dx^4} + \frac{d^4u}{dy^4} + \frac{d^4u}{dz^4} + 2\frac{d^4u}{dx^2 dy^2} + 2\frac{d^4u}{dy^2 dz^2} + 2\frac{d^4u}{dz^2 dx^2} = 0.$$

36. If $z = \dfrac{x}{x^2 + y^2}$, and $\phi = \tan^{-1}\left(\dfrac{y}{x}\right)$, prove that

$$\frac{d^n z}{dx^n} = (-1)^n \frac{1 \cdot 2 \cdot 3 \ldots n \cdot \cos(n+1)\phi \cdot \cos^{n+1}\phi}{x^{n+1}},$$

$$\frac{d^{2n} z}{dy^{2n}} = (-1)^n \frac{1 \cdot 2 \cdot 3 \ldots 2n \cdot \cos(2n+1)\phi \cdot \cos^{2n-1}\phi}{x^{2n+1}}$$

$$\frac{d^{2n+1} z}{dy^{2n+1}} = (-1)^{n+1} \frac{1 \cdot 2 \cdot 3 \ldots (2n+1) \sin(2n+2)\phi \cdot \cos^{2n+2}\phi}{x^{2n+2}}.$$

37. If u be a homogeneous function of the nth degree in x, y, z, and u_1, u_2, u_3, denote its differential coefficients with regard to x, y, z, respectively, while u_{11}, u_{12}, &c., in like manner denote its second differential coefficients; prove that

$$\begin{vmatrix} u_{11} & u_{12} & u_{13} & u_1 \\ u_{21} & u_{22} & u_{23} & u_2 \\ u_{31} & u_{32} & u_{33} & u_3 \\ u_1 & u_2 & u_3 & 0 \end{vmatrix} = -\frac{nu}{n-1} \begin{vmatrix} u_{11} & u_{12} & u_{13} \\ u_{21} & u_{22} & u_{23} \\ u_{31} & u_{32} & u_{33} \end{vmatrix}.$$

38. If u be a homogeneous function of the nth degree in x, y, z, w, show that for all values of the variables which satisfy the equation $u = 0$ we have

$$\begin{vmatrix} u_{11}, & u_{12}, & u_{13}, & u_1 \\ u_{21}, & u_{22}, & u_{23}, & u_2 \\ u_{31}, & u_{32}, & u_{33}, & u_3 \\ u_1, & u_2, & u_3, & 0 \end{vmatrix} = \frac{w^2}{(n-1)^2} \begin{vmatrix} u_{11}, & u_{12}, & u_{13}, & u_{14} \\ u_{21}, & u_{22}, & u_{23}, & u_{24} \\ u_{31}, & u_{32}, & u_{33}, & u_{34} \\ u_{41}, & u_{42}, & u_{43}, & u_{44} \end{vmatrix}.$$

39. Show that the equation

$$\frac{d}{d\mu}\left\{(1-\mu^2)\frac{dP}{d\mu}\right\} + \frac{1}{1-\mu^2}\frac{d^2P}{d\theta^2} + 6P = 0,$$

is satisfied if P is any of the quantities

$$\frac{1}{3} - \mu^2, \quad (1-\mu^2)\cos 2\theta, \quad (1-\mu^2)\sin 2\theta, \quad \mu\sqrt{1-\mu^2}\cos\theta, \quad \mu\sqrt{1-\mu^2}\sin\theta,$$

or any linear function of them.

40. If $x + \lambda$ be substituted for x in the quantic

$$a_0 x^n + n a_1 x^{n-1} + \frac{n(n-1)}{1 \cdot 2} a_2 x^{n-2} + \&c. + a_n,$$

and if $a'_0, a'_1, \ldots a'_r \ldots$ denote the corresponding coefficients in the new quantic; prove that

$$\frac{da'_r}{d\lambda} = r a'_{r-1}.$$

It is easily seen that in this case we have

$$a'_r = a_r + r a_{r-1} \lambda + \frac{r(r-1)}{1 \cdot 2} a_{r-2} \lambda^2 + \&c. \ldots + a_0 \lambda^r; \therefore \&c.$$

41. If ϕ be any function of the differences of the roots of the quantic in the preceding example, prove that

$$\left(a_0 \frac{d}{da_1} + 2a_1 \frac{d}{da_2} + 3a_2 \frac{d}{da_3} + \ldots + n a_{n-1} \frac{d}{da_n}\right)\phi = 0.$$

This result follows immediately, since any function of the differences of the roots remains unaltered when $x + \lambda$ is substituted for x, and accordingly $\frac{d\phi}{d\lambda} = 0$ in this case.

42. Being given

$$u = xy + \sqrt{1 - x^2 - y^2 + x^2 y^2}, \quad v = x\sqrt{1-y^2} + y\sqrt{1-x^2},$$

prove that

$$\frac{du}{dx}\frac{dv}{dy} - \frac{dv}{dx}\frac{du}{dy} = 0,$$

and explain the meaning of the result.

43. Find the minimum value of

$$\frac{\sin A}{\sin B \sin C} + \frac{\sin B}{\sin C \sin A} + \frac{\sin C}{\sin A \sin B}, \text{ where } A + B + C = 180°.$$

44. Prove that

$$\phi\left(x\frac{d}{dx}\right)f(ax) \equiv \phi\left(a\frac{d}{da}\right)f(ax),$$

where $\phi(x)$ is a rational function of x.

45. Show that the reciprocal polar to the evolute of the ellipse

$$\frac{x^2}{a^2} + \frac{y^2}{b^2} = 1,$$

with respect to the circle described on the line joining the foci as diameter, has for its equation

$$\frac{a^2}{x^2} + \frac{b^2}{y^2} = 1.$$

46. If the second term be removed from the quantic

$$(a_0, a_1, a_2, \ldots a_n)(x, y)^n$$

by the substitution of $x - \dfrac{a_1}{a_0} y$, instead of x, and if the new quantic be denoted by $(A_0, 0, A_2, A_3, \ldots A_n)(x, y)$; show that the successive coefficients $A_2, A_3, \ldots A_n$ are obtained by the substitution of a_1 for x and $-a_0$ for y in the series of quantics

$$(a_0, a_1, a_2)(x, y), \quad (a_0, a_1, a_2, a_3)(x, y), \ldots (a_0, a_1, \ldots a_n)(x, y).$$

47. Distinguish the maxima and minima values of

$$\frac{1 + 2x \tan^{-1} x}{1 + x^2}.$$

48. If $y = \dfrac{a'x^2 + 2b'x + c'}{ax^2 + 2bx + c}$, prove that

$$\frac{1}{2}\frac{dy}{dx} = \frac{(ac - b^2)y^2 + (ac' + a'c - 2bb')y + a'c' - b'^2}{(ab')x^2 - (ca')x + (bc')}.$$

49. If $lX + mY + nZ$, $l'X + m'Y + n'Z$, $l''X + m''Y + n''Z$, be substituted for x, y, z, in the quadratic expression $ax^2 + by^2 + cz^2 + 2dyz + 2ezx + 2fxy$; and if a', b', c', d', e', f', be the respective coefficients in the new expression; prove that

$$\begin{vmatrix} a', & f', & e', \\ f', & b', & d', \\ e', & d', & c', \end{vmatrix} = 0 \text{ whenever } \begin{vmatrix} a, & f, & e \\ f, & b, & d \\ e, & d, & c \end{vmatrix} = 0.$$

50. If the transformation be *orthogonal*, i.e. if

$$x^2 + y^2 + z^2 = X^2 + Y^2 + Z^2,$$

prove that the preceding determinants are equal to one another.

51. Prove that the maximum and minimum values of the expression

$$ax^4 + 4bx^3 - 6cx^2 + 4dx + e$$

are the roots of the cubic

$$a^3 z^3 - 3(a^2 I - 3H^2) z^2 + 3(aI^2 - 18HJ) z - \Delta = 0,$$

where $\quad H = ac - b^2, \quad I = ae - 4bd + 3c^2,$

$$J = \begin{vmatrix} a, & b, & c \\ b, & c, & d \\ c, & d, & e \end{vmatrix}, \text{ and } \Delta = I^3 - 27J^2.$$

By Art. 138 it is evident that the equation in z is obtained by substituting $e - z$ instead of e in the discriminant of the biquadratic; accordingly we have for the resulting equation

$$(I - az)^3 = 27(J - zH)^2,$$

since the discriminant of the biquadratic is

$$I^3 - 27J^2 = 0.$$

In general, the equation in z whose roots are the $n - 1$ maximum and minimum values of a given function of n dimensions in x, can be got from the discriminant of the function, by substituting in it, instead of the absolute term, the absolute term *minus* z.

It is evident that the discriminant of the function in x is, in all cases, the absolute term in the equation in z.

52. If Δ be the product of the squares of the differences of the roots of

$$x^3 - px^2 + qx - r = 0,$$

find an expression in terms of the roots for $\dfrac{d\Delta}{dr}$, by solving from three equations of the form

$$\frac{d\Delta}{da} = \frac{d\Delta}{dp}\frac{dp}{da} + \frac{d\Delta}{dq}\frac{dq}{da} + \frac{d\Delta}{dr}\frac{dr}{da}.$$

Ans. $2(\beta + \gamma - 2\alpha)(\gamma + \alpha - 2\beta)(\alpha + \beta - 2\gamma)$.

53. If $X + Y\sqrt{-1}$ be a function of $x + y\sqrt{-1}$, prove that X and Y satisfy the equations

$$\frac{d^2X}{dx^2} + \frac{d^2X}{dy^2} = 0, \text{ and } \frac{d^2Y}{dx^2} + \frac{d^2Y}{dy^2} = 0.$$

54. If the three sides of a triangle are a, $a + \alpha$, $a + \beta$, where α and β are infinitesimals, find the three angles, expressed in circular measure.

Ans. $\dfrac{\pi}{3} - \dfrac{\alpha+\beta}{a\sqrt{3}},\ \dfrac{\pi}{3} + \dfrac{2\alpha-\beta}{a\sqrt{3}},\ \dfrac{\pi}{3} + \dfrac{2\beta-\alpha}{a\sqrt{3}}.$

55. If $y = x + \alpha x^3$, where α is an infinitesimal, find the order of the error in taking $x = y - \alpha y^3$.

56. The sides a, b, c, of a right-angled triangle become $a + \alpha$, $b + \beta$, $c + \gamma$, where α, β, γ are infinitesimals; find the change in the right angle.

Ans. $\dfrac{c\gamma - a\alpha - b\beta}{ab}$.

57. If a curve be given by the equations

$$2x = \sqrt{t^2 + 2t} + \sqrt{t^2 - 2t},$$

$$2y = \sqrt{t^2 + 2t} - \sqrt{t^2 - 2t},$$

find the radius of curvature in terms of t.

58. In the curve whose equation is $y = e^{-x^2}$, determine all the cases where the tangent is parallel to the axis of x.

If θ be the greatest angle which any of its tangents makes with the axis of x, prove that $\tan \theta = \sqrt{\dfrac{2}{e}}$.

59. In a curve traced on a sphere, prove the following formula for the radius of curvature at any point:

$$\tan \rho = \frac{\sin r\, dr}{\cos p\, dp}.$$

60. Apply this form to show that in a spherical ellipse $\sin p \sin p' = \text{const.}$, where p and p' are the perpendiculars from the foci on any great circle touching the ellipse.

61. Prove the following relation between (ρ, ρ'), the radii of curvature at corresponding points of two reciprocal polar curves:

$$\rho\rho' = \frac{k^2}{\cos^3 \psi},$$

where ψ is the angle between the radius vector and normal.

62. If AB, BC, CD, ... be the sides of an equilateral polygon inscribed in any curve, and if AD be produced to meet BC in P; prove that, when the sides of the polygon are diminished indefinitely, $BP = 3\dfrac{\rho^2}{\rho'}$, where ρ and ρ' are the radii of curvature at B and at the corresponding point of the evolute.

63. If
$$U = \frac{\sqrt{(1-x)(1+y+y^2)} + \sqrt{(1-y)(1+x+x^2)}}{x-y},$$

and
$$V = \left(\frac{\sqrt{1-x^3} - \sqrt{1-y^3}}{x-y}\right)^2 + x + y,$$

find the value of
$$\frac{dU}{dx}\frac{dV}{dy} - \frac{dV}{dx}\frac{dU}{dy}.$$

64. If
$$V = x^n + \frac{1}{x^n}, \quad \text{and} \quad z = x + \frac{1}{x},$$

prove that
$$(z^2 - 4)\frac{d^2V}{dz^2} + z\frac{dV}{dz} - n^2 V = 0$$

65. Determine b and k so that the curve

$$(x^2 + y^2)(x \cos \alpha + y \sin \alpha - a) = k^2(x \cos \beta + y \sin \beta - b)$$

may have a cusp; α, β, and a being given and the coordinates being rectangular.

Prove that in this case the cuspidal tangent makes equal angles with the asymptote and with the line drawn from the cusp to the origin.

66. Find the coordinates of the two real finite points of inflexion on the curve $y^2 = (x-2)^2(x-5)$, and show that they subtend a right angle at the double point.

67. If x, y, z, be given in terms of three new variables u, v, w, by the following equations: $x = Pu$, $y = (P-b)v$, $z = (P-c)w$, where

$$P = \frac{1 + bv^2 + cw^2}{u^2 + v^2 + w^2};$$

it is required to prove that $dx^2 + dy^2 + dz^2 = L^2 du^2 + M^2 dv^2 + N^2 dw^2$, and to determine the actual values of L, M, N.

68. If $x + y = X$, $y = XY$, prove that

$$x\frac{d^2u}{dx^2} + y\frac{d^2u}{dx\,dy} + \frac{du}{dx} = X\frac{d^2u}{dX^2} - Y\frac{d^2u}{dX\,dY} + \frac{du}{dX}.$$

69. Being given $x = u^3 - 3uv^2$, $y = 3u^2v - v^3$, find what $\dfrac{xdy - ydx}{xdx + ydy}$ becomes in terms of u, v, du, dv.

$$\text{Ans. } \frac{udv - vdu}{udu + vdv}.$$

70. If the polar equation of a curve be $r = a \sec^2\dfrac{\theta}{2}$, find an expression for its radius of curvature at any point.

71. Show that the differential $\dfrac{dx}{\sqrt{x^4 - 3x^2 + 3}}$ is transformed into

$$\frac{l\,dy}{\sqrt{(1 + y^2 \tan^2\lambda)(1 + y^2 \cot^2\lambda)}},$$

by assuming $x = \sqrt[4]{3\dfrac{1-y}{1+y}}$, and find the value of λ.

$$\text{Ans. } \lambda = 7° 30'.$$

72. If $y^4 + xy = 1$, prove that

$$y^2\frac{d^2y}{dx^2} + 3x\frac{dy^3}{dx^3} + y\frac{dy^2}{dx^2} = 0.$$

73. The pair of curves represented by the equation

$$r^2 - 2rF(\omega) + c^2 = 0$$

may be regarded as the envelope of a series of circles whose centres lie on a certain curve, and which cut orthogonally the circle whose radius is c, and whose centre is the origin (Mannheim, *Journal de Math.*, 1862).

74. A chord PQ cuts off a constant area from a given oval curve; show that the radius of curvature of its envelope will be $\tfrac{1}{4}PQ (\cot\theta + \cot\phi)$, θ and ϕ being the angles at which PQ cuts the curve.

75. In the polar equations of two curves,

$$F(r, \omega) = 0, \quad f(r, \omega) = 0,$$

if $R^{\frac{1}{n}}$ be substituted for r, and $n\Omega$ for ω, prove that the curves represented by the transformed equations intersect at the same angle as the original curves.

(Mr. W. Roberts, *Liouville's Journal*, Tome 13, p. 209).

Miscellaneous Examples. 441

This result follows immediately from the property that $\frac{rd\omega}{dr}$ is unaltered by the transformation in question.

76. A system of concentric and similarly situated equilateral hyperbolas is cut by another such system having the same centre, under a constant angle, which is double that under which the axes of the two systems intersect.
Ibid., p. 210.

77. In a triangle formed by three arcs of equilateral hyperbolas, having the same centre (or by parabolas having the same focus), the sum of the angles is equal to two right angles. *Ibid.*, p. 210.

78. Being given two hyperbolic tangents to a conic, the arc of any third hyperbolic tangent, which is intercepted by the two first, subtends a constant angle at the focus. *Ibid.*, p. 212.

An equilateral hyperbola which touches a conic, and is concentric with it, is called a *hyperbolic tangent* to the conic.

79. A system of confocal cassinoids is cut orthogonally by a system of equilateral hyperbolas passing through the foci and concentric with the cassinoids.
Ibid., p. 214.

The student will find a number of other remarkable theorems, deduced by the same general method, in Mr. Roberts' Memoir. This method is an extension of the method of inversion.

80. If P_n be the coefficient of x^n in the expansion of $(1 - 2ax + x^2)^{-\frac{1}{2}}$, prove the two following equations:

$$(a^2 - 1)\frac{dP_n}{da} = naP_n - nP_{n-1},$$

$$nP_n = (2n - 1)aP_{n-1} - (n - 1)P_{n-2}.$$

81. If at each point on a curve a right line be drawn making a constant angle with the radius vector drawn to a fixed point, prove that the envelope of the line so drawn is a curve which is similar to the *negative pedal* of the given curve, taken with respect to the fixed point as pole.

82. If $\quad 2U \equiv ax^2 + 2bxy + cy^2, \quad 2V \equiv a'x^2 + 2b'xy + c'y^2$,

and

$$\begin{vmatrix} \frac{dU}{dx}, & \frac{dU}{dy} \\ \frac{dV}{dx}, & \frac{dV}{dy} \end{vmatrix}^2 \equiv AU^2 + 2BUV + CV^2, \text{ find } A, B, C.$$

83. Prove that the values of the diameters of curvature of the curve $y^2 = f(x)$ where it meets the axis of x are $f'(a), f'(\beta), \ldots$ if $a, \beta, \ldots$ be the roots of $f(x) = 0$.

Hence find the radii of curvature of $y^2 = (x^2 - m^2)(x - a)$ at such points.

84. A constant length PQ is measured along the tangent at any point P on a curve; give, by aid of Art. 290, a geometrical construction for the centre of curvature of the locus of the point Q.

85. In same case, if PQ' be measured equal to PQ, in the opposite direction along the tangent, prove that the point P, and the centres of curvature of the loci of Q, and Q', lie *in directum*.

86. A framework is formed by four rods jointed together at their extremities; prove that the distance between the middle points of either pair of opposite sides is a maximum or a minimum when the other rods are parallel, being a maximum when the rods are uncrossed, and a minimum when they cross.

87. At each point of a closed curve are formed the rectangular hyperbola, and the parabola, of closest contact; show that the arc of the curve described by the centre of the hyperbola will exceed the arc of the oval by twice the arc of the curve described by the focus of the parabola; provided that no parabola has five-pointic contact with the curve. (*Camb. Math. Trip.* 1875.)

88. A curve rolls on a straight line, determine the nature of the motion of one of its involutes. (*Prof. Crofton.*)

89. Prove the following properties of the three-cusped hypocycloid:—

(1). The segment intercepted by any two of the three branches on any tangent to the third is of constant length. (2). The locus of the middle point of the segment is a circle. (3). The tangents to these branches at its extremities intersect at right angles on the inscribed circle. (4). The normals corresponding to the three tangents intersect in a common point, which lies on the circumscribed circle.

Definition.—The right line joining the feet of the perpendiculars drawn to the sides of a triangle from any point on its circumscribed circle is called the *pedal line of the triangle* relative to the point.

90. Prove that the envelope of the pedal line of a triangle is a three-cusped hypocycloid, having its centre at the centre of the nine-point circle of the triangle. (Steiner, *Ueber eine besondere curve dritter klasse, und vierten grades*, Crelle, 1857.)

This is called *Steiner's Envelope*, and the theorem can be demonstrated, geometrically, as follows:—

Let P be any point on the circumscribed circle of a triangle ABC, of which D is the intersection of the perpendiculars; then it can be shown without difficulty, that the pedal line corresponding to P passes through the middle point of DP. Let Q denote this middle point, then Q lies on the nine-point circle of the triangle ABC. If O be the centre of the nine-point circle, it is easily seen that, as Q moves round the circle, the angular motion of the pedal line is half that of OQ, and takes place in the opposite direction. Let R be the other point in which the pedal line cuts the nine-point circle, and, by drawing a consecutive position of the moving line, it can be seen immediately that the corresponding point T on the envelope is obtained by taking $QT = QR$. Hence it can be readily shown that the locus of T is a three-cusped hypocycloid.

This can also be easily proved otherwise by the method of Art. 295 (*a*).

91. The envelope of the tangent at the vertex of a parabola which touches three given lines is a three-cusped hypocycloid.

92. The envelope of the parabola is the same hypocycloid.

For fuller information on Steiner's Envelope, and the general properties of the three-cusped hypocycloid, the student is referred, amongst other memoirs, to Cremona, Crelle, 1865. Townsend, *Educ. Times. Reprint.* 1866. Ferrers, *Quar. Jour. of Math.*, 1866. Serret, *Nouv. Ann.*, 1870. Painvin, *ibid.*, 1870. Cahen, *ibid.*, 1875.

ON THE FAILURE OF TAYLOR'S THEOREM.

As no mention has been made in Chapter III. of the cases when Taylor's Series becomes inapplicable, or what is usually called the failure of Taylor's Theorem, the following extract from M. Navier's *Leçons d'Analyse* is introduced for the purpose of elucidating this case:—

On the Case when, for certain particular Values of the Variable, Taylor's Series does not give the Development of the Function.—The existence of Taylor's Series supposes that the function $f(x)$ and its differential coefficients $f'(x)$, $f''(x)$, &c., do not become infinite for the value of x from which the increment h is counted. If the contrary takes place, the series will be inapplicable.

Suppose, for example, that $f(x)$ is of the form $\dfrac{F(x)}{(x-a)^m}$, m being any positive number, and $F(x)$ a function of x which does not become either zero or infinite when $x = a$.

If, conformably to our rules, $\dfrac{F(x+h)}{(x+h-a)^m}$ be developed in a series of positive powers of h, all the terms would become infinite when we make $x = a$. At the same time the function has then a determinate value, viz.: $\dfrac{F(a+h)}{h^m}$. But as the development of this value according to powers of h must necessarily contain negative powers of h, it cannot be given by Taylor's Series.

Taylor's Series naturally gives indeterminate results when, the proposed function $f(x)$ containing radicals, the particular value attributed to x causes these radicals to disappear in the function and in its differential coefficients. In order to understand the reason, we remark that a radical of the form $(x-a)^{\frac{p}{q}}$, p and q denoting whole numbers, which forms part of a function $f(x)$, gives to this function q different values, real or imaginary. As this same radical is reproduced in the differential coefficients of the function, these coefficients also present a number, q, of values. But, if the particular value a be attributed to x, the radical will disappear from all the terms of the series, while it remains always in the function, where it becomes $h^{\frac{p}{q}}$. Therefore the series no longer represents the function, because the latter has many values, while the series can have but one. The analysis solves this contradiction by giving infinite values to the terms of the series, which consequently does not any longer represent a determined result.

The development of $f(x)$ ought, in the case with which we are occupied, to contain terms of the form $h^{\frac{p}{q}}$. We should obtain the development by making $x = a + h$ in the proposed function.

On the Failure of Taylor's Theorem.

Fractional powers of h would appear in the latter development: for example, suppose

$$f(x) = 2ax - x^2 - a\sqrt{x^2 - a^2};$$

this gives

$$f'(x) = 2(a - x) + \frac{ax}{\sqrt{x^2 - a^2}};$$

$$f''(x) = -2 + \frac{a}{\sqrt{x^2 - a^2}} - \frac{ax^2}{(x^2 - a^2)^{\frac{3}{2}}}.$$

On making $x = a$, we have $f(x) = a^2$, and all the differential coefficients become infinite. This circumstance indicates that the development of $f(x + h)$ ought to contain fractional powers of h when $x = a$: in fact the function becomes then

$$f(a + h) = a^2 - h^2 + a\sqrt{2ah + h^2},$$

of which the development according to powers of h would contain $h^{\frac{1}{2}}$, $h^{\frac{3}{2}}$, $h^{\frac{5}{2}}$, &c.

It should be remarked that a radical contained in the function $f(x)$ may disappear in two different ways when a particular value is attributed to the variable x, that is, 1°, when the quantity contained under the radical vanishes: 2°, when a factor with which the radical may be affected vanishes.

In the former case the development according to Taylor's Theorem can never agree with the function $f(x + h)$ for the particular value of x in question, for the reason already indicated.

But it is not the same in the latter case, because the factor with which the radical is affected, and which becomes zero in the function, may cease to affect the radical in the differential coefficients of higher orders; in fact it may not disappear at all, and the series may in consequence present the necessary number of values.

For example, let the proposed function be

$$f(x) = (x - a)^m \sqrt{x - b},$$

m being a positive integer.

Here we have

$$f'(x) = m(x - a)^{m-1}\sqrt{x - b} + \frac{m(x - a)^m}{2\sqrt{x - b}},$$

$$f''(x) = m(m - 1)(x - a)^{m-2}\sqrt{x - b} + \frac{m(x - a)^{m-1}}{\sqrt{x - b}} - \frac{(x - a)^m}{4(x - b)^{\frac{3}{2}}}.$$

Each differentiation causes one of the factors of $(x - a)^m$ to disappear in the first term. After m differentiations these factors would entirely disappear; and consequently the supposition $x = a$, in causing the first m-derived functions to vanish, will leave the radical $\sqrt{x - b}$ to remain in all the others.

On the Conditions for a Maximum or Minimum of a Function of any Number of Variables (Art. 163).

The conditions for a maximum or a minimum in the case of two or of three variables have been given in Chapter X.

It can be readily seen that the mode of investigation, and the form of the conditions there given, admit of extension to the case of any number of independent variables.

We shall commence with the case of four independent variables. Proceeding as in Art. 162, it is obvious that the problem reduces to the consideration of a quadratic expression in four variables which shall preserve the same sign for all real values of the variable.

Let the quadratic be written in the form

$$a_{11}x_1^2 + a_{22}x_2^2 + a_{33}x_3^2 + a_{44}x_4^2 + 2a_{12}x_1x_2 + 2a_{13}x_1x_3 + 2a_{14}x_1x_4 + 2a_{23}x_2x_3,$$

$$+ 2a_{24}x_2x_4 + 2a_{34}x_3x_4, \qquad (1)$$

in which a_{11}, a_{12}, a_{22}, &c., represent the respective second differential coefficients of the function, as in Art. 162.

We shall first investigate the conditions that this expression shall be always a positive quantity; in this case a_{11} evidently is necessarily positive: again, multiplying by a_{11}, the expression may be written in the following form:—

$$(a_{11}x_1 + a_{12}x_2 + a_{13}x_3 + a_{14}x_4)^2 + (a_{11}a_{22} - a_{12}^2)x_2^2 + (a_{11}a_{33} - a_{13}^2)x_3^2$$

$$+ (a_{11}a_{44} - a_{14}^2)x_4^2 + 2(a_{11}a_{23} - a_{12}a_{13})x_2x_3 + 2(a_{11}a_{24} - a_{12}a_{14})x_2x_4$$

$$+ 2(a_{11}a_{34} - a_{13}a_{14})x_3x_4. \qquad (2)$$

Also, in order that the part of this expression after the first term shall be always positive, we must have, by the Article referred to, the following conditions:—

$$a_{11}a_{22} - a_{12}^2 > 0, \qquad (3)$$

$$(a_{11}a_{22} - a_{12}^2)(a_{11}a_{33} - a_{13}^2) - (a_{11}a_{23} - a_{12}a_{13})^2 > 0, \qquad (4)$$

and

$$\begin{vmatrix} a_{11}a_{22} - a_{12}^2, & a_{11}a_{23} - a_{12}a_{13}, & a_{11}a_{24} - a_{12}a_{14} \\ a_{11}a_{23} - a_{12}a_{13}, & a_{11}a_{33} - a_{13}^2, & a_{11}a_{34} - a_{13}a_{14} \\ a_{11}a_{24} - a_{12}a_{14}, & a_{11}a_{34} - a_{13}a_{14}, & a_{11}a_{44} - a_{14}^2 \end{vmatrix} > 0. \qquad (5)$$

To express this determinant in a simpler form, we write it as follows:—

$$\frac{1}{a_{11}} \begin{vmatrix} a_{11}, & a_{12}, & a_{13}, & a_{14} \\ 0, & a_{11}a_{22} - a_{12}^2, & a_{11}a_{23} - a_{12}a_{13}, & a_{11}a_{24} - a_{12}a_{14} \\ 0, & a_{11}a_{23} - a_{12}a_{13}, & a_{11}a_{33} - a_{13}^2, & a_{11}a_{34} - a_{13}a_{14} \\ 0, & a_{11}a_{24} - a_{12}a_{14}, & a_{11}a_{34} - a_{13}a_{14}, & a_{11}a_{44} - a_{14}^2 \end{vmatrix}. \qquad (6)$$

Next, to form a new determinant, multiply the first row by a_{12}, a_{13}, a_{14}, successively, and add the resulting terms to the 2nd, 3rd, and 4th rows, respectively; then, since each term in the rows after the first contains a_{11} as a factor, the determinant is evidently equivalent to

$$a_{11}^2 \begin{vmatrix} a_{11}, & a_{12}, & a_{13}, & a_{14} \\ a_{12}, & a_{22}, & a_{23}, & a_{24} \\ a_{13}, & a_{23}, & a_{33}, & a_{34} \\ a_{14}, & a_{24}, & a_{34}, & a_{44} \end{vmatrix}. \qquad (7)$$

In like manner the relation in (4) is at once reducible to the form

$$a_{11} \begin{vmatrix} a_{11}, & a_{12}, & a_{13} \\ a_{12}, & a_{22}, & a_{23} \\ a_{13}, & a_{23}, & a_{33} \end{vmatrix} > 0.$$

Hence we conclude that whenever the following conditions are fulfilled, viz.:

$$a_{11} > 0, \quad \begin{vmatrix} a_{11}, & a_{12} \\ a_{12}, & a_{22} \end{vmatrix} > 0, \quad \begin{vmatrix} a_{11}, & a_{12}, & a_{13} \\ a_{12}, & a_{22}, & a_{23} \\ a_{13}, & a_{23}, & a_{33} \end{vmatrix} > 0, \quad \begin{vmatrix} a_{11}, & a_{12}, & a_{13}, & a_{14} \\ a_{12}, & a_{22}, & a_{23}, & a_{24} \\ a_{13}, & a_{23}, & a_{33}, & a_{34} \\ a_{14}, & a_{24}, & a_{34}, & a_{44} \end{vmatrix} > 0, \quad (8)$$

the quadratic expression (1) is *positive* for all real values of x_1, x_2, x_3, x_4.

Accordingly the conditions are the same as in the case (Art. 162) of three variables, x_1, x_2, x_3; with the addition that the determinant (7) shall be also positive.

In like manner it can be readily seen that if the second and fourth of the preceding determinants be positive, and the two others negative, the quadratic expression (1) is negative for all real values of the variables.

The last determinant in (8) is called the *discriminant* of the quadratic function, and the preceding determinant is derived from it by omitting the extreme row and column, and the other is derived from that in like manner.

When the discriminant vanishes, it can be seen without difficulty that the expression (1) is reducible to the sum of *three* squares.

It can be easily proved by *induction* that the preceding principle holds in general, and that in the case of n variables the conditions can be deduced from the *discriminant* in the manner indicated above.

According as the number of rows in a determinant is even or odd, the determinant is said to be one of an even or of an odd order.

Conditions of Maxima and Minima in General.

If the notation already adopted be generalized, the coefficient of x_r^2 is denoted by a_{rr}, and that of $x_r x_m$, by $2a_{rm}$. In this case the discriminant of the quadratic function in n variables is

$$\begin{vmatrix} a_{11}, & a_{12}, & a_{13}, & \ldots & a_{1n} \\ a_{12}, & a_{22}, & a_{23}, & \ldots & a_{2n} \\ a_{13}, & a_{23}, & a_{33}, & \ldots & a_{3n} \\ \cdot & \cdot & \cdot & & \cdot \\ a_{1n}, & a_{2n}, & a_{3n}, & \ldots & a_{nn} \end{vmatrix}, \qquad (9)$$

and the conditions that the quadratic expression shall be always *positive* are, that the determinant (9) and the series of determinants derived in succession by erasing the outside row and column shall be all positive.

To establish this result, we multiply the quadratic function by a_{11}, and it is evident that it may be written in the form

$$(a_{11}x_1 + a_{12}x_2 + \ldots a_{1n}x_n)^2 + (a_{11}a_{22} - a_{12}^2) x_2^2 + \ldots + (a_{11}a_{nn} - a_{1n}^2) x_n^2$$
$$+ 2(a_{11}a_{23} - a_{12}a_{13}) x_2 x_3 + \&c. + (2a_{11}a_{rm} - a_{1r}a_{1n}) x_r x_n + \ldots$$

In order that this should be always positive, it is necessary that the part after the first term should be always positive. This is a quadratic function of the $n-1$ variables $x_2, x_3, \ldots x_n$. Accordingly, assuming that the conditions in question hold for it, its discriminant must be positive, as also the series of determinants derived from it. But the discriminant is

$$\begin{vmatrix} a_{11}a_{22} - a_{12}^2, & a_{11}a_{23} - a_{12}a_{13}, & \ldots & a_{11}a_{2n} - a_{12}a_{1n} \\ a_{11}a_{23} - a_{12}a_{13}, & a_{11}a_{33} - a_{13}^2, & \ldots & a_{11}a_{3n} - a_{13}a_{1n} \\ a_{11}a_{24} - a_{12}a_{14}, & a_{11}a_{34} - a_{13}a_{14}, & \ldots & a_{11}a_{4n} - a_{14}a_{1n} \\ \cdot & \cdot & & \cdot \\ a_{11}a_{2n} - a_{12}a_{1n}, & a_{11}a_{3n} - a_{13}a_{1n}, & \ldots & a_{11}a_{nn} - a_{1n}^2 \end{vmatrix}. \qquad (10)$$

Writing this as in (6), and proceeding as before, it is easily seen that the determinant becomes

$$a_{11}^{n-2} \begin{vmatrix} a_{11}, & a_{12}, & a_{13}, & \ldots & a_{1n} \\ a_{12}, & a_{22}, & a_{23}, & \ldots & a_{2n} \\ a_{13}, & a_{23}, & a_{33}, & \ldots & a_{3n} \\ \cdot & \cdot & \cdot & & \cdot \\ a_{1n}, & a_{2n}, & a_{3n}, & \ldots & a_{nn} \end{vmatrix}, \qquad (11)$$

i.e. the discriminant of the function multiplied by a_{11}^{n-2}.

Hence we infer, that if the principle in question hold for $n-1$ variables it holds for n. But it has been shown to hold in the cases of 3 and 4 variables; consequently it holds for any number.

We conclude finally that the quadratic expression in n variables is always positive, whenever the series of determinants

$$a_{11},\ \begin{vmatrix} a_{11}, & a_{12} \\ a_{12}, & a_{22} \end{vmatrix},\ \begin{vmatrix} a_{11}, & a_{12}, & a_{13} \\ a_{12}, & a_{22}, & a_{23} \\ a_{13}, & a_{23}, & a_{33} \end{vmatrix},\ \ldots\ \begin{vmatrix} a_{11}, & a_{12}, & \ldots & a_{1n} \\ a_{12}, & a_{22}, & \ldots & a_{2n} \\ \cdot & \cdot & & \cdot \\ \cdot & \cdot & & \cdot \\ a_{1n}, & a_{2n}, & \ldots & a_{nn} \end{vmatrix},\quad (12)$$

are all positive.

Again, if the series of determinants of an even order be all positive, and those of an odd order, commencing with a_{11}, be all negative, the quadratic expression is negative for all real values of the variables.

Hence we infer that the number of *independent* conditions for a maximum or a minimum in the case of n variables is $n-1$, as stated in Art. 163.

It is scarcely necessary to state that similar results hold if we interchange any two of the suffix numbers; *i.e.* if any of the coefficients, a_{22}, a_{33}, $\ldots a_{nn}$, be taken instead of a_{11} as the leading term in the series of determinants.

If the determinants in (12) be denoted by Δ_1, Δ_2, Δ_3, $\ldots \Delta_n$, it can be proved without difficulty that, whenever none of these determinants vanishes, the quadratic expression under consideration may be written in the form

$$\Delta_1 U_1^2 + \frac{\Delta_2}{\Delta_1} U_2^2 + \frac{\Delta_3}{\Delta_2} U_3^2 + \ldots + \frac{\Delta_n}{\Delta_{n-1}} U_n^2. \quad (13)$$

Hence, in general, when the quadratic is transformed into a sum of squares, the number of positive squares in the sum depends on the number of continuations of signs in the series of determinants in (12).

It is easy to see *independently* that the series of conditions in (12) are necessary in order that the quadratic function under consideration should be always positive; the preceding investigation proves, however, that they are not only *necessary*, but that they are *sufficient*.

Again, since these results hold if any two or more of the suffix numbers be interchanged, we get the following theorem in the theory of numbers: that if the series of determinants given in (12) be all positive, then every determinant obtained from them by an interchange of the suffix numbers is also necessarily positive.

Also, since, when a quadratic expression is reduced to a sum of squares, the number of positive and negative squares in the sum is fixed (Salmon's *Higher Algebra*, Art. 162), we infer that the number of variations of sign in any series of determinants obtained from (12) by altering the suffix numbers is the same as the number of variations of sign in the series (12).

As already stated, a quadratic expression can be transformed in an infinite number of ways by linear transformations into the sum of a number of squares multiplied by constant coefficients; there is, however, one mode that is unique, **viz.**, what is styled the orthogonal transformation.

Conditions of Maxima and Minima in General.

In this case, if $X_1, X_2, X_3, \ldots X_n$ denote the new linear functions, we have

$$V = x_1^2 + x_2^2 + \ldots + x_n^2 = X_1^2 + X_2^2 + \&c. + X_n^2;$$

and also, denoting the coefficients of the squares in the transformed expression by $a_1, a_2, \ldots a_n$,

$$U = a_{11}x_1^2 + a_{22}x_2^2 + \ldots + a_{nn}x_n^2 + \ldots + 2a_{12}x_1x_2 + 2a_{1r}x_1x_r + \ldots$$

$$= a_1 X_1^2 + a_2 X_2^2 + \ldots a_n X_n^2.$$

Hence, equating the discriminants of $U - \lambda V$ for the two systems, we get

$$\begin{vmatrix} a_{11} - \lambda, & a_{12}, & \ldots & a_{1n} \\ a_{12}, & a_{22} - \lambda, & \ldots & a_{2n} \\ a_{13}, & a_{23}, & \ldots & a_{3n} \\ \cdot & \cdot & \cdot & \cdot \\ a_{1n}, & a_{2n}, & \ldots & a_{nn} - \lambda \end{vmatrix} = (a_1 - \lambda)(a_2 - \lambda) \ldots (a_n - \lambda). \quad (14)$$

Accordingly, the coefficients $a_1, a_2, \ldots a_n$, are the roots of the determinant at the left-hand side of equation (14).

Moreover, in order that the function U should be always positive or always negative for all real values of the variables $x_1, x_2, \ldots x_n$, the coefficients $a_1, a_2, \ldots a_n$, must be all positive in the former case, and all negative in the latter; and consequently, in either case, the roots of the determinant in (14) must all have the same sign.

The application of this result to the determination of the conditions of maxima and minima is easily seen; however, as the conditions thus arrived at are clumsy and complicated in comparison with those given in (12), it is not considered necessary to enter into their discussion here.

INDEX.

Acnode, 259.
Approximations, 42.
    further trigonometrical applications of, 130–8.
Arbogast's method of derivations, 88.
Arc of plane curve, differential expressions for, 220, 223.
Archimedes, spiral of, 301, 303.
Asymptotes, definition of, 242, 249.
    method of finding, 242, 245.
    number of, 243.
    parallel, 247.
    of cubic, 249, 325.
    in polar coordinates, 250.
    circular, 252.

Bernoulli's numbers, 93.
    series, 70.
Bertrand, on limits of Taylor's series, 77.
Bobillier's theorem, 368, 374.
Boole, on transformation of coordinates, 412.
Brigg's logarithmic system, 26.
Burnside, on covariants, 412.

Cardioid, 297, 372.
Cartesian oval, or Cartesian, 233, 375.
    third focus, 376.
    tangent to, 379.
    confocals intersect orthogonally, 381.
Casey, on new form of tangential equation, 339.
    on cycloid, 373.
    on Cartesians, 382.
Cassini, oval of, 233, 333.
Catenary, 288, 321.
Cayley, 259, 266.
Centre of curve, 237.

Centrode, 363.
Change of single independent variable, 399.
    of two independent variables, 403, 410.
Chasles, on envelope of a carried right line, 356.
    construction for centre of instantaneous rotation, 359.
    generalization of method of drawing normals to a roulette, 360.
    on epicycloids, 373.
    on Cartesian oval, 376.
    on cubics, 418.
Circle of inflexions in motion of a plane area, 354, 358, 367, 374.
Conchoid of Nicomedes, 332, 361.
    centre of curvature of, 370.
Concomitant functions, 411.
Condition that $Pdx + Qdy$ is a total differential, 146.
Conjugate points, 259.
Contact, different orders of, 304.
Convexity and concavity, 278.
Crofton, on Cartesian oval, 378, 379, 380.
Crunode, 259.
Cubics, 262, 281, 323, 334.
Curvature, radius of, 286, 287, 295, 297, 301.
    chord of, 296.
    at a double point, 310.
    at a cusp, 311, 313.
    measure of, on a surface, 209.
Cusps, 259, 266, 315.
    curvature at, 311.
Cycloid, 335, 356.
    equation of, 335, 336.
    radius of curvature, and evolute, 337.
    length of arc, 338.

Descartes, on normal to a roulette, 336.
  ovals of, 375.
Differential coefficients, definition, 5.
  successive, 34.
Differentiation, of a product, 13, 14.
  a quotient, 15.
  a power, 16, 17.
  a function of a function, 17.
  an inverse function, 18.
  trigonometrical functions, 19, 20.
  circular functions, 21, 22.
  logarithm, 25.
  exponential functions, 26.
  functions of two variables, 115.
    three or more variables, 117.
  an implicit function, 120.
  partial, 113, 406.
    of a function of two variables, 115.
    of three or more variables, 115.
    applications in plane trigonometry, 130.
      in spherical trigonometry, 133.
    successive, 144.
  of $\phi(x + at, y + \beta t)$ with respect to t, 148.
Discriminant of a ternary quadratic expression, 129, 194, 196.
  of any quadric, 447.
Double points, 258, 261.

Elimination, of constants, 384.
  of transcendental functions, 386.
  of arbitrary functions, 387, 396.
Envelope, 270.
  of $La^2 + 2Ma + N = 0$, 272.
  of a system of confocal conics, Ex. 8, p. 276.
  of a carried curve, 355.
    centre of curvature of, 357.
Epicyclics, 363.
  are epi- or hypo-trochoids, 366.
Epicycloids and hypocycloids, 339, 356.
  radius of curvature of, 342.
  cusps in, 341.
  double generation of, 343.
  evolute of, 344.
  length of arc, 345.
  pedal, 346, 372.
  regarded as envelope, 347.

Epitrochoids and hypotrochoids, 347.
  ellipse as a case of, 348, 363.
  centre of curvature of, 351.
  double generation of, 367.
Equation of tangent to a plane curve, 212, 218.
  normal, 215.
Errors in trigonometrical observation, 135.
Euler, formulæ for sin x and cos x, 69.
  theorem on homogeneous functions, 123, 127, 148, 162.
  on double generation of epi- and hypocycloids, 344.
Evolute, 297.
  of parabola, 298.
  of ellipse, 299, 308; as an envelope, 297.
  of equiangular spiral, 300.
Expansion of a function by Taylor's series, 61.
  of $\phi(x + h, y + k)$, 156.
  of $\phi(x + h, y + k, z + l)$, 159.

Family of curves, 270.
Ferrers, on Bobillier's theorem, 369.
  on Steiner's envelope, 442.
Folium of Descartes, 333.
Functions, elementary forms of, 2.
  continuous, 3.
  derived, 3.
  successive, 34.
  examples of, 46.
  partial derived, 113.
  elliptic, illustrations of, 136, 138.

Graves, on a new form of tangential equation, 339.

Harmonic polar of point of inflexion on a cubic, 281.
Huygens, approximation to length of circular arc, 66.
Hyperbolic branches of a curve, 246.
Hypocycloid, *see* epicycloid.
Hypotrochoid, *see* epitrochoid.

Indeterminate forms, 96.
  treated algebraically, 96–9.
  treated by the calculus, 99, *et seq*.
Infinitesimals, orders of, 36.
  geometrical illustration, 57.
Inflexion, points of, 279, 281.
  in polar coordinates, 303.

Intrinsic equation of a curve, 304.
of a cycloid, 338.
of an epicycloid, 350.
of the involute of a circle, 301.
Inverse curves, 225.
tangent to, 225.
radius of curvature, 295.
conjugate Cartesians as, 378.
Involute, 297.
of circle, 300, 358, 374.
of cycloid, 356.
of epicycloid, 357.

Jacobians, 415-27.

Lagrange, on derived functions, 4, *note*.
on limits of Taylor's series, 76.
on addition of elliptic integrals, 136.
theorem on expansion in series, 151.
on Euler's theorem, 163.
condition for maxima and minima, 191, 197, 199, 202.
La Hire, circle of inflexions, 354.
on cycloid, 373.
Landen's transformation in elliptic functions, 133.
Laplace's theorem on expansion in series, 154.
Legendre, on elliptic functions, 137.
on rectification of curves, 233.
Leibnitz, on the fundamental principle of the calculus, 40.
theorem on the nth derived function of a product, 51.
on tangents to curves in vectorial coordinates, 234.
Lemniscate, 259, 277, 296, 329, 333.
Limaçon, is inverse to a conic, 227, 331, 334, 349, 361, 372.
Limiting ratios, algebraic illustration of, 5.
trigonometrical illustration, 7.
Limits, fundamental principles as to, 11.

Maclaurin, series, 65, 81.
on harmonic polar for a cubic, 282.
Mannheim, construction for axes of an ellipse, 374.
Maxima or minima, 164.
geometrical examples, 164, 183.

algebraic examples, 166.
of $\dfrac{ax^2 + 2bxy + cy^2}{a'x^2 + 2b'xy + c'y^2}$, 166, 177.
condition for, 169, 174.
problem on area of section of a right cone, 181.
for implicit functions, 185.
quadrilateral of given sides, 186.
for two variables, 191; Lagrange's condition, 191, 197.
for functions of three variables, 198.
of n variables, 199, 447.
application to surfaces, 200.
undetermined multipliers applied to, 204.
Multiple points on curves, 256, 265, 367.
Multipliers, method of undetermined, 204.

Napier, logarithmic system, 25.
Navier, geometrical illustration of fundamental principles of the calculus, 8.
on Taylor's theorem, 443.
Newton's definition of fluxion, 10.
prime and ultimate ratios, 40.
expansions of $\sin x$, $\cos x$, $\sin^{-1} x$, &c., 64, 69.
by differential equations, 85.
method of investigating radius of curvature, 291.
on evolute of epicycloid, 345.
Nicomedes, conchoid of, 332.
Node, 259.
Normal, equation of, 215.
number passing through a given point, 220.
in vectorial coordinates, 233.

Orthogonal transformations, 409, 414, 449.
Osc-node, 259.
Osculating curves, 309.
circle, 291, 306.
conic, 317.
Oscul-inflexion, point of, 314, 317.

Parabola, of the third degree, 262, 288.
osculating, 318.
Parabolic branches of a curve, 246.
Parameter, 270.

Partial differentiation, 113, 406.
Pascal, limaçon of, 227.
Pedal, 227.
    tangent to, 227.
    examples of, 230.
    negative, 227.
Plücker, on locus of cusps of cubics having given asymptotes, 265.
Points, de rebroussement, 266.
    of inflexion, 279.
Polar conic of a point, 219.
Proctor, definition of epi- and hypo-cycloids, 399.
    epicyclics, 366.
Ptolemy, epicyclics, 366.

Quetelet, on Cartesian oval, 376, 381.

Radius of curvature, 286.
    in Cartesian coordinates, 287, 289.
    in r, p coordinates, 295.
    in polar coordinates, 301.
    at singular points, 310.
    of envelope of a moving right line, 358.
Reauleaux, on centrodes of moving areas, 363.
Reciprocal polars, 228, 230.
Remainder in series, Taylor's, 76, 79.
    Maclaurin's, 81.
Resultant of concurrent lines, 234.
Roberts, W., extension of method of inversion, 429.
Rotation, of a plane area, 359.
    centre of instantaneous, 360, 364.
    of a rigid body, 371.
Roulettes, 335.
    normal to, 336.
    centre of curvature, 352; Savary's construction, 352.
    circle of inflexions of, 354.
    motion of a plane figure reduced to, 362.
    spherical, 370.

Savary's construction for centre of curvature of roulette, 353.
Series, Taylor's, 61, 70, 76.
    binomial, 63, 82.
    logarithmic, 63, 82.
    for $\sin x$ and $\cos x$, 64, 66, 81.
    Maclaurin's, 64, 81.
    exponential, 65, 81.
    Bernoulli's, 70.
    convergent and divergent, 72, 75.
    for $\sin^{-1} x$, 68, 85.
    for $\tan^{-1} x$, 68, 84.
    for $\sin mx$ and $\cos mx$, 87.
    Arbogast's, 88.
    Lagrange's, 151.
Spinode, 259.
Stationary, points, 266.
    tangents, 282.
Subtangent and subnormal, 215.
    polar, 223.
Symbols, separation of, 53.
    representation of Taylor's theorem by, 70, 160.

Tacnode, 266.
Tangent to curve, 212, 218, 258.
    number through a point, 219.
    expression for perpendicular on, 217, 224.
    expression for intercept on, 232.
Taylor's series, 61.
    symbolic form of, 70.
    Lagrange on limits of, 76.
    extension to two variables, 156.
    to three variables, 159.
    symbolic form of, 160.
    on inapplicability of, 443.
Three-cusped hypocycloid, 350, 372, 430, 442.
Tracing of curves, 322, 328.
Transformations, linear, 408.
    orthogonal, 409, 449.
Trisectrix, 332.
Trochoids, 339.

Ultimate intersection, locus of, 271.
    for consecutive normals, 290.
Undetermined multipliers, application to maxima and minima, 204.
    applied to envelope, 273.
Undulation, points of, 280.

Variables, dependent and independent, 1.
Variations of elements of a triangle, plane, 130; spherical, 133.
Vectorial coordinates, 233.

Whewell, on intrinsic equation, 304.

THE END.